METHODS IN MOLECULAR BIOLOGY

For further volumes:
http://www.springer.com/series/7651

For over 35 years, biological scientists have come to rely on the research protocols and methodologies in the critically acclaimed *Methods in Molecular Biology* series. The series was the first to introduce the step-by-step protocols approach that has become the standard in all biomedical protocol publishing. Each protocol is provided in readily-reproducible step-by-step fashion, opening with an introductory overview, a list of the materials and reagents needed to complete the experiment, and followed by a detailed procedure that is supported with a helpful notes section offering tips and tricks of the trade as well as troubleshooting advice. These hallmark features were introduced by series editor Dr. John Walker and constitute the key ingredient in each and every volume of the *Methods in Molecular Biology* series. Tested and trusted, comprehensive and reliable, all protocols from the series are indexed in PubMed.

Prokaryotic Gene Regulation

Methods and Protocols

Edited by

Eveline Peeters and Indra Bervoets

Research Group of Microbiology, Vrije Universiteit Brussel, Brussels, Belgium

Humana Press

Editors
Eveline Peeters
Research Group of Microbiology
Vrije Universiteit Brussel
Brussels, Belgium

Indra Bervoets
Research Group of Microbiology
Vrije Universiteit Brussel
Brussels, Belgium

ISSN 1064-3745 ISSN 1940-6029 (electronic)
Methods in Molecular Biology
ISBN 978-1-0716-2412-8 ISBN 978-1-0716-2413-5 (eBook)
https://doi.org/10.1007/978-1-0716-2413-5

This Humana imprint is published by the registered company Springer Science+Business Media, LLC, part of Springer Nature.
The registered company address is: 1 New York Plaza, New York, NY 10004, U.S.A.

Preface

The fitness and evolutionary success of prokaryotic microorganisms can be largely attributed to a swift response and adaption of their physiology, metabolism, and behavior to continuously changing environmental conditions. An intricate network of gene regulatory mechanisms lies at the basis of this adaptive response, impacting on different levels of information processing (transcriptional, posttranscriptional, and posttranslational). Although more advanced omics techniques have become available to researchers to assess gene regulation from a system-level global perspective, many knowledge gaps remain with respect to the involved regulatory elements (whether DNA, RNA, or protein based) and the underlying molecular mechanisms. To this end, it can be insightful to combine proven classical biochemical techniques with newly developed methodologies. This volume represents a collection of descriptions on versatile methodologies to investigate prokaryotic gene regulation, focused on different levels of information processing and useful for various model organisms, whether archaeal or bacterial, or both. Most chapters cover experimental methodologies with a step-by-step explanation of needed materials and reagents, the procedure to be followed, and useful tips on how to avoid pitfalls or to perform troubleshooting.

This book, *Prokaryotic gene regulation*, is divided into different sections. The first section covers methodologies that enable the study of the structure of the bacterial/archaeal chromosome, the main template for all gene regulatory processes, and its epigenetic modification. Although not yet well understood, especially in archaea, both aspects presumably affect gene expression. A widely used technique for the sequence-specific study of chromosome architecture are chromosome conformation capture (3C) assays (Chap. 1), while micrococcal nuclease assays enable the investigation of fundamental structural units in histone-containing archaeal chromosomes (Chap. 2). Genome-wide patterns of DNA methylation can be revealed using a SMRT sequencing approach (Chap. 3). The next section provides a selection of methodologies enabling to reach higher levels of the understanding of transcription initiation, a key step in information processing. A synergetic combination of bioinformatic and experimental studies often enables to formulate research hypotheses and initiate a study or to reach deeper levels of understanding, but also to generate parts for synthetic biology engineering of gene regulation, for example, by designing customized promoters for the model bacterium *Escherichia coli* (Chap. 4). In a next step of the information processing flow, the basal transcription machinery is assembled, for which computational approaches can provide insights into protein dynamics, as exemplified for the assembly of the *Mycobacterium tuberculosis* RNA polymerase (Chap. 5). To generate a deeper insight into the functioning of the basal transcription machinery, in vitro transcription approaches are useful. These are typically optimized for each model organism, with the purification of host-specific components of the basal transcription machinery. In this book, this technique is elaborated for archaeal species belonging to *Sulfolobales* (Chap. 6). The third section is devoted to the investigation of regulatory transcription factors, which are often considered as the main players in gene regulation in prokaryotic cells. The identification of transcription factors could be done using a bioinformatic prediction approach (Chap. 7) or by an experimental screening (Chap. 8). A further study of sequence-specific interactions between transcription factors and DNA, a molecular interaction at the core of gene regulatory processes, could be achieved with various experimental methodologies.

These methodologies are also relevant for the study of protein-DNA interactions in the context of other systems besides transcription factors (e.g., replication and recombination). Described methodologies range from the use of electron microscopy to study protein-DNA complexes in *E. coli* (Chap. 9), high-speed atomic force microscopy (Chap. 10), and electrophoretic mobility shift assays and in-gel footprinting (Chap. 11) to chemical protection and premodification binding interference assays (Chap. 12) and isothermal titration calorimetry (Chap. 13). The final section of the book is devoted to a next stage of information processing at which gene regulation occurs, namely the RNA-based level. It is established that the secondary structure of RNA molecules can affect their function, posttranscriptional regulation, and/or translation, with in vitro SHAPE being an elegant technique to unravel this structure (Chap. 14). For the study of interactions between non-coding RNAs and their targets, microscale thermophoresis is described (Chap. 15), while toeprint assays are presented for the study of RNA-protein interactions (Chap. 16).

We would like to thank all authors for sharing their valuable knowledge in this diverse collection of chapters and are convinced that the book will become a valuable resource for researchers in the field of prokaryotic gene regulation.

Brussels, Belgium ***Indra Bervoets***

Brussels, Belgium ***Eveline Peeters***

Contents

Contributors

PAUL BABITZKE • *Department of Biochemistry and Molecular Biology, Center for RNA Molecular Biology, The Pennsylvania State University, University Park, PA, USA*

RANI BAES • *Research Group of Microbiology, Department of Bioengineering Sciences, Vrije Universiteit Brussel, Brussel, Belgium*

AMBER JOKA BERNAUW • *Research Group of Microbiology, Department of Bioengineering Sciences, Vrije Universiteit Brussel, Brussels, Belgium*

INDRA BERVOETS • *Research Group of Microbiology, Department of Bioengineering Sciences, Vrije Universiteit Brussel, Brussels, Belgium*

SNEHA BHEEMIREDDY • *Molecular Biophysics Unit, Indian Institute of Science, Bengaluru, Karnataka, India*

DANIEL CHARLIER • *Research Group of Microbiology, Department of Bioengineering Sciences, Vrije Universiteit Brussel, Brussels, Belgium; Research Group of Microbiology, Department of Bioengineering Sciences, Vrije Universiteit Brussel, Brussels, Belgium*

JIM CLAUWAERT • *KERMIT, Department of Data Analysis and Mathematical Modelling, Ghent University, Ghent, Belgium*

MOHEA COUTURIER • *Carl R. Woese Institute for Genomic Biology, University of Illinois at Urbana-Champaign, Urbana, IL, USA; Research Group of Microbiology, Department of Bioengineering Sciences, Vrije Universiteit Brussel, Brussels, Belgium*

REMUS T. DAME • *Macromolecular Biochemistry, Leiden Institute of Chemistry, Leiden University, Leiden, The Netherlands; Centre for Microbial Cell Biology, Leiden University, Leiden, The Netherlands; Centre for Interdisciplinary Genome Research, Leiden University, Leiden, The Netherlands*

LAURIEN DETMAR • *Macromolecular Biochemistry, Leiden Institute of Chemistry, Leiden University, Leiden, The Netherlands*

SAN HADŽI • *Department of Physical Chemistry, Faculty of Chemistry and Chemical Technology, University of Ljubljana, Ljubljana, Slovenia*

RAFAEL HERNANDEZ-GUERRERO • *Instituto de Investigaciones en Matemáticas Aplicadas y en Sistemas, Universidad Nacional Autónoma de México, Unidad Académica Yucatán, Mé rida, Yucatán, México*

BRITTA JORDAN • *Institute for General Microbiology, Christian-Albrechts-University, Kiel, Germany*

SANDOR KASAS • *International Joint Research Group VUB-EPFL BioNanotechnology & NanoMedicine, Brussels, Belgium; Laboratory of Biological Electron Microscopy, Ecole Polytechnique Fédérale de Lausanne (EPFL), Lausanne, Switzerland; Centre Universitaire Romand de Médecine Légale, UFAM, Université de Lausanne, Lausanne, Switzerland*

VEERKE DE KOCK • *Research Group of Microbiology, Department of Bioengineering Sciences, Vrije Universiteit Brussel, Brussels, Belgium*

YURII F. KRUPYANSKII • *Department of Structure of Matter, Semenov Federal Research Center of Chemical Physics, RAS, Moscow, Russia*

JURIJ LAH • *Department of Physical Chemistry, Faculty of Chemistry and Chemical Technology, University of Ljubljana, Ljubljana, Slovenia*

LEONARDO LEDESMA • *Posgrado en Ciencia e Ingeniería de la Computación, Universidad Nacional Autónoma de México, Ciudad de México, México*

NATALIYA LOIKO • *Department of Microbiology, Federal Research Center 'Fundamentals of Biotechnology' RAS, Moscow, Russia*
JO MAERTENS • *Centre for Synthetic Biology (CSB), Department of Biotechnology, Ghent University, Ghent, Belgium*
HUGO MARUYAMA • *Department of Bacteriology, Osaka Dental University, Hirakata, Japan*
FRIEDERIKE MEY • *Centre for Synthetic Biology (CSB), Department of Biotechnology, Ghent University, Ghent, Belgium*
MARJAN DE MEY • *Centre for Synthetic Biology (CSB), Department of Biotechnology, Ghent University, Ghent, Belgium*
ANDREY MOISEENKO • *Department of Biology, Faculty of Biology, Lomonosov Moscow State University, Moscow, Russia; Department of Structure of Matter, Semenov Federal Research Center of Chemical Physics, RAS, Moscow, Russia*
LISA NICKEL • *Institute for General Microbiology, Christian-Albrechts-University, Kiel, Germany*
EVELINE PEETERS • *Research Group of Microbiology, Department of Bioengineering Sciences, Vrije Universiteit Brussel, Brussels, Belgium*
ERNESTO PEREZ-RUEDA • *Instituto de Investigaciones en Matemáticas Aplicadas y en Sistemas, Universidad Nacional Autónoma de México, Unidad Académica Yucatán, Mé rida, Yucatán, México*
FATEMA-ZAHRA M. RASHID • *Macromolecular Biochemistry, Leiden Institute of Chemistry, Leiden University, Leiden, The Netherlands; Centre for Microbial Cell Biology, Leiden University, Leiden, The Netherlands; Centre for Interdisciplinary Genome Research, Leiden University, Leiden, The Netherlands*
RUTH A. SCHMITZ • *Institute for General Microbiology, Christian-Albrechts-University, Kiel, Germany*
OLGA S. SOKOLOVA • *Department of Biology, Faculty of Biology, Lomonosov Moscow State University, Moscow, Russia; Department of Biology, Shenzhen MSU-BIT University, Shenzhen, China*
NARAYANASWAMY SRINIVASAN • *Molecular Biophysics Unit, Indian Institute of Science, Bengaluru, Karnataka, India*
MICHIEL STOCK • *KERMIT, Department of Data Analysis and Mathematical Modelling, Ghent University, Ghent, Belgium*
DAVID SYBERS • *Research Group of Microbiology, Department of Bioengineering Sciences, Vrije Universiteit Brussel, Brussels, Belgium*
CHRISTIAN TELLGREN-ROTH • *Department of Immunology, Genetics and Pathology, National Genomics Infrastructure, Science for Life Laboratory, Uppsala University, Uppsala, Sweden*
MAARTEN VAN BREMPT • *Centre for Synthetic Biology (CSB), Department of Biotechnology, Ghent University, Ghent, Belgium*
WILLEM WAEGEMAN • *KERMIT, Department of Data Analysis and Mathematical Modelling, Ghent University, Ghent, Belgium*
RONNIE G. WILLAERT • *Research Group Structural Biology Brussels, Alliance Research Group VUB-UGent NanoMicrobiology (NAMI), Brussels, Belgium; International Joint Research Group VUB-EPFL BioNanotechnology & NanoMedicine, Brussels, Belgium*
HELEN YAKHNIN • *Department of Biochemistry and Molecular Biology, Center for RNA Molecular Biology, The Pennsylvania State University, PA, USA*

Chapter 1

Chromosome Conformation Capture in Bacteria and Archaea

Fatema-Zahra M. Rashid, Laurien Detmar, and Remus T. Dame

Abstract

The three-dimensional structure of the chromosome is encoded within its sequence and regulates activities such as replication and transcription. This necessitates the study of the spatial organization of the chromosome in relation to the underlying sequence. Chromosome conformation capture (3C) techniques are proximity ligation-based approaches that simplify the three-dimensional architecture of the chromosome into a one-dimensional library of hybrid ligation junctions. Deciphering the information contained in these libraries resolves chromosome architecture in a sequence-specific manner. This chapter describes the preparation of 3C libraries for bacteria and archaea. It details how the three-dimensional architecture of local chromatin can be extracted from the 3C library using qPCR (3C-qPCR), and it summarizes the processing of 3C libraries for next-generation sequencing (3C-Seq) for a study of global chromosome organization.

Key words 3C-Seq, 3C-qPCR, *Escherichia coli*, *Saccharolobus solfataricus*

1 Introduction

In prokaryotes, the genome sequesters into a phase-separated nucleoid in the cytoplasm. Within this compartment, chromosomes organize into domains—physically insulated regions of the chromosome in which the chromatin interacts more frequently with itself than with flanking chromatin. Domain organization is hierarchical, with larger domains formed from the preferential interaction of two or more smaller domains [1–3].

Three-dimensional chromosome organization is determined by the physical properties of the DNA polymer and by the architectural properties of nucleoid-associated proteins (NAPs) bound to the macromolecule. These proteins maintain chromosome structure and remodel it in response to environmental stimuli, directly affecting genome processes. In vitro studies of *Escherichia coli* H-NS demonstrate that the protein responds to changes in

Eveline Peeters and Indra Bervoets (eds.), *Prokaryotic Gene Regulation: Methods and Protocols*,
Methods in Molecular Biology, vol. 2516, https://doi.org/10.1007/978-1-0716-2413-5_1,

osmolarity by switching between its DNA-bridging and DNA-stiffening modes [4]. At low osmolarity, H-NS bridges a pair of DNA duplexes to form a DNA:H-NS:DNA bridge that functions as a transcription road-block [4, 5]. At higher osmolarities, H-NS preferentially forms filaments along DNA that are conducive to transcription [4, 5]. In *Bacillus subtilis*, a pair of hairpins formed by the directional movement of SMC-ScpAB complexes along the chromatin brings distantly placed DnaA boxes in close proximity to OriC to initiate chromosome replication [1]. In *Sulfolobus acidocaldarius* and *Sulfolobus islandicus*, the SMC-like coalescin protein (ClsN) separates transcriptionally repressed B-compartments from transcriptionally active A-compartments, and maintains silencing in the B-compartment [6, 7]. Conversely, genome processes affect chromosome structure. In *Caulobacter crescentus*, long, highly-transcribed genes form domain boundaries by generating extended supercoil diffusion barriers that physically separate flanking DNA [8]. Furthermore, gene transcription in the Ter > Ori direction slows the movement of SMC complexes toward the chromosome terminus [9, 10]. In *S. acidocaldarius*, the transcription inhibitory drug actinomycin D weakens loop formation in the chromosome and, chromosome compartmentalization into transcriptionally active and repressed chromatin [7]. In *Haloferax volcanii*, actinomycin D treatment weakens the organization of the chromosome into CIDs [11].

The intricate interplay between chromosome architecture and its processes is encoded within the genome sequence as, for instance, replication origins and termini, loading and offloading sites for motor proteins that track along the DNA, binding sites of varying affinity for the repertoire of NAPs available within a cell, and transcription initiation and termination sites of differing strengths (Reviewed in [12]). The inherent sequence dependence of chromosome architecture calls for describing chromosome structure in terms of its sequence. To that end, chromosome conformation capture (3C) techniques detect the physical proximity of chromosome segments in three-dimensional space [13].

The architecture of local chromatin can be gleaned using 3C-qPCR, a one-to-one technique that combines proximity ligation with quantitative PCR to determine the relative interaction frequency between a pair of loci [14]. With the limitation of a lower throughput, the technique allows local three-dimensional chromosome architecture to be studied at the resolution of individual restriction fragments [14].

3C-Seq is a high-throughput approach that combines proximity ligation with next-generation sequencing to probe physical interactions amongst all segments of the chromosome in an unbiased manner [15]. The technique, therefore, describes chromosome structure in relation to genome sequence at a global scale. As an ensemble technique, 3C-Seq determines the probability that two regions of the chromosome lie in physical proximity

[15]. This highlights common architectural features of an organism's chromosome such as its domains and domain boundaries, compartmentalization into hetero- and eu-chromatin in eukaryotes, and, in some prokaryotes, the presence of inter-arm interactions [1–3, 6, 8]. As a single-cell technique, 3C-Seq—by extrapolation of the observations from its comparable Hi-C variant—could identify the mutual inclusivity or exclusivity of these features [16–18].

This chapter builds on our earlier work [19] and describes protocols for the preparation of ensemble 3C libraries compatible with quantitative PCR (3C-qPCR) and for sequencing on an Illumina® NGS platform (3C-Seq) from *E. coli*, and *Saccharolobus solfataricus* cultures. The design, experimentation, and data analysis for 3C-qPCR studies are described in detail using the *proVWX* operon of *Escherichia coli* as a model.

2 Materials

2.1 Cell Fixation

1. 37% w/v formaldehyde solution, stored at room temperature away from direct sunlight.
2. 2.5 M glycine solution filter-sterilized using a sterile, non-pyrogenic 0.2 μm filter (*see* **Note 1**).
3. 1× TE Buffer: 10.0 mM Tris-HCl, 1.0 mM EDTA·Na_2, pH 8.0. Store the buffer at room temperature and cool it to 4 °C prior to use.
4. Liquid nitrogen.
5. Refrigerated benchtop centrifuge with rotors for 50 mL, 15 mL, and 1.5 mL tubes.
6. Roller bench.
7. −80 °C freezer.
8. 50 mL and 15 mL conical tubes.
9. Serological pipettes.
10. Pipette pump.
11. Elongated Pasteur pipette [19].
12. 2.5 μL, 20 μL, 200 μL, and 1000 μL micropipettes.
13. Low-retention micropipette tips.
14. Low-retention 1.5 mL microfuge tubes.
15. **For *E. coli*:** Methanol, ACS reagent ≥99.8%. Store the reagent at room temperature and cool it to 4 °C prior to use.
16. **For *E. coli*:** 1× PBS solution: 137.0 mM NaCl, 2.7 mM KCl, 4.3 mM Na_2HPO_4, 1.47 mM KH_2PO_4, pH 7.4. Store 1× PBS at room temperature and cool it to 4 °C prior to use.
17. **For *S. solfataricus*:** Thermomixer with a thermoblock for 15 mL conical tubes.

2.2 3C Library Preparation

1. 1× TE Buffer: 10.0 mM Tris-HCl, 1.0 mM EDTA·Na_2, pH 8.0.
2. **For *E. coli*:** 40,000 U/μL Ready-Lyse Lysozyme (Epicentre).
3. **For *E. coli*:** 1.0 M NaCl.
4. **For *S. solfataricus*:** Glass beads, 0.5 mm in diameter.
5. Nuclease-free water.
6. 10% w/v sodium dodecyl sulfate (SDS) solution: Dissolve 1.0 g of SDS (ACS reagent, ≥99.0%) in nuclease-free water to a final volume of 10.0 mL.
7. 10% v/v Triton X-100 solution: Dilute 1.0 mL of Triton X-100 (molecular biology grade) with 9.0 mL of nuclease-free water. Store the solution at room temperature (20–25 °C) in the dark.
8. Restriction enzyme: 10 U/μL NlaIII (*see* **Note 2**).
9. 10× restriction enzyme buffer, as supplied with the selected restriction enzyme (*see* **Note 2**).
10. Qubit® dsDNA HS Assay Kit and a Qubit® fluorometer from ThermoFisher Scientific (*see* **Note 3**).
11. 10× T4 DNA Ligase buffer: 0.5 M Tris-HCl, 0.1 M $MgCl_2$, 10.0 mM ATP, 0.1 M DTT, pH 7.5.
12. 20.0 mg/mL bovine serum albumin (BSA).
13. 2000 U/μL T4 DNA Ligase.
14. 0.5 M EDTA, pH 8.0.
15. 10.0 mg/mL RNase A.
16. 20.0 mg/mL proteinase K.
17. 5.0 M NaCl.
18. 25:24:1 phenol:chloroform:isoamyl alcohol solution (*see* **Note 4**).
19. Chloroform (*see* **Note 4**).
20. 1.0 M sodium acetate (NaOAc), pH 8.0.
21. 5.0 mg/mL glycogen.
22. 100% and 70% (v/v) ethanol. Keep the solutions cold at −20 °C.
23. 10.0 mM Tris, pH 8.0.
24. Thermomixer with a thermoblock for 1.5 mL microfuge tubes.
25. Benchtop centrifuge.
26. Refrigerated benchtop centrifuge.
27. Vortex mixer.
28. Agarose gel electrophoresis set-up.
29. −20 °C freezer.

30. 2.5 μL, 20 μL, 200 μL, and 1000 μL micropipettes.
31. Low-retention micropipette tips.
32. Low-retention 1.5 mL microfuge tubes.
33. 15 mL conical tubes.

2.3 3C-qPCR

1. 3C-qPCR primers at a concentration of 100.0 μM. The design of the primers is described in Subheading 3.3.1.
2. DNA oligonucleotides for the synthesis of the 3C-qPCR control library at a concentration of 10.0 μM. The design of the oligonucleotides is detailed in Subheading 3.3.1.
3. PCR kit.
4. TaqMan quantitative PCR master mix (e.g., PrimeTime® Gene Expression Master Mix from Integrated DNA technologies).
5. TaqMan probe (e.g., PrimeTime® probe 5′ 6-FAM/ZEN/3' IBFQ from Integrated DNA technologies). The design of the TaqMan probe is described in Subheading 3.3.1.
6. Nuclease-free water.
7. PCR tubes.
8. 96-well PCR plates with white wells.
9. PCR plate sealing films.
10. Thermal cycler with a heated lid.
11. Real-Time PCR detection system and analysis software (e.g., CFX96 Touch Real-Timc PCR Detection System and the CFX Maestro software for CFX Real-Time PCR Instruments from Bio-Rad).
12. Agarose gel electrophoresis set-up.
13. Agarose gel DNA extraction and purification kit.
14. Qubit® dsDNA HS Assay Kit and a Qubit® fluorometer from ThermoFisher Scientific (*see* **Note 3**).
15. 2.5 μL, 20 μL, 200 μL, and 1000 μL micropipettes.
16. **Preferred:** Multi-channel pipette for 0.5–10 μL.
17. Low-retention micropipette tips.
18. Low-retention 1.5 mL microfuge tubes.

2.4 3C-Seq

1. Solid Phase Reverse Immobilization (SPRI) beads. Store the beads as 1.0 mL aliquots at 4 °C. Ensure that the beads are equilibrated to 25 °C and vortexed before use.
2. Nuclease-free water.
3. 10.0 mM Tris, pH 8.0.
4. 10.0 mM Tris, pH 8.0 + 0.1% Tween 20.

5. Qubit® dsDNA HS Assay Kit and a Qubit® fluorometer from ThermoFisher Scientific.
6. Illumina® library preparation kit (e.g., KAPA HyperPlus kit from KAPA Biosystems).
7. Illumina®-compatible paired-end sequencing adapters at a concentration of 15.0 μM (e.g., KAPA Single-Indexed Adapter Sets A and B from KAPA Biosystems). The sequencing adapters should be barcoded to allow multiplexing during NGS.
8. Magnetic racks for PCR tubes and 1.5 mL microfuge tubes.
9. Thermomixer with a thermoblock for 1.5 mL microfuge tubes.
10. Thermal cycler with a heated lid.
11. Vortex mixer.
12. PCR tubes.
13. Agarose gel electrophoresis set-up.
14. 2.5 μL, 20 μL, 200 μL, and 1000 μL micropipettes.
15. Low-retention micropipette tips.
16. Low-retention 1.5 mL microfuge tubes.

3 Methods

The protocols described here are optimized for 3C in *Escherichia coli* and *Saccharolobus solfataricus*. To process other prokaryotes using these protocols, optimize the cell lysis and solubilization steps by adjusting the treatment duration—and, if relevant, the concentrations—of Ready-Lyse Lysozyme, SDS, and physical shearing with glass beads.

3.1 Cell Fixation

3.1.1 Escherichia coli

1. Transfer an aliquot of an *Escherichia coli* culture equivalent to ~5.0×10^9 cells into a chilled 50 mL conical tube.
2. Harvest the cells by centrifugation at 3000 × *g* for 5 min at 4 °C.
3. Decant the supernatant and carefully remove the remaining liquid with a micropipette. Resuspend the cell pellet in 4.0 mL of 1× PBS.
4. Add 16.0 mL of methanol to the cell suspension (*see* **Note 5**). Mix well by inversion and incubate the cells at 4 °C for 10 min on a roller bench at 30 rpm.
5. Collect the cells at 3000 × *g* for 5 min at 4 °C.
6. Discard the supernatant and resuspend the cells in 20.0 mL of 1× PBS to wash.
7. Centrifuge the cell suspension at 3000 × *g* for 5 min at 4 °C.

8. During the centrifugation, prepare 40.0 mL of 3% formaldehyde in 1× PBS by adding 3.2 mL of 37% formaldehyde solution to 36.8 mL of cold 1× PBS in a 50 mL conical tube. Mix well by inversion and place the solution on ice or at 4 °C.
9. Discard the supernatant of the centrifugation performed at **step** 7 of this section. Resuspend the cell pellet in 40.0 mL of 3% formaldehyde and incubate the cell suspension at 4 °C for 1 h on a roller bench at 30 rpm.
10. Add 6.4 mL of 2.5 M glycine to the fixation reaction (to a final concentration of 0.375 M) to quench excess formaldehyde. Incubate the cells for 15 min at 4 °C on a roller bench at 30 rpm.
11. Collect the fixed cells by centrifugation at 3000 × *g* for 10 min at 4 °C.
12. Decant the supernatant and carefully remove the remaining liquid with a micropipette. Resuspend the pellet in 5.0 mL of 1× TE and divide the suspension into five aliquots of 1.0 mL in sterile 1.5 mL microfuge tubes. Each aliquot contains approximately 10^9 cells.
13. Pellet the fixed cells by centrifugation at 10,000 × *g* for 2 min at 4 °C.
14. Remove the supernatant with an elongated Pasteur pipette.
15. Proceed immediately with 3C library preparation or flash freeze the cell pellet in liquid nitrogen. Frozen pellets can be stored at −80 °C.

3.1.2 Saccharolobus solfataricus

1. Heat up a 15 mL conical tube to 80 °C in a Thermomixer.
2. Transfer an aliquot of a *Saccharolobus solfataricus* culture (grown at 80 °C) equivalent to ~5.0 × 10^9 cells to the conical tube at 80 °C.
3. Add 37% formaldehyde solution to the culture to a final concentration of 1%. This corresponds to 28.8 μL of 37% formaldehyde per mL of culture. Mix the reaction by inversion and return the conical tube to the Thermomixer.
4. Change the set temperature of the Thermomixer to 75 °C and allow the fixation reaction to cool to 75 °C (*see* **Note 6**). This takes approximately 6 min with the Eppendorf Thermomixer® C.
5. Add 2.5 M glycine to the fixation reaction to a final concentration of 0.125 M. This corresponds to 54.1 μL of 2.5 M glycine per mL of culture. Mix well by inversion and incubate the reaction at 4 °C on a roller bench at 30 rpm for 15 min.
6. Collect the cells by centrifugation at 3000 × *g* for 5 min at 4 °C.

7. Discard the supernatant and resuspend the cell pellet in 10.0 mL of 1× TE to wash.
8. Centrifuge the cells at 3000 × *g* for 5 min at 4 °C.
9. Pour out the supernatant and carefully remove the remaining liquid with a micropipette. Resuspend the cell pellet in 5.0 mL of 1× TE and separate the suspension of fixed cells into five 1.0 mL aliquots in low-retention 1.5 mL microfuge tubes.
10. Pellet the fixed cells by centrifugation at 10,000 × *g* for 2 min at 4 °C.
11. Remove the supernatant with an elongated Pasteur pipette.
12. Use the fixed cells immediately for 3C library preparation, or flash freeze the pellets in liquid nitrogen and store the cells at −80 °C.

3.2 3C Library Preparation

3.2.1 Lysis and Solubilization: Escherichia coli

1. Dilute 0.5 μL of 40,000 U/μL Ready-Lyse Lysozyme (Epicentre) with 4.0 μL of 1× TE and 0.5 μL of 1.0 M NaCl. Place the dilution on ice.
2. Resuspend a pellet of fixed *Escherichia coli* cells in 50.0 μL of 1× TE and add 0.5 μL of the diluted lysozyme (2000 U) to the cell suspension. Mix by pipetting and incubate the tube at 37 °C in a Thermomixer at 1000 rpm for 15 min.
3. Add 2.5 μL of 10% SDS to the lysis reaction and mix by pipetting. Incubate the tube for 15 min at 37 °C in a Thermomixer at 1000 rpm.
4. Add 45.0 μL of 1× TE to a 5.0 μL aliquot of the lysed and solubilized cells. Store the preparation at −20 °C as 'Chromatin'. This control is used to qualitatively examine the efficiency of cell lysis and solubilization (*see* **steps 2** and **3** of Subheading 3.2.1; Fig. 1). Continue with processing Chromatin in Subheading 3.2.6.
5. Make up the volume of the lysed and solubilized cells to 53.0 μL with 5.0 μL of 1× TE and continue with 3C library preparation (*see* Subheading 3.2.3).

3.2.2 Lysis and Solubilization: Saccharolobus solfataricus

1. Resuspend a pellet of fixed cells in 60.0 μL of 1× TE. Add 25–30 glass beads to the cell suspension and vortex vigorously for 5–7 min to shear the cells. Ensure that the beads flow with the cell suspension on the sides of the tube during vortexing.
2. Centrifuge the tube for 5 s to spin down the disrupted cells and sediment the glass beads.
3. Add 2.8 μL of 10% SDS to the lysate and mix by briefly vortexing the tube. Incubate the tube for 15 min at 37 °C in a Thermomixer at 1000 rpm.

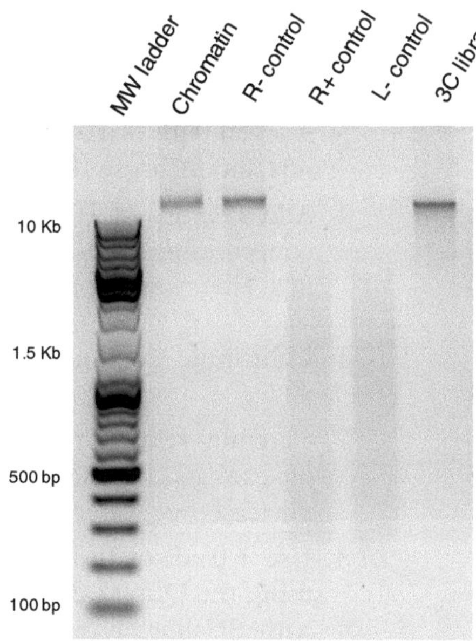

Fig. 1 Agarose gel electrophoresis of a 3C preparation. **Chromatin:** Fixed chromatin runs as a high molecular weight band on an agarose gel, often with a smear toward higher molecular weights. Chromatin preparations that show low molecular weight smears are degraded, hence unreliable for use in a 3C-Seq/3C-qPCR study. **R- control:** Fixed chromatin incubated in the restriction digestion buffer without the restriction enzyme shows a profile similar to the Chromatin. The appearance of low molecular weight DNA in R- is indicative of degradation, in which case, the digestion observed in R+ cannot be relied on for 3C-based experiments. **R+ control:** Fixed chromatin treated with a restriction enzyme shows a smear of digestion products. An R+ profile that is similar to C and R- indicates inefficient digestion. **L- control:** Digested chromatin incubated in the ligation buffer in the absence of T4 DNA ligase shows a profile similar to R+. Degradation may be inferred from a shift in the DNA profile toward lower molecular weights. **3C Library:** Proximity ligation of restriction digested, fixed chromatin results in a DNA profile that is indistinguishable from Chromatin and R-. Nevertheless, a 'shift-up' of the DNA profile toward higher molecular weights is sufficient to infer successful ligation

4. Mix 5.0 μL of the lysed and solubilized cells with 45.0 μL of 1× TE in a 1.5 mL low-retention microfuge tube. Store the preparation at −20 °C as 'Chromatin'. Continue with processing this sample in Subheading 3.2.6.
5. Transfer 53.0 μL of the lysed and solubilized cells into a 1.5 mL low-retention microfuge tube to continue with 3C library preparation (*see* Subheading 3.2.3).

3.2.3 Restriction Digestion

1. Add 136.0 μL of nuclease-free water, 25.0 μL of 10% Triton X-100, and 25.0 μL of 10× restriction digestion buffer to 53.0 μL of lysed and solubilized cells (*see* **step 5** of either Subheadings 3.2.1 or 3.2.2).

2. Mix by inversion and incubate the sample at 37 °C for 10 min in a Thermomixer at 1000 rpm to sequester SDS in the cell lysate with Triton X-100.
3. Add 10.0 μL of 10 U/μL NlaIII (*see* **Note 2**) to the digestion mix and incubate the reaction at 37 °C for 3 h at 450 rpm.
4. Add 12.5 μL of 10% SDS to the digestion reaction to a final concentration of 0.5%. Incubate the sample for 20 min at 37 °C in a Thermomixer at 1000 rpm to terminate the reaction.

3.2.4 Fractionation

1. Centrifuge the digested chromatin at 25,000 × *g* for 1 h at 4 °C.
2. Carefully remove the supernatant with an elongated Pasteur pipette and resuspend the gel-like pellet in 200.0 μL of nuclease-free water by vigorous pipetting and vortexing.
3. Use 1.0 μL of the sample to determine DNA concentration using the Qubit® dsDNA HS Assay Kit (*see* **Note 3**). Proceed with 3C library preparation if at least 1200 ng of digested chromatin is available.
4. Keep a 200–500 ng aliquot aside as the R+ control to visualize the efficiency of restriction digestion (*see* Fig. 1). Make up the volume of the control to 50.0 μL using 1× TE and store it at −20 °C. Continue with processing the R+ control as described in Subheading 3.2.6.

3.2.5 Ligation

1. Adjust the volume of an aliquot of digested chromatin containing between 1.0–3.0 μg of DNA to 895.0 μL with nuclease-free water.
2. Add 100.0 μL of 10× T4 DNA Ligase Buffer, 5.0 μL of 20.0 mg/mL BSA, and 2.0 μL of 2000 U/μL T4 DNA Ligase to the chromatin and mix well by inversion. Multiple reactions can be prepared to use up all the fractionated chromatin.
3. Incubate the reaction(s) in a Thermomixer for 16 h at 16 °C at 0 rpm, and for 1 h at 25 °C at 450 rpm.
4. Terminate the ligation reaction(s) with 20.5 μL of 0.5 M EDTA.

3.2.6 Reverse Cross-Linking and DNA Purification

1. Add 16.6 μL of 10.0 mg/mL RNase A to the ligation reaction (s). For Chromatin and R+ (*see* **step 4** of Subheadings 3.2.1, 3.2.2, and 3.2.4), use 0.8 μL of 10.0 mg/mL RNase A. Mix by pipetting and incubate the samples for 30 min at 37 °C in a Thermomixer at 450 rpm.
2. Add 12.5 μL of 20.0 mg/mL Proteinase K and 120.8 μL of 5.0 M NaCl to the ligation reaction(s). Use 0.64 μL of 20.0 mg/mL Proteinase K and 6.2 μL of 5.0 M NaCl for Chromatin and R+ (*see* **step 4** of Subheadings 3.2.1, 3.2.2, and 3.2.4). Incubate the samples for 16 h at 65 °C in a Thermomixer at 450 rpm.

3. Transfer the ligation sample into a 15 mL conical tube. Pool corresponding samples if applicable (*see* **step 2** of Subheading 3.2.5).
4. Add 1.0 volume of 25:24:1 phenol:chloroform:isoamyl alcohol to ligation sample and controls (*see* **Note 4**). Vortex vigorously for 10 s.
5. Centrifuge the samples for 10 min at 13,000 × *g* at 4 °C.
6. Transfer the top aqueous layer into sterile 1.5 mL or 15 mL tubes, as necessary (*see* **Note 4**).
7. Repeat **steps 4–6**.
8. Add 1.0 volume of chloroform to the ligation sample and controls and vortex vigorously for 10 s (*see* **Note 4**).
9. Centrifuge the tubes for 10 min at 13,000 × *g* at 4 °C.
10. Transfer the top aqueous layer into sterile 1.5 mL or 15 mL tubes, as necessary (*see* **Note 4**).
11. Add 0.1 volumes of 1.0 M sodium acetate, pH 8.0, 0.025 volumes of 5.0 mg/mL glycogen, and 2.5 volumes of cold 100% ethanol to the ligation sample and controls. Vortex well and incubate the solutions overnight at −20 °C.
12. Centrifuge the solutions for 20 min at 25,000 × *g* at 4 °C.
13. Remove the supernatant with an elongated Pasteur pipette. A white pellet should be visible at the bottom of the tubes.
14. Add 500.0 μL of cold 70% ethanol to the tubes to wash the pellets.
15. Centrifuge the tubes at 25,000 × *g* for 5 min at 4 °C.
16. Remove the supernatant with an elongated Pasteur pipette.
17. Repeat **steps 14–16**.
18. Dry the pellets in a 50 °C incubator.
19. Dissolve each pellet in 31.0 μL of nuclease-free water in a Thermomixer at 60 °C for 20 min at 1000 rpm.
20. Use 1.0 μL of the controls and the purified ligation sample, hereafter referred to as the 3C library, to determine DNA concentration using the Qubit® dsDNA HS Assay Kit (*see* **Note 3**).
21. Visualize between 100 and 500 ng of each sample on a 1.5% agarose gel to qualitatively assess 3C library preparation. Typically, the Chromatin migrates as a heavy >10 kb band on a 1.5% agarose gel, R+ runs as a smear of digestion products, and a well-ligated 3C library shows a shift of the R+ profile to higher molecular weights (*see* Fig. 1). Ensure that at least 1.0 μg of the 3C library is available for the next steps of the protocol.

3.3 3C-qPCR

3C-qPCR determines the relative interaction frequency between a pair of loci by quantifying the hybrid ligation junction formed between the target loci during **step 3** of Subheading 3.2.5. Ideally, PCR cycles at which the reagents and enzymes are not limiting are associated with an exponential increase in the quantity of a target fragment. In TaqMan probe-based qPCR, this increase is associated with an increase in fluorescence, since the TaqMan probe—designed to anneal to the target—hydrolyzes during amplification across the target region, resulting in the separation of the fluorophore positioned at one end of the probe from its quencher positioned at the other. The fluorescence measured after each qPCR cycle is, therefore, a measure of the quantity of the target in the reaction. In qPCR runs, the *Starting Quantity* (*SQ*) of the target is calculated from the *Cycle of quantitation* (*Cq*)—the cycle at which the fluorescence of a reaction rises above the background signal.

3.3.1 Experimental Design

1. Identify the digestion sites of the restriction enzyme used to prepare the 3C library at the locus to be studied with 3C-qPCR. For instance, at the region encompassing the promoter of the *proVWX* operon in *Escherichia coli* K-12 MG1655 (GenBank: NC_000913.3; chr:2804122-2804995), NlaIII digests the chromatin at three positions: chr:2804700, chr:2804817, and chr:2804994.
2. Design primers that hybridize ~50 bp upstream of the restriction digestion site for each fragment of interest (*see* **Note 7**) (*see* Fig. 2). The melting temperature (T_m) of the primers should be ~55 °C with no secondary structures having a T_m higher than 40 °C forming in the oligonucleotide. The change in Gibbs free energy (ΔG) for the formation of any primer homodimers should be less than 15% of the ΔG of complete hybridization. For example, primers for NlaIII-digested chromatin at NC_000913.3 chr:2804122-2804995 are (Table 1):
3. Design a TaqMan probe with a T_m of >60 °C for the anchor fragment between the primer annealing coordinates and the restriction digestion site. The probe should hybridize to the strand complementary to the primer-annealing strand (*see* **Note 8**) (*see* Fig. 2). For instance, in the case of the promoter region of the *proVWX* operon, fragment ProVWX_2 that carries the σ^{70}-dependent promoter may be selected as the anchor. The TaqMan probe for this fragment, then, is (Table 2):
4. Design primers and a TaqMan probe for the Internal control (*see* **Note 9**) as described in **steps 2** and **3** of Subheading 3.3.1. In our experience, the interaction between fragments Chr:2804701-2804817 (ProVWX_2) and Chr:2805902-2806570 serves as an internal control in NlaIII-digested 3C libraries of exponential phase *Escherichia coli* cultures grown at different osmolarities. Therefore, in the example described

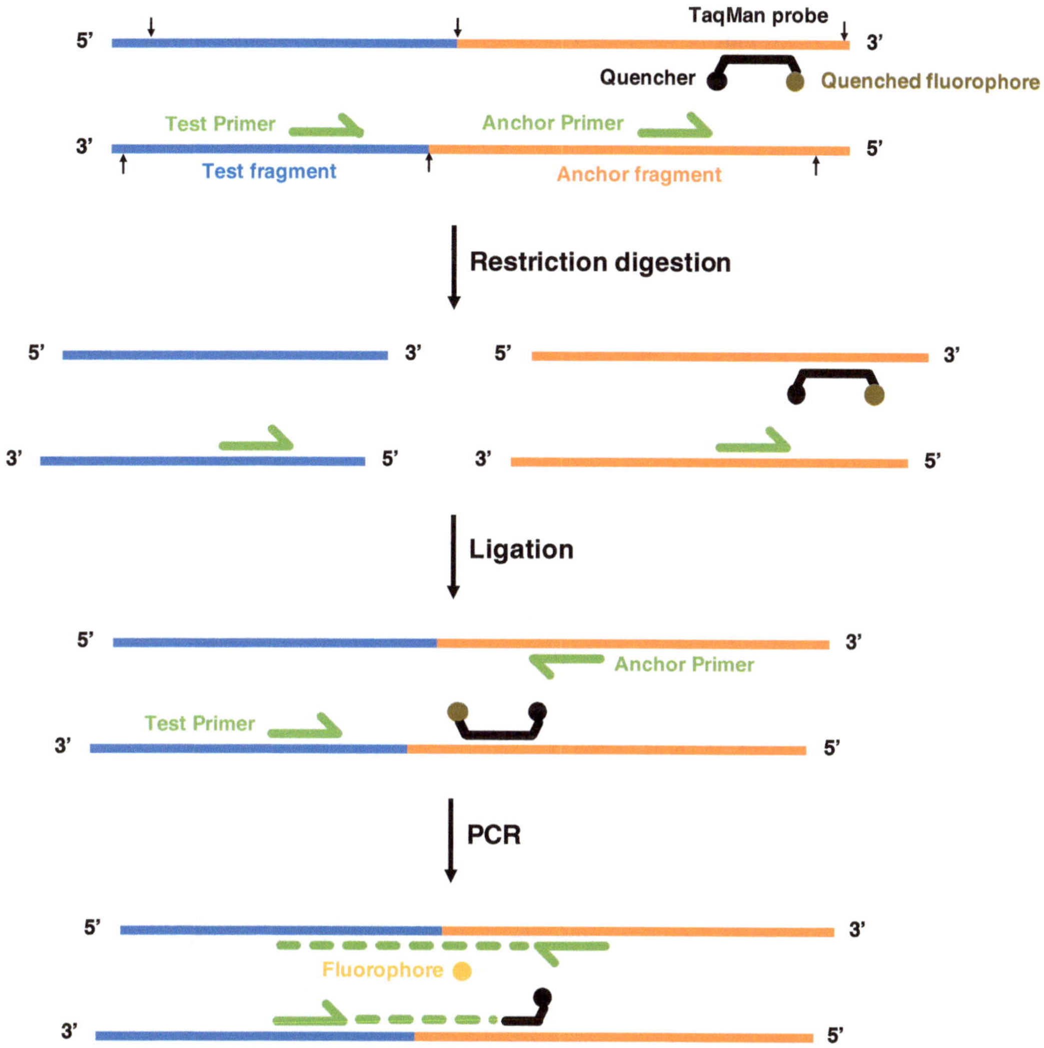

Fig. 2 3C-qPCR design [14]. 3C-qPCR primers (green) for the test (blue) and anchor (orange) fragments are designed to anneal along the same DNA strand, such that the primers are oriented uni-directionally. The TaqMan probe (black) for the anchor fragment is designed to hybridize to the strand complementary to the primer-annealing strand and is positioned between the anchor primer and restriction digestion site (short black arrows) of the anchor fragment. This design ensures that a successful PCR only occurs if the DNA molecule is digested and if the fragments re-ligate in a manner that re-orients only one of the fragments. Such a junction orients the primers to extend toward each other. The hybridization of the anchor primer and the TaqMan probe to opposite strands affords specificity to the signal. The design only allows the detection of a hybrid junction formed between the anchor and test fragment, since the TaqMan probe is only hydrolyzed, and hence, fluorescence is only detected, when elongation from the test primer proceeds across the ligation junction toward the anchor fragment

Table 1
3C-qPCR primers for NlaIII-digested chromatin at NC_000913.3 chr:2804122-2804995

Fragment	Fragment position	Primer sequence	T_m	ΔG of complete hybridization	ΔG for homodimer formation
ProVWX_1	Chr:2804123-2804700	cgatttgctctcagccctt	55.3 °C	−38.36 kcal/mol	−4.74 kcal/mol
ProVWX_2	Chr:2804701-2804817	gcctcagattctcagtatgttagg	54.6 °C	−40.88 kcal/mol	−4.67 kcal/mol
ProVWX_3	Chr:2804818-2804994	aaactgggctatcgcttgg	55.3 °C	−38.99 kcal/mol	−3.61 kcal/mol

Table 2
The TaqMan probe for fragment ProVWX_2

Probe	ProVWX_2.probe
Sequence	5′-/6-FAM/AACATACTG/ZEN/AGAATCTGAGGCAACCCCTGA/IBFQ/-3′
Annealing position	Chr:2804713-2804742
T_m	62.3 °C

Table 3
IC_Primer for the Internal control fragment

Fragment	Internal control
Fragment position	Chr:2805902-2806570
Primer name	IC_primer
Primer sequence	GGATATGGCTGGCGAGAAG
T_m	55.5 °C
ΔG of complete hybridization	−38.38 kcal/mol
ΔG for homodimer formation	−3.91 kcal/mol

here, the primers for the internal control are the ProVWX_2 primer and the IC_primer (Table 3). Since ProVWX_2 is an anchor fragment, a new TaqMan probe does not need to be designed in this case.

5. Determine the DNA sequences of the interactions to be tested by 3C-qPCR. For the example used in this section, the interactions are (Table 4):

Table 4
DNA sequences of ProVWX interactions

Interaction	Sequence
ProVWX_2-ProVWX_1	gcctcagattctcagtatgttagggtagaaaaaag tgactatttccattgggtaatatatcgacatagac aaataaaggaatctttctattgcatgaatattgatgcga gcaaatcgccggaatttcccgtgatataagggctgagagcaaatcg
ProVWX_2-ProVWX_3	gcctcagattctcagtatgttagggtagaaaaaagtgactatttccattgggtaatatat cgacatagacaaataaaggaatctttctattgcatgatgacaaatatctcgccttct tcaatggccagactggcgtcttttacgccaagcgatagcccagttt
ProVWX_2-InternalControl	gcctcagattctcagtatgttagggtagaaaaaagtgactatttccattgggtaatatatcga catagacaaataaaggaatctttctattgcatggcatcaagcagtggacgaataattttcgc cgctcgcggacttctcgccagccatatcc

6. Design oligonucleotides to synthesize the Interaction fragments by overlap extension PCR (oePCR) (*see* **Note 10**) according to the scheme provided in Fig. 3. In the case of the example presented in this text, the oligonucleotides are (Table 5):
7. Synthesize the anchor, test, and internal control fragments (Tables 4 and 5) separately with the following reaction (Table 6) and program (Table 7):
8. Purify the anchor and test fragments synthesized in **step** 7 of Subheading 3.3.1 using the GFX™ PCR DNA and Gel Band Purification Kit and elute the molecules in 10.0 μL of 10.0 mM Tris, pH 8.0.
9. Join the test and anchor fragments (Table 5) to synthesize the interaction fragments (Table 4) with the reaction (Table 8) and program (Table 9) described below (*see* **Note 10**):
10. Resolve the complete volume of the reaction performed in **step 9** of Subheading 3.3.1 on a 3% agarose gel. Purify the interaction fragments using the GFX™ PCR DNA and Gel Band Purification Kit, and elute the fragments in 15.0 μL of 10.0 mM Tris, pH 8.0.
11. Use the Qubit® dsDNA HS Assay Kit to determine the concentration of the Interaction fragments (*see* **Note 3**).
12. For each Interaction fragment, determine the mass in ng that corresponds to 1.0 pmol of the molecule. The NEBioCalculator® made available online by New England Biolabs is a suitable program for mass to mole calculations for double-stranded DNA on the basis of the length of the molecule in base pairs. In the case of the ProVWX Interaction fragments (Table 10):

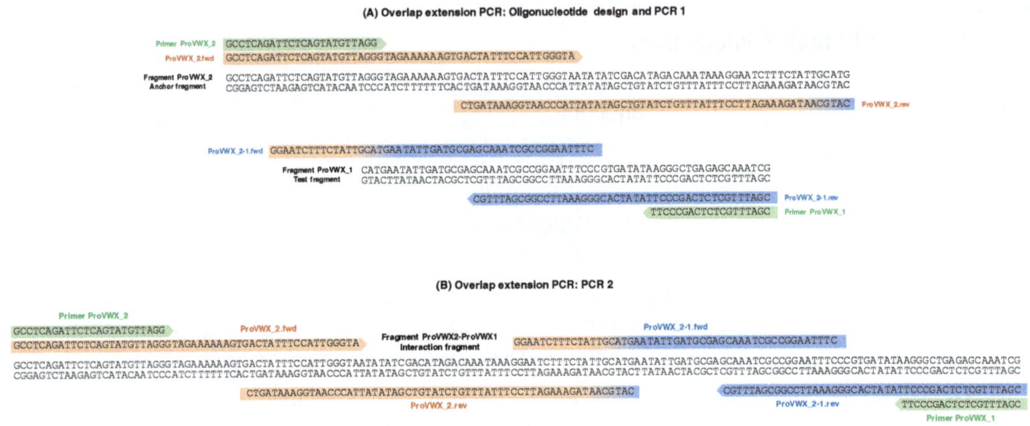

Fig. 3 Overlap extension PCR (oePCR) (*see* **Note 10**) for the synthesis of ProVWX Interaction fragments. **Panel A:** To synthesize short DNA fragments, such as the anchor and test fragments, design a pair of oligonucleotides that are perfectly complementary to each other over a ~20 bp-long region at the 3′ end. During oePCR 1, the oligonucleotides hybridize to each other at the 3′ ends. The non-hybridized 5′ regions act as templates for the extension of the 3′ ends to synthesize a merged double-stranded DNA (dsDNA) construct. Note that the 5′ end of the forward oligonucleotide of the test fragment carries as extension (blue oligonucleotide with an orange extension) that is complementary to the 5′ end of the reverse primer of the anchor fragment (orange oligonucleotide with blue shading). As a result, the test fragment will carry a 20 bp-extension that is identical to part of the anchor fragment. **Panel B:** The anchor and test fragments are merged in oePCR 2—a reaction that uses two double-stranded DNA templates (see Table 8)—to synthesize the Interaction fragment. In the first cycle of the PCR program, the DNA templates are melted during denaturation at 98 °C, upon lowering the temperature to 63 °C, a fraction of the single-stranded molecules anneal such that the extension of the test fragment hybridizes to its complementary site on the single-stranded anchor fragment. Given that the template molecules are double-stranded, the complementary annealing may involve a pair of 5′ ends, or a pair of 3′ ends. In the latter case, the 3′ ends are extended to synthesize a merged template. In subsequent cycles, primers hybridize to the ends of the merged template to amplify the longer Interaction fragment construct

13. Prepare a 10.0 fmol/μL 3C-qPCR control library (in triplicate) by pooling 1.0 pmol of each Interaction fragment and adjusting the final volume of the pool to 100.0 μL with 10.0 mM Tris, pH 8.0.
14. Serially dilute the 3C-qPCR control libraries prepared in **step 13** of Subheading 3.3.1 with 10.0 mM Tris, pH 8.0 to concentrations of 1.0 fmol/μL, 0.1 fmol/μL, 0.01 fmol/μL, and 0.001 fmol/μL.
15. Store the 3C-qPCR control library preparations at 4 °C to avoid repeated freezing and thawing.

Table 5
Oligonucleotides to synthesize ProVWX Interaction fragments by overlap extension PCR (oePCR)

Fragment	Oligonucleotides	Oligonucleotide sequences
ProVWX_2 (Anchor)	ProVWX_2.fwd ProVWX_2.rev	gcctcagattctcagtatgttagggtagaaaaaagtgactatttccattgggta catgcaatagaaagattcctttatttgtctatgtcgatatattacccaatggaaatagtc
ProVWX_2-1 (Test)	ProVWX_2-1.fwd ProVWX_2-1.rev	ggaatctttctattgcatgaatattgatgcgagcaaatcgccggaatttc cgatttgctctcagcccttatatcacgggaaattccggcgatttgc
ProVWX_2-3 (Test)	ProVWX_2-3.fwd ProVWX_2-3.rev	ggaatctttctattgcatgatgacaaatatctcgccttcttcaatggccagactg aaactgggctatcgcttggcgtaaaagacgccagtctggccattgaagaaggcga
ProVWX_2-InternalControl (Internal Control)	ProVWX_2-InternalControl.fwd ProVWX_2-InternalControl.rev	ggaatctttctattgcatggcatcaagcagtggacgaataattttcgccgctc ggatatggctggcgagaagtccgcgagcggcgaaaattattcgtccactg

Table 6
oePCR mix for the synthesis of the ProVWX anchor, test, and internal control fragments

Reagent	Volume
Nuclease free water	5.9 μL
5× HF Buffer	2.0 μL
2.0 mM dNTPs	1.0 μL
10.0 μM Forward primer	0.5 μL
10.0 μM Reverse primer	0.5 μL
Phusion polymerase	0.1 μL

Table 7
oePCR program for the synthesis of the ProVWX anchor, test, and internal control fragments

Step	Temperature	Duration (mm:ss)	Number of cycles
Initial denaturation	98 °C	01:00	1
Denaturation Annealing/extension	98 °C 63 °C	00:15 00:20	30
Final extension	72 °C	05:00	1
Hold	4 °C	∞	1

Table 8
oePCR mix for the synthesis of the ProVWX Interaction fragments

Reagent	Volume
Nuclease free water	27.5 μL
5× HF Buffer	10.0 μL
2.0 mM dNTPs	5.0 μL
Anchor fragment	1.0 μL
Test fragment (or Internal Control fragment)	1.0 μL
10.0 μM Forward primer of the Anchor fragment	2.5 μL
10.0 μM Reverse primer of the Test fragment (or the Internal Control fragment)	2.5 μL
Phusion polymerase	0.5 μL

Table 9
oePCR program for the synthesis of the ProVWX Interaction fragments

Step	Temperature	Duration (mm:ss)	Number of cycles
Initial denaturation	98 °C	01:00	1
Denaturation Annealing/extension	98 °C 63 °C	00:15 00:20	30
Final extension	72 °C	05:00	1
Hold	4 °C	∞	1

Table 10
The mass in ng that corresponds to 1.0 pmol of the ProVWX Interaction fragments

Interaction fragment	Length (bp)	ng/pmol
ProVWX2-ProVWX1	155	95.82
ProVWX2-ProVWX3	163	100.8
ProVWX2-InternalControl	154	95.20

3.3.2 Experimentation

1. To study the relative interaction frequency between the anchor fragment and a test fragment, set up the following samples in a 96-well plate:
   - Four biological replicates of a 3C library (*see* **Note 11**). Test each biological replicate in triplicate. (Count: 12 samples)
   - The relative interaction frequency between anchor and test fragments may be affected by cell culture conditions. To detect such a change, include four biological replicates of

Table 11
Reaction setup for 3C-qPCR

Component	Volume
Template/sample (*see* **step 1** of Subheading 3.3.2)	0.5 μL
[a]2X PrimeTime® Gene Expression Master Mix	5.0 μL
[a]Anchor primer (10.0 μM)	0.5 μL
[a]Test primer (10.0 μM)	0.5 μL
[a]TaqMan probe (10.0 μM)	0.15 μL
[a]Nuclease free water	3.35 μL

[a] Assemble these components into a master mix for the required number of reactions plus a 10% excess. Add 9.5 μL of the master mix to each sample well in a single pipetting step

Table 12
3C-qPCR program

Step	Temperature	Duration (mm:ss)	Repeats
Polymerase activation	95 °C	03:00	1
Denaturation Annealing/extension Plate reading (SYBR/FAM channel)	95 °C 60 °C N/A	00:15 01:00 N/A	45

3C libraries prepared from cells cultured under a different environmental condition. Test each biological replicate in triplicate. (Count: 12 samples)

- 3C-qPCR control libraries at concentrations of 1.0 fmol/μL, 0.1 fmol/μL, 0.01 fmol/μL, and 0.001 fmol/μL (*see* **step 14** of Subheading 3.3.1). (Count: 12 samples)
- No template controls. (Count: 3 samples)

2. Set up the reaction for each sample as described in Table 11:
3. Seal the 96-well plate with a clear film.
4. Centrifuge the plate briefly to collect the reactions at the bottom of each well and to remove air bubbles.
5. Place the 96-well plate in a CFX96 Touch Real-Time PCR Detection System and run the following program (Table 12):
6. Repeat the experiment for all anchor–test pairs and for the Internal Control.

3.3.3 Data Analysis

1. Export the data file of the qPCR run (in its *.zpcr format) from the CFX96 Touch Real-Time PCR Detection System. Open the file with the CFX Maestro software for CFX Real-Time PCR Instruments from Bio-Rad and save it in its *.pcrd form.
2. Use Plate Setup>View/Edit Plate to visualize the plan of the 96-well plate.
   - Highlight the wells that were left empty during the 3C-qPCR run and select the 'Exclude Wells in Analysis' option.
   - Assign the 'Sample Type' of the wells that contained 3C libraries as 'Unknown'.
   - Assign the 'Sample Type' of the wells that contained the 3C-qPCR control library as 'Standard'. Using the 'Load' option, input the concentration of the control library for each of the Standard wells. This is the *Starting Quantity* (*SQ*) of the sample and enables the program to plot a Standard Curve of the *Cq* value against the logarithm of the *SQ* for quantifying 'Unknown' samples (*see* **Note 12**). In a reliable run, the R^2 value of the Standard Curve is higher than 0.980.
   - Assign the 'Sample Type' of the no-template control wells of the 3C-qPCR run as 'NTC'. In a reliable run, these samples should not have a *Cq* value.
   - Add the 'Target Name' and 'Sample Name' for each well. The 'Target Name' is the interaction being tested for. In the example presented here, the 'Target Names' are: ProVWX2-ProVWX1, ProVWX2-ProVWX3, ProVWX2-CrosslinkCtrl. The 'Sample Name' corresponds to the identity of the template.
3. Export the 3C-qPCR data to a *.xlsx format using Export>Export All Data Sheets> Excel 2007 (*.xlsx).
4. The 'Quantification Cq Results' datasheet lists the *SQ* (Column K) (*see* **Notes 12** and **13**) of the interaction being tested for in all wells included in the analysis (Column B). Recall that each 3C library is tested in triplicate (*see* **step 1** of Subheading 3.3.2). Use the *SQ* values computed for these reactions to calculate the average *SQ* (SQ_{avg}) for each 3C library and the error in the measurement.
5. For each 3C library, divide the SQ_{avg} of the test interaction with the SQ_{avg} of the Internal Control to calculate the normalized relative interaction frequency (RIF_{norm}) for the test interaction (*see* **Notes 9** and **12**).
6. Use the RIF_{norm} values of biological replicates to determine the average relative interaction frequency (RIF_{avg}) for the test interaction and the error in the measurement.

7. The RIF_{avg} values and errors of all test interactions can be used to plot a graph of the relative interaction frequency between the anchor and test fragments against the position of the test fragment in the chromosome. The RIF_{norm} values and errors can also be plotted in the same graph.

3.4 3C-Seq

3.4.1 NGS Library Preparation

1. Remove glycogen and excess salt from the 3C library by adding 60.0 μL of KAPA HyperPure beads with 20.0 μL of the 3C library in a low-retention 1.5 mL microfuge tube (*see* **Note 14**). Vortex to mix. Incubate the sample at 25 °C in a Thermomixer for 5 min.
2. Wash the beads according to the manufacturer's instructions and elute the 3C library in 36.0 μL of 10.0 mM Tris, pH 8.0 (*see* **Note 15**).
3. Use 1.0 μL of the eluted library to determine DNA concentration using the Qubit® dsDNA HS Assay Kit. Proceed to the next step if more than 500 ng of the 3C library is available.
4. Add 5.0 μL of 10× KAPA Frag buffer and 10.0 μL of the KAPA Frag enzyme to the 3C library on ice. Vortex gently to mix and incubate the tube in a Thermomixer set at 37 °C for 45 min.
5. Transfer the fragmentation reaction to ice and set the temperature of the Thermomixer to 65 °C.
6. Mix 7.0 μL of the End Repair and A-Tailing Buffer and 3.0 μL of the End Repair and A-Tailing Enzyme Mix on ice (supplied with the KAPA HyperPlus kit). Transfer the mix (10.0 μL) to the fragmentation reaction and vortex gently.
7. Place the reaction in the Thermomixer at 65 °C for 30 min.
8. Transfer the reaction to ice and set the temperature of the Thermomixer to 20 °C.
9. Add the following to 60.0 μL of end-repaired and A-tailed DNA: 5.0 μL of 15.0 μM Illumina® paired-end sequencing adapter (barcoded), 5.0 μL of nuclease-free water, 30.0 μL of ligation buffer (KAPA HyperPlus), and 10.0 μL of DNA Ligase (KAPA HyperPlus). To sequence several 3C libraries in a single Illumina® lane, ligate each library to a different sequencing adapter.
10. Vortex gently and incubate the reaction at 20 °C for 2 h.
11. Clean up the adapter-ligated 3C library (hereafter referred to as the 3C-Seq library) with 88.0 μL of KAPA HyperPure beads (*see* **Note 14**). Wash the beads according to the manufacturer's instructions and elute the library in 50.0 μL of 10.0 mM Tris, pH 8.0 for double-sided size selection.
12. Add 35.0 μL of KAPA HyperPure beads to the 3C-Seq library. Incubate the mix at 25 °C for 5 min to allow fragments longer than 450 bp to immobilize onto the SPRI beads (*see* **Note 14**).

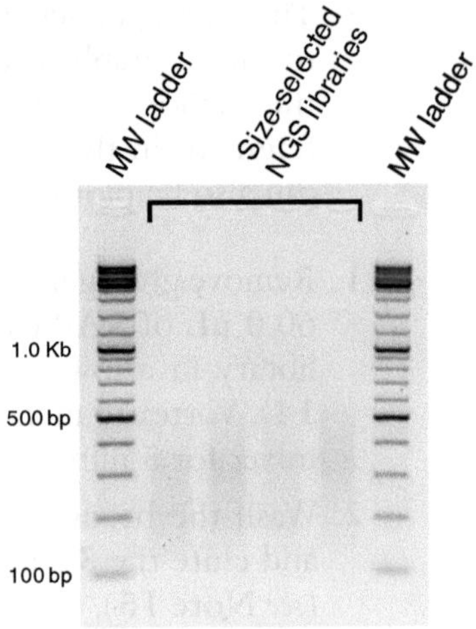

Fig. 4 Typical run of a size-selected NGS library on a 2% agarose gel. A size-selected NGS library runs as a smear of DNA fragments between 150 and 600 bp

13. Place the tube on a magnetic rack until the solution clears.
14. Transfer the clear supernatant, now containing fragments shorter than ~450 bp, into a 1.5 mL low-retention microfuge tube.
15. Add 10.0 μL of KAPA HyperPure beads to the supernatant to immobilize fragments longer than ~250 bp to the SPRI beads (*see* **Note 14**). Place the sample at 25 °C for 5 min.
16. Wash the beads according to the manufacturer's instructions and elute the size-selected 3C-Seq library in 20.0 μL of 10.0 mM Tris, pH 8.0. The eluted 3C-Seq library contains molecules between ~250 and ~450 bp in length.
17. Use 1.0 μL of the 3C-Seq library to determine DNA concentration using the Qubit® dsDNA HS Assay Kit. The concentration of the library should be between 5 and 20 ng/μL.
18. Assess the quality of the 3C-Seq library by determining the size distribution of the DNA fragments using agarose gel electrophoresis (*see* **Note 16**) (*see* Fig. 4). 1.0–2.0 μL of the 3C-Seq library is sufficient for this purpose.
19. Libraries that are marked with distinct Illumina® paired-end sequencing adapters can be pooled for sequencing. The ratio at which the libraries are pooled determines the fraction of total reads available for each library.

3.4.2 Data Analysis and Generation of Contact Maps

Reads from Illumina® paired-end sequencing of a 3C library are mapped to restriction fragments of the reference genome. Ligation junctions are assigned to a pair of 'bins' by the midpoint coordinate of the corresponding restriction fragments. In this context, a 'bin' refers to a segment of a computationally partitioned genomic sequence that spans between several bps and several kbs. For instance, the 4,641,652 bp-long chromosome of *E. coli* K-12 MG1655 (GenBank: NC_000913.3) can be partitioned into 4641 bins spanning 1 kb and an additional shorter bin that is 652 bp long. The interaction data is contained in a symmetric square matrix where every value represents the interaction frequency between a pair of bins. The final, normalized contact map is obtained after balancing the matrix by equalizing the sum of all its rows and columns. Detailed methods for processing 3C-based data are available in [20].

The resolution of a chromosome contact map is determined by its bin size. Smaller bins increase the resolution of the map at the cost of a decreased signal-to-noise ratio. Thus, the selected bin size is a trade-off between the two factors and is heavily influenced by read depth. In our experience, 15 million valid reads of ligation junctions are sufficient to generate a normalized contact map of the *E. coli* chromosome with a resolution of 5 kb. The bin size is further limited by the choice of restriction enzyme. Ideally, the length of all the fragments that are generated by the restriction digestion of the chromosome is shorter than the chosen bin size. The occurrence of larger fragments produces empty bins in the contact map that result in the misrepresentation of chromosome contact profiles.

4 Notes

1. 2.5 M glycine should be stored at 20–25 °C. Glycine crystallizes out of solution at lower temperatures.
2. The restriction enzyme selected for 3C studies should function optimally at 37 °C (or at an even lower temperature—for instance, the optimal temperature for CviAII activity is 25 °C) as restriction digestion at higher temperatures leads to premature reverse cross-linking of DNA-protein complexes. To ensure a good digestion efficiency during library preparation, it is important that the activity of the selected enzyme is not blocked by DNA methylation. Restriction enzymes define the resolution of the 3C library. For 3C-qPCR it is necessary to ensure that the enzyme digests the target region sufficiently well that it separates predicted regulatory elements from each other and digests the target region into fragments of an optimal size to be able to resolve structural changes. For 3C-Seq, digestion of the fixed chromosome by the restriction enzyme should, generally, produce fragments that are shorter than the chosen 3C map resolution; long restriction fragments introduce biases in the contact matrix.

3. Conventional UV spectroscopy determines the concentration of dsDNA in solution based on the absorbance of the sample at 260 nm. The measurement requires a negative control of the buffer in which the dsDNA is dissolved. As such, the technique cannot be used to measure dsDNA concentration in 3C-Seq and 3C-qPCR samples that contain the cell lysate (*see* **step 3** of Subheading 3.2.4), or glycogen and ethanol at an unknown concentration (*see* **step 20** of Subheading 3.2.6, and **step 11** of Subheading 3.3.1). The use of the Qubit® dsDNA HS Assay Kit and the Qubit® fluorometer from ThermoFisher Scientific overcomes this limitation. The kit is based on the use of a fluorophore that exhibits specificity to dsDNA and shows fluorescence enhancement upon interaction with dsDNA. It can, therefore, be used to accurately determine the concentration of dsDNA in 3C samples.
4. 25:24:1 phenol:chloroform:isoamyl alcohol and chloroform are neurotoxic organic solvents. Perform all steps requiring these solvents under a fume hood with necessary personal protective equipment.
5. Ensure that the *E. coli* cells are completely resuspended in 1× PBS prior to the addition of methanol. In an 80% methanol solution, even vigorous pipetting and vortexing only break a bacterial pellet into large clumps.
6. Formaldehyde fixation of *Saccharolobus solfataricus* requires the reaction to be carried out at a temperature as close as possible to 80 °C—the optimal growth temperature of the organism. This is meant to prevent modifications to chromosome structure as a result of temperature stress. However, the fixation reaction is inefficient at higher temperatures. To strike a balance between the conflicting constraints, fixation of *Saccharolobus solfataricus* is initiated at 80 °C with the temperature being gradually decreased to 75 °C.
7. 3C-qPCR primers are designed such that the primers anneal along the same strand of the chromosome and are oriented uni-directionally (*see* Fig. 2) [14]. This design ensures that a fluorescent signal will only be observed in a 3C-qPCR experiment if, firstly, the (local) chromosome is efficiently digested and, secondly, the fragments are re-ligated in a manner that re-orients one of two, and hence, its primer. This enables the primer pair to extend toward each other as required for a successful PCR (*see* Fig. 2) [14]. The 3C-qPCR primer design described here does not detect re-ligations where the fragments orient in a manner that causes the primers to extend away from each other, or in the same direction [14].

8. The TaqMan probe is designed to hybridize to the strand complementary to the primer-annealing strand. This way, the probe is only hydrolyzed when elongation from the test primer proceeds across the ligation junction toward the anchor fragment. The design ensures that any fluorescence signal detected during 3C-qPCR arises as a result of a ligation between the anchor and test fragment (*see* Fig. 2) [14].
9. Differences in the efficiencies of restriction digestion and ligation between 3C libraries and variation in the concentration of the template between reactions necessitate that the data obtained from 3C-qPCR runs are normalized with an Internal Control. The Internal Control is an interaction that is expected to occur at the same relative interaction frequency across all the libraries evaluated in a 3C-qPCR study. The *Starting Quantity* of the Internal Control can, therefore, be used to normalize the *Starting Quantities* of the test interactions.
10. Overlap extension PCR (oePCR) enables the merging of a pair of DNA molecules, with the requirements that the molecules share complementarity at the 3′ end of the single-stranded template (*see* Fig. 3), and that these ends anneal specifically to each other at the temperature used for the annealing/extension steps of the PCR program. In oePCR, the non-hybridized 5′ region of the single-stranded molecules act as templates for the extension of the 3′ ends to synthesize a merged double-stranded DNA (dsDNA) construct. For the synthesis of short DNA fragments up to ~130 bp in length, a pair of oligonucleotides that share a ~20 bp-long complementary region at the 3′ end can be extended (*see* Fig. 3, panel A; **steps 6** and 7 of Subheading 3.3.1). A conventional template is not necessary for such a reaction (*see* **step** 7 of Subheading 3.3.1). In the case of longer DNA constructs, the setup requires that the dsDNA molecules to be merged are added to the reaction. In conventional terms, two templates are used for the reaction (*see* **step 9** of Subheading 3.3.1). The first cycle of the PCR program involves melting of dsDNA molecules, the annealing of complementary 3′ ends of the single-stranded templates, and the extension of the 3′ ends to synthesize a merged template. During subsequent cycles, primers that hybridize to the ends of the merged templates amplify the long DNA construct (*see* Fig. 3, panel B; **step 9** of Subheading 3.3.1).
11. 3C libraries are considered as biological replicates when the libraries are prepared from cells that are cultured, fixed, and processed for 3C under identical experimental conditions, but prepared independently of each other.
12. The CFX Maestro software does not assign a unit to the *Starting Quantity* of the 'Standard' and the 'Unknown' samples (*see*

steps 2 and **4** of Subheading 3.3.3). As a user, be mindful that identical quantification units are used in the 3C-qPCR experiments.

13. The Starting Quantity (*SQ*) for each well can also be calculated using the formula [14]:

$$SQ = 10^{((Cq - \text{Intercept})/\text{Slope})}$$

where,

Cq is the cycle of quantitation exported in Column H of the 'Quantification Cq Results' file.

Intercept and Slope correspond to the Y-Intercept and Slope of the Standard Curve are exported in Column E and Column D, respectively, of the 'Standard Curve Results' datasheet.

14. Solid Phase Reverse Immobilization (SPRI) beads size-selectively, and reversibly bind DNA molecules. The immobilization of DNA onto the surface of the beads is dependent on the concentration of polyethylene glycol (PEG) in the DNA-beads solution. Heavier DNA molecules immobilize onto the bead surface at low PEG concentrations. Shorter molecules require higher concentrations. Suspending DNA-bound SPRI beads in a buffer that lacks PEG elutes DNA from the bead surface. Since SPRI beads are supplied in a PEG-containing storage buffer, the cut-off for the length of the molecules that bind to the bead surface is determined by the volumetric ratio of the bead solution to the DNA solution. At a 3.0× volumetric ratio of the beads to DNA (*see* **step 1** of Subheading 3.4.1), DNA fragments of all lengths are expected to immobilize onto the bead surface. In **step 11** of Subheading 3.4.1, the adapter-ligated 3C library is immobilized onto the beads at a 0.8× volumetric ratio of the bead solution to the DNA solution. In this case, however, the molecular weight cut-off cannot be clearly defined since the ligation buffer of the DNA solution contains PEG6000 that interferes with size selection. The double-sided selection of the adapter-ligated 3C library performed at **steps 12–16** of Subheading 3.4.1, extracts DNA molecules between 250 and 450 bp in length. In the first step, a 0.7× volumetric ratio of beads to DNA retains fragments shorter than 450 bp in the supernatant. Next, separating the supernatant fraction and supplementing it with the beads solution to a final PEG buffer:DNA solution ratio of 0.9, immobilizes fragments between 250 and 450 bp in length onto the bead surface. The size-selected library is then eluted from the beads in 10.0 mM Tris, pH 8.0.

15. Elution of the 3C library in nuclease-free water at **step 2** of Subheading 3.4.1 reduces the efficiency of library fragmentation (*see* **step 4** of Subheading 3.4.1).
16. On a 2% agarose gel, a purified, size-selected NGS library runs as a smear between 150 and 600 bp (*see* Fig. 4). A 'laddered' pattern in the smear or a strong DNA signal at ~120 bp indicate the presence of PCR duplicates and adapter dimers, respectively, both of which consume a significant proportion of the read depth and reduce the quality of the sequencing run.

Acknowledgments

This work was supported by grants from the Netherlands Organization for Scientific Research [VICI 016.160.613 and OCENW. GROOT.2019.012], the FOM Foundation for Fundamental Research on Matter program 'Crowd management: The physics of genome processing in complex environments and the Human Frontier Science Program (HFSP) [RGP0014/2014].

References

1. Marbouty M, Le GA, Cattoni DI et al (2015) Condensin- and replication-mediated bacterial chromosome folding and origin condensation revealed by Hi-C and super-resolution imaging. Mol Cell 59:588–602
2. Le TBK, Imakaev MV, Mirny LA et al (2013) High-resolution mapping of the spatial organization of a bacterial chromosome. Science (80-) 342:731–734
3. Lioy VS, Cournac A, Marbouty M et al (2018) Multiscale structuring of the E. coli chromosome by nucleoid-associated and condensin proteins. Cell 172:771–783.e18
4. Valk RA van der, Vreede J, Qin L et al (2017) Mechanism of environmentally driven conformational changes that modulate H-NS DNA-bridging activity. elife 6:e27369
5. Kotlajich MV, Hron DR, Boudreau BA et al (2015) Bridged filaments of histone-like nucleoid structuring protein pause RNA polymerase and aid termination in bacteria. elife 2015: e04970
6. Takemata N, Samson RY, Bell SD (2019) Physical and functional compartmentalization of archaeal chromosomes. Cell 179: 165–179.e18
7. Takemata N, Bell SD (2021) Multi-scale architecture of archaeal chromosomes. Mol Cell 81: 473–487.e6
8. Le TB, Laub MT (2016) Transcription rate and transcript length drive formation of chromosomal interaction domain boundaries. EMBO J 35:1582–1595
9. Tran NT, Laub MT, Le TBK (2017) SMC progressively aligns chromosomal arms in Caulobacter crescentus but is antagonized by convergent transcription. Cell Rep 20:2057–2071
10. Wang X, Brandão HB, Le TBK et al (2017) Bacillus subtilis SMC complexes juxtapose chromosome arms as they travel from origin to terminus. Science (80-) 355:524–527
11. Cockram C, Thierry A, Gorlas A et al (2021) Euryarchaeal genomes are folded into SMC-dependent loops and domains, but lack transcription-mediated compartmentalization. Mol Cell 81:459–472.e10
12. Dame RT, Rashid FZM, Grainger DC (2020) Chromosome organization in bacteria: mechanistic insights into genome structure and function. Nat Rev Genet 21:227–242

13. Dekker J, Rippe K, Dekker M et al (2002) Capturing chromosome conformation. Science (80-) 95:1306–1311
14. Hagege H, Klous P, Braem C et al (2007) Quantitative analysis of chromosome conformation capture assays (3c-qpcr). Nat Protoc 2:1722–1733
15. Lieberman-Aiden E, Berkum NL Van, Williams L et al (2009) Comprehensive mapping of long-range interactions reveals folding principles of the human genome. Science ('80-) 326: 289–293
16. Nagano T, Lubling Y, Várnai C et al (2017) Cell-cycle dynamics of chromosomal organization at single-cell resolution. Nature 547: 61–67
17. Nagano T, Lubling Y, Stevens TJ et al (2013) Single-cell Hi-C reveals cell-to-cell variability in chromosome structure. Nature 502:59–64
18. Stevens TJ, Lando D, Basu S et al (2017) 3D structures of individual mammalian genomes studied by single-cell Hi-C. Nature 544:59–64
19. Crémazy FG, Rashid FZM, Haycocks JR et al (2018) Determination of the 3D genome organization of bacteria using Hi-C. In: Methods in molecular biology, pp 3–18
20. Hofmann A, Heermann DW (2018) Processing and analysis of Hi-C data on bacteria. In: Methods in molecular biology, pp 19–31

Chapter 2

Micrococcal Nuclease Digestion Assays for the Analysis of Chromosome Structure in Archaea

Hugo Maruyama

Abstract

The digestion of chromosomes using micrococcal nuclease (MNase) enables the analysis of their fundamental structural units. For example, the digestion of eukaryotic chromatin using MNase results in laddered DNA fragments (~150 bp increment), which reflects the length of the DNA wrapped around regularly spaced nucleosomes. Here, we describe the application of MNase to examine the chromosome structure in Archaea. We used *Thermococcus kodakarensis*, a hyperthermophilic euryarchaeon that encodes proteins homologous to eukaryotic histones. Methods for chromosome extraction and agarose gel electrophoresis of MNase-digested DNA including small fragments (~30 bp) are also described.

Key words Archaea, Chromosome, Histone, Nucleosome, Micrococcal nuclease

1 Introduction

Micrococcal nuclease (MNase) is an endonuclease as well as an exonuclease that has been widely used to analyze chromatin structure in eukaryotic cells [1]. When acting on chromatin, it preferentially digests the linker DNA between nucleosomes rather than the DNA wrapped around a histone octamer within a nucleosome. As MNase requires calcium for its activity, the presence of EDTA can inhibit digestion [2]. MNase digestion of eukaryotic chromatin followed by agarose gel electrophoresis resulted in a laddered pattern of DNA with a minimum of ~150 bp and steps of 150–200 bp, which reflects the DNA length within a nucleosome core and a regularly spaced eukaryotic nucleosome [2].

Archaea is a domain that is phylogenetically distinct from Bacteria and Eukarya [3]. Although they are prokaryotes, archaeal species use sets of DNA-binding proteins that are different from those of bacteria [4]. In addition, Archaea constitute a variety of lineages that use different types of chromosomal proteins. For example, most species in the phylum Euryarchaeota encode either

Eveline Peeters and Indra Bervoets (eds.), *Prokaryotic Gene Regulation: Methods and Protocols*,
Methods in Molecular Biology, vol. 2516, https://doi.org/10.1007/978-1-0716-2413-5_2,

archaeal histones (most species in Euryarchaeota) or HTa (species in Thermoplasmatales), a protein homologous to the bacterial nucleoid-associated protein HU, whereas species in Crenarchaeota utilize other DNA-binding proteins such as Alba and Cren7 to build the chromosome structure [5, 6]. The MNase digestion assay has revealed a fundamental difference between Euryarchaeota and Crenarchaeota with respect to the way they organize their genomic DNA into a chromosome at the most fundamental level. Chromosomal DNA in Euryarchaeota is wrapped around the DNA-binding protein (either histone or HTa), whereas that in Crenarchaeota is not [7–13]. In this chapter, the application of MNase digestion assays in analyzing the chromosome structure of archaeal species will be described in detail. In particular, *Thermococcus kodakarensis*, a hyperthermophilic euryarchaeon, is used as an example.

2 Materials

2.1 Cell Culture

1. Archaeal strain: *T. kodakarensis* KOD1 (*see* **Note 1**).
2. Anaerobic culture tubes or bottles (*see* **Note 2**).
3. Anaerobic chamber (*see* **Note 3**).
4. 0.8x Artificial seawater (ASW): 16 g/L NaCl, 2.4 g/L $MgCl_2{\cdot}6H_2O$, 4.8 g/L $MgSO_4{\cdot}7H_2O$, 0.8 g/L $(NH_4)_2SO_4$, 0.16 g/L $NaHCO_3$, 0.24 g/L $CaCl_2{\cdot}2H_2O$, 0.4 g/L KCl, 0.34 g/L KH_2PO_4, 40 mg/L NaBr, 16 mg/L $SrCl_2{\cdot}6H_2O$, 8 mg/L $Fe(NH_4)$ citrate [14] (*see* **Note 4**).
5. ASW-YT-S^0: 0.8x ASW, 5.0 g/L yeast extract, 5.0 g/L tryptone, 0.8 mg/L resazurin, 0.2% elemental sulfur. Prior to inoculation, add $Na_2S{\cdot}9H_20$ to the medium until the red color of resazurin disappeared (*see* **Note 5**).

2.2 Chromosome Extraction

1. Extraction buffer: 25 mM HEPES (pH 7.0), 15 mM $MgCl_2$, 100 mM NaCl, 0.4 M sorbitol, and 0.5% Triton X-100 (*see* **Note 6**)

2.3 MNase Digestion

1. MNase buffer: 20 mM Tris-HCl (pH 7.4), 5 mM NaCl, and 2.5 mM $CaCl_2$
2. Micrococcal nuclease (*see* **Note 7**)
3. Bovine pancreatic ribonuclease (RNase A)
4. 10x Stop solution: 2% SDS, 100 mM EDTA, 100 mM EGTA
5. Proteinase K (20 mg/mL)
6. Phenol:chloroform:isoamyl alcohol mixture (25:24:1)
7. 100% and 70% ethanol

2.4 Agarose Gel Electrophoresis

1. Low-melting-temperature agarose (*see* **Note 8**)
2. 1x TBE buffer: 89 mM Tris, 89 mM boric acid, and 2 mM EDTA (*see* **Note 9**)
3. Low-molecular-weight DNA marker that covers the size of DNA fragments as small as ~30 bp
4. Ethidium bromide (*see* **Note 10**)

3 Methods

3.1 Cell Culture

Cells can be cultured in any desired medium (such as ASW-YT-S^0 or ASW-YT-pyruvate) and harvested at any growth phase. Here is an example using ASW-YT-S^0. Typical growth curves of *T. kodakarensis* are shown in Fig. 1.

1. Cultivate *T. kodakarensis* KOD1 under anaerobic conditions at 85 °C in the nutrient-rich medium ASW-YT-S^0 (*see* **Note 11**).
2. Harvest the cells at desired growth phase.
3. Centrifuge the cell culture for 5 min at 3000 × *g* at 4 °C.
4. Discard the supernatant.
5. Wash the cells with 0.8 × ASW of approximately 5–10% of the original culture volume.
6. Centrifuge the cells for 5 min at 3000 × *g* at 4 °C and discard the supernatant (*see* **Note 12**).
7. Resuspend the cells in 0.8 × ASW.
8. Divide the cell solution into 1.5 mL tubes so that each tube contains cells equivalent to approximately 30 mL of the cell culture medium (*see* **Note 13**).

3.2 Chromosome Extraction

This protocol was optimized for the chromosome extraction of *T. kodakarensis* (and other species in the order Thermococcales). To extract chromosomes from other archaea, the extraction buffer and method should be determined empirically.

1. Resuspend the fresh or thawed cell pellet derived from 30 mL culture in a small volume of 0.8 × ASW. The volume should be approximately 10% of the reaction volume in the next step (*see* Fig. 2a) (*see* **Note 14**).
2. Disrupt the cells by adding the extraction buffer to achieve a final concentration of ~10^{10} cells/mL (*see* **Note 15**).
3. Incubate for 10 min at 4 °C with occasional pipetting (*see* **Note 16**).
4. Centrifuge the tube at 14,000 × *g* for 20 min.

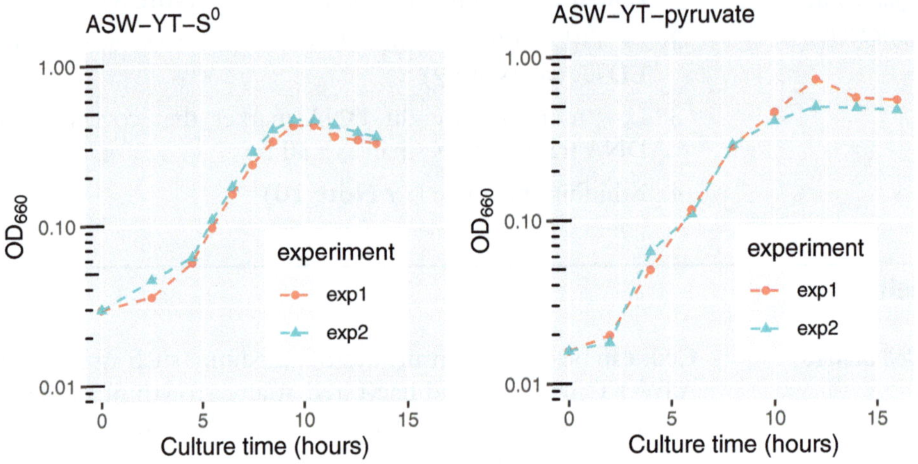

Fig. 1 Typical growth curves of *T. kodakarensis*. Cells were cultured at 85 °C in ASW-YT-S^0 (left) or ASW-YT-pyruvate (right) medium. Curves obtained from two experiments for each medium are shown. It usually takes about 10 h to reach the stationary phase, although the duration of the lag phase depends on the initial dilution rate at the time of inoculation from the preculture

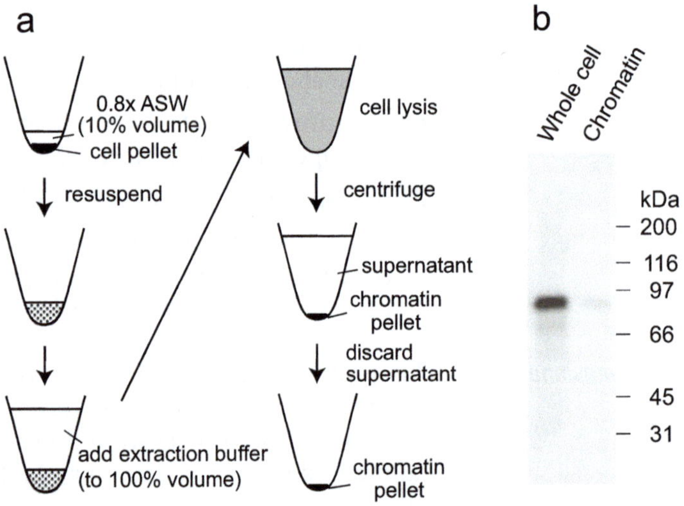

Fig. 2 Chromosome extraction. (**a**) Schematic representation of the steps in this procedure. For efficient cell lysis, it is important to resuspend cells in 0.8x ASW (~10% of the total volume) before adding extraction buffer. (**b**) Western blotting with antibody against Lon_{TK}, a membrane-bound protease, shows that chromosomes and membrane proteins are effectively separated. Proteins from whole-cell extract (left lane) and chromatin fraction (right lane) obtained from the same number of cells were loaded

5. Discard the supernatant (or save it for analysis).
6. Wash the insoluble chromatin fraction containing DNA and chromatin-associated protein with the extraction buffer (*see* **Note 17**).

3.3 MNase Digestion

1. For each tube containing chromatin derived from 30 mL of the original cell culture, add MNase buffer containing RNase A (final concentration, 10 μg/mL) and varying amounts (units) of MNase (e.g., 0, 0.3, 1, 3, 10, and 30 units) in a total reaction volume of 100 μL (*see* **Notes 18** and **19**).
2. Incubate for 30 min at 37 °C.
3. Occasionally, during the incubation, gently pipette the reaction mixture to ensure efficient digestion.
4. To stop the digestion reaction, add approximately 11 μL (11% of the reaction volume) of 10x Stop solution (to make it 1x). Add Proteinase K to a final concentration of 0.2 mg/mL.
5. Incubate for 3 h at 37 °C.
6. Add MNase buffer to bring the volume of the solution to 400 μL (for convenience in the next DNA extraction step).
7. Purify the DNA using general phenol:chloroform:isoamyl alcohol extraction and ethanol precipitation (*see* **Note 20**).

3.4 Agarose Gel Electrophoresis

For the best resolution of small DNA fragments, 4% NuSieve GTG agarose in 1x TBE buffer is used (*see* Fig. 3a). It is also possible to use 2.5% agarose (1 × or 0.5 × TBE), but this results in a lower resolution of small DNA (*see* Fig. 3b).

1. Add chilled 1x TBE buffer to a beaker that is 2–4 times the volume of the solution (*see* **Note 21**).
2. Slowly sprinkle the NuSieve GTG agarose powder while rapidly stirring the solution.
3. Soak the agarose in the buffer for 15 min before heating (*see* **Note 22**).
4. Heat the solution using a microwave oven (*see* **Note 23**).
5. Mix thoroughly and cool the solution to 50–60 °C before casting.
6. Measure the concentration of the DNA. Apply 2–5 μg of DNA samples and run through the agarose gel along with a low-molecular-weight DNA marker (*see* **Note 24**).
7. Stain the gel with ethidium bromide to visualize the DNA bands (*see* **Note 10**).

3.5 Interpretation of the Results

In the case of *T. kodakarensis*, the minimum size of DNA protected by MNase was ~60 bp (*see* Fig. 3a), which corresponds to the size of DNA in contact with a histone tetramer. Furthermore, in addition to the 60 bp DNA, a ladder-like pattern of DNA in ~30 bp increments (i.e., ~90, ~120, ~150, etc.) was observed (*see* Fig. 3a), which correspond to the size of DNA in contact with a histone hexamer, an octamer, and a decamer, where ~30 bp corresponds to

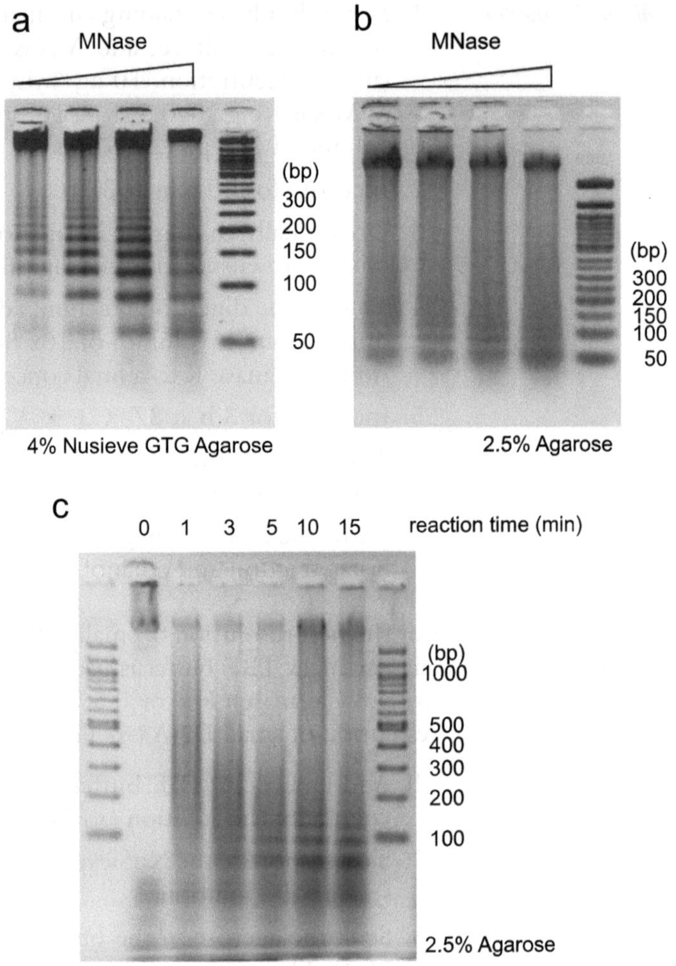

Fig. 3 Agarose gel electrophoresis of chromatin and cells of *T. kodakarensis* digested with MNase. Chromatin extracted from stationary phase cells was digested with increasing amounts of MNase (0.3, 1, 3, and 10 units in 100 μL reaction) in the presence of RNase A (10 μg/mL). DNA were separated through (**a**) 4% NuSieve GTG agarose in 1x TBE, or (**b**) 2.5% agarose in 1x TBE. (**c**) MNase digestion was performed directly on cells (instead of extracted chromatin) with a fixed MNase concentration and varying reaction time at 37 °C. DNA was separated on 2.5% agarose gel. Molecular weight of the DNA ladder is shown

the size of DNA in direct contact with a histone dimer [11]. Such flexibility in histone multimer formation is predicted to be widespread among archaea [5].

4 Notes

1. Each archaeal species requires a specific culture medium. Information on the composition of the media can be found on the websites of several resource organizations for microorganisms [e.g., American Type Culture Collection (ATCC) and Japan Collection of Microorganisms (JCM)].
2. For example, glass tubes with rubber caps are impermeable to gasses. Glass bottles with inner rubber caps can also be used. For large-scale culture, a 500 mL glass bottle can be used.
3. It is convenient to perform cell inoculation in an anaerobic chamber. However, it is possible to inoculate cells without an anaerobic chamber. For example, *T. kodakarensis* cells cultured in sealed bottles with a rubber cap can be transferred using a needle attached to a syringe from one tube to another containing a fresh medium.
4. Commercially available artificial seawater can also be used instead of ASW.
5. Mix ASW, yeast extract, tryptone, and resazurin. After autoclaving, add Na_2S and elemental sulfur powder separately.
6. The extraction buffer was originally developed for *Pyrococcus abysii* [15]. It can also be used to isolate chromosomes of Thermococcales species (*Thermococcus* and *Pyrococcus*). It should be noted that this buffer may not be applicable to other lineages of Archaea. A method to disrupt cells while maintaining chromosome integrity should be developed for each archaeon.
7. MNases are endonucleases and exonucleases derived from *Staphylococcus aureus* [1].
8. For example, NuSieve GTG agarose (Lonza) is suitable for the separation of low-molecular-weight DNA. Normal agarose can also be used but with the less efficient separation of small DNA fragments. Refer to the instructors' manuals for details.
9. TBE is suitable for electrophoresis of small (<1 kb) DNA.
10. Other DNA staining reagents, such as SYBR Green or GelRed, may also be used.
11. For inoculation, start with 1/100th dilution (volume) of a preculture at the stationary phase. The dilution ratio can be adjusted to harvest cells at the desired growth phase.
12. The washing step can be repeated more than once if necessary.
13. Aliquots are prepared for convenience in the next chromatin extraction step. If necessary, the aliquots (cell pellets) can be frozen at this stage for later use. The optimal culture volume to be divided in each aliquot depends on the growth phase of the

cells. The 30 mL described in this protocol is a good starting point for *T. kodakarensis* culture at the stationary phase, where the optical density at 660 nm (OD_{660}) reaches 0.3–0.4. As cell culture in log phase contains fewer cells, preparing an aliquot of cells derived from a larger culture volume (e.g., 60 mL) is a good choice. The maximum cell density also depends on the culture medium used. If a medium such as ASW-YT-pyruvate is used instead of ASW-YT-S^0, the cell density will be greater (*see* Fig. 1) [16], and the culture volume for each aliquot should be smaller and determined empirically.

14. This step is important for achieving homogeneous cell disruption. Without this step, only the cells at the periphery of the pellet will be lysed, thereby clogging the pellet and leaving the cells inside the pellet unlysed.
15. Based on a typical cell number of ~2×10^8 cells/ mL in ASW-YT-S^0 medium at the stationary phase ($OD_{660} = 0.4$), the cells derived from 30 mL stationary phase culture (6×10^9 cells) should be lysed in 0.6 mL of extraction buffer to achieve 10^{10} cells/mL. Other investigators have calculated the cell density of *T. kodakarensis* in ASW-YT-pyruvate medium in the early stationary phase ($OD_{600} = 1.4$) to be ~9×10^8 cells/mL [16].
16. Every few minutes, tap the tube gently or pipette the solution up and down to unclog aggregated cells in order to achieve efficient cell disruption.
17. Here, soluble proteins (supernatant) and the insoluble fraction containing a large amount of chromosomal DNA (pellet) are separated. If necessary, western blotting can be performed using antibodies against proteins with specific localization (chromatin, cytoplasm, membrane, etc.) to confirm the efficiency of the separation. For example, western blotting confirmed that Lon_{TK}, a membrane-bound protease, was expelled from the chromatin fraction of *T. kodakarensis* prepared by this method (*see* Fig. 2b) [9]. Chromatin pellets can be frozen at −20 °C until use.
18. The amount of MNase should ideally be in the range that achieves "no digestion" to "complete digestion." To cover this range, five or six different concentrations, including zero, should be sufficient. The concentration of MNase should be determined empirically, as the activity of commercial MNase may vary among manufacturers. It may be easier to find and set the concentrations of MNase such that DNA digestion reaches a plateau after 30 min of incubation at 37 °C. It is also possible to fix one concentration of MNase and vary the reaction time (*see* Fig. 3c). However, in this case, care should be taken to stop the reaction completely after each time point.

19. It is also possible to perform MNase digestion directly on cell lysate instead of the extracted chromatin (*see* Fig. 3c). If chromosome extraction (purification) is difficult in the target archaeal species, this option can also be considered.

20. Other equivalent methods can be used to purify the DNA. Select a method that can retain DNA fragments of approximately 30 bp.

21. It is important to use a chilled buffer to avoid partial melting of the low-melting-temperature agarose.

22. This reduces the tendency of the agarose solution to form foam during heating.

23. When using a low-melting-temperature agarose, it is important not to melt the agarose at once. Instead, heat the solution on medium power for ~2 min, then allow it to sit at room temperature for ~15 min, and repeat the heating process. Refer to the manufacturer's instructions for further details.

24. For histone-encoding archaea such as *T. kodakarensis*, the smallest DNA fragment can be as small as ~30 bp. DNA markers should be selected so that the size of the MNase-digested DNA fragment can be estimated.

References

1. Zaret K (2005) Micrococcal nuclease analysis of chromatin structure. Curr Protoc Mol Biol 69:21.1.1–21.1.17. https://doi.org/10.1002/0471142727.mb2101s69
2. Clark DJ (2010) Nucleosome positioning, nucleosome spacing and the nucleosome code. J Biomol Struct Dyn 27(6):781–793. https://doi.org/10.1080/073911010010524945
3. Woese CR, Kandler O, Wheelis ML (1990) Towards a natural system of organisms: proposal for the domains Archaea, Bacteria, and Eucarya. Proc Natl Acad Sci U S A 87(12): 4576–4579
4. Peeters E, Driessen RP, Werner F, Dame RT (2015) The interplay between nucleoid organization and transcription in archaeal genomes. Nat Rev Microbiol 13(6):333–341. https://doi.org/10.1038/nrmicro3467
5. Henneman B, van Emmerik C, van Ingen H, Dame RT (2018) Structure and function of archaeal histones. PLoS Genet 14(9): e1007582. https://doi.org/10.1371/journal.pgen.1007582
6. Driessen Rosalie PC, Dame Remus T (2011) Nucleoid-associated proteins in Crenarchaea. Biochem Soc Trans 39(1):116–121. https://doi.org/10.1042/bst0390116
7. Nalabothula N, Xi L, Bhattacharyya S, Widom J, Wang JP, Reeve JN, Santangelo TJ et al (2013) Archaeal nucleosome positioning in vivo and in vitro is directed by primary sequence motifs. BMC Genomics 14:391. https://doi.org/10.1186/1471-2164-14-391
8. Mattiroli F, Bhattacharyya S, Dyer PN, White AE, Sandman K, Burkhart BW, Byrne KR et al (2017) Structure of histone-based chromatin in Archaea. Science 357(6351):609–612. https://doi.org/10.1126/science.aaj1849
9. Maruyama H, Shin M, Oda T, Matsumi R, Ohniwa RL, Itoh T, Shirahige K et al (2011) Histone and TK0471/TrmBL2 form a novel heterogeneous genome architecture in the hyperthermophilic archaeon Thermococcus kodakarensis. Mol Biol Cell 22(3):386–398. https://doi.org/10.1091/mbc.E10-08-0668
10. Maruyama H, Prieto EI, Nambu T, Mashimo C, Kashiwagi K, Okinaga T, Atomi H et al (2020) Different proteins mediate step-wise chromosome architectures in Thermoplasma acidophilum and Pyrobaculum calidifontis. Front Microbiol 11:1247. https://doi.org/10.3389/fmicb.2020.01247
11. Maruyama H, Harwood JC, Moore KM, Paszkiewicz K, Durley SC, Fukushima H,

Atomi H et al (2013) An alternative beads-on-a-string chromatin architecture in Thermococcus kodakarensis. EMBO Rep 14(8):711–717. https://doi.org/10.1038/embor.2013.94

12. Hocher A, Rojec M, Swadling JB, Esin A, Warnecke T (2019) The DNA-binding protein HTa from Thermoplasma acidophilum is an archaeal histone analog. elife 8:e52542. https://doi.org/10.7554/eLife.52542
13. Ammar R, Torti D, Tsui K, Gebbia M, Durbic T, Bader GD, Giaever G et al (2012) Chromatin is an ancient innovation conserved between Archaea and Eukarya. elife 1:e00078. https://doi.org/10.7554/eLife.00078
14. Robb FT, Place AR (1995) Media for thermophiles. In: Robb FT, Place AR (eds) Archaea: a laboratory manual -thermophiles. Cold Spring Harbor Laboratory Press, Cold Sprig Harbor, pp 167–168
15. Matsunaga F, Forterre P, Ishino Y, Myllykallio H (2001) In vivo interactions of archaeal Cdc6/Orc1 and minichromosome maintenance proteins with the replication origin. Proc Natl Acad Sci U S A 98(20): 11152–11157. https://doi.org/10.1073/pnas.191387498
16. Spaans SK, van der Oost J, Kengen SW (2015) The chromosome copy number of the hyperthermophilic archaeon Thermococcus kodakarensis KOD1. Extremophiles 19(4): 741–750. https://doi.org/10.1007/s00792-015-0750-5

Chapter 3

Detecting DNA Methylations in the Hyperthermoacidophilic Crenarchaeon *Sulfolobus acidocaldarius* Using SMRT Sequencing

Christian Tellgren-Roth and Mohea Couturier

Abstract

DNA methylations are one of the most well-known epigenetic modifications along with histone modifications and noncoding RNAs. They are found at specific sites along the DNA in all domains of life, with 5-mC and 6-mA/4-mC being well-characterized in eukaryotes and bacteria respectively, and they have not only been described as contributing to the structure of the double helix itself but also as regulators of DNA-based processes such as replication, transcription, and recombination. Different methods have been developed to accurately identify and/or map methylated motifs to decipher the involvement of DNA methylations in regulatory networks that affect the cellular state.

Although DNA methylations have been detected along archaeal genomes, their involvement as regulators of DNA-based processes remains the least known. To highlight the importance of DNA methylations in the control of key cellular mechanisms and their dynamics in archaea cells, we have used single-molecule real-time (SMRT) sequencing. This sequencing technology allows the identification and direct mapping of the methylated motifs along the genome of an organism. In this chapter, we present a step-by-step protocol for detecting DNA methylations in the hyperthermophilic crenarchaeon *Sulfolobus acidocaldarius* using SMRT sequencing. This protocol can easily be adapted to other prokaryotes.

Key words Single-molecule real-time (SMRT) sequencing, DNA methylome, Epigenetics, Hyperthermophile, *Sulfolobus acidocaldarius*, *Archaea*

1 Introduction

DNA methylations, one of the well-known epigenetic modifications, are found at specific sites along the DNA in all domains of life and are carried out by DNA methyltransferases (MTases). These enzymes catalyze the transfer of the methyl group from the universal donor S-adenosyl-L-methionine to the exogenous nitrogen of the adenine or cytosine forming N^6-methyl-adenine (6-mA) or N^4-methyl-cytosine (4-mC) modifications, respectively, or directly to the endogenous carbon leading to a 5-methyl-cytosine (5-mC) modification. Initially, prokaryotic DNA MTases were described

Eveline Peeters and Indra Bervoets (eds.), *Prokaryotic Gene Regulation: Methods and Protocols*,
Methods in Molecular Biology, vol. 2516, https://doi.org/10.1007/978-1-0716-2413-5_3,

as part of restriction-modification (R-M) systems allowing discrimination between self and foreign DNA. Later, orphan DNA MTases were identified and were reported to be involved in embryonic development and genomic imprinting in eukaryotes along with maintenance of genome integrity and regulation of cell cycle in bacteria [1, 2]. The influences of DNA methylations on key cellular processes can be explained by their impact on the intrinsic mechanical properties of the double-stranded DNA, which, by consequence, may affect the spatial organization and the dynamics of the genome. In vitro studies have reported that DNA methylations modulate the local stability of the DNA by favoring the fusion of the two strands [3], decreasing the flexibility [4, 5], or changing the curvature [6] of the double helix. In addition, methyl groups, being added onto the double helix, modulate the interface between DNA and DNA-binding proteins, which not only control the action of prokaryotic endonucleases from R-M systems but also control gene expression by influencing the binding of transcriptional regulators [7, 8]. Indeed, it has been suggested, in bacteria and eukaryotes, that structural alterations of the interface between DNA-binding proteins and DNA via DNA methylations regulate gene expression by directly inhibiting, or recruiting, the binding of transcription factors [9–12] or indirectly by changing the DNA compaction level [13–15]. To further investigate the physiological roles that DNA methylations might have in biological regulatory mechanisms, efforts have been made to better detect and map them along cellular and viral genomes.

As part of R-M systems in prokaryotes, DNA methylations are known to be site-specific and associated with methylation-sensitive restriction enzymes, which are perfect characteristics to investigate the methylation state of specific motifs [16, 17]. Although this detection method is advantageous to specifically study the methylation state of a DNA motif, the use of restriction enzymes is limited to known motifs. The detection of DNA methylations can also be done by antibodies raised against a specific type of DNA methylation [13, 18–21]. While this method is useful for quickly checking the presence/absence of a specific type of methylation, with commercially available antibodies only raised against 5-mC and 6-mA, the identity of motif(s) associated with the DNA methylation and the corresponding location along the genome have to be determined by sequencing. Advances made in developing next-generation sequencing for metagenomics purposes have also been used to improve the DNA methylation detection: (*a*) Single-molecule real-time (SMRT) sequencing works by detecting the fluorescence signal emitted during the incorporation of a fluorescent nucleotide when a circular DNA template is replicated in a tiny well. DNA modifications on the template create distinctive kinetic patterns and can be used to identify the type of these modifications [22, 23]; (*b*) Nanopore sequencing works by monitoring the changes in electrical current as a DNA template is passed through a protein nanopore. Modified bases leave a different trace than standard bases [24, 25]. Both technologies read native molecules and accurately reveal the position of each methylated motif. The advantage of using SMRT sequencing is that an advanced analysis pipeline is available, reporting the type of DNA methylation, especially for 6-mA and 4-mC, and corresponding motifs without any prior knowledge.

Among the three domains of life, the archaeal domain remains the least known. However, this domain offers the opportunity to study specific bacterial and eukaryotic features as well as unique ones. Indeed, archaea are a mosaic of microorganisms sharing similarities with Bacteria such as cell size, metabolism, genome organization, and the absence of a nucleus. On the other hand, archaeal informational processes (replication, transcription, and recombination) are related to their eukaryotic counterparts. In archaea, DNA methylations, suggested to be involved in R-M systems, were identified a long time ago by using restriction assays [26, 27] and, more recently, using next-generation sequencing [28–30]. However, the study of epigenetics seems left behind in the third domain of life, Archaea, as opposed to Eukarya and Bacteria. Indeed, few studies identified the DNA MTases and/or characterized these systems in archaea as opposed to bacterial and eukaryotic systems. In the halophilic euryarchaeon, *Haloferax volcanii*, and in the hyperthermoacidophilic crenarchaeon, *Sulfolobus acidocaldarius*, DNA methylations have been suggested to be part of DNA metabolism [31], or cell cycle regulation [30], respectively, without being linked to R-M systems. These two studies testify for the presence of orphan DNA MTases in archaea, but the significance of DNA methylations as epigenetic regulatory mechanisms remains largely uncharacterized in this domain of life. In this chapter, we describe a step-by-step protocol on how to detect DNA methylations using SMRT sequencing in the hyperthermoacidophilic model organism *Sulfolobus acidocaldarius*.

2 Materials

2.1 Sulfolobus acidocaldarius *DSM639 Culture*

All media contain ultrapure water with 18.2 MΩ.cm resistivity at 25 °C, which has been sterilized by autoclaving. All stock solutions should be sterile filtered through a 0.22 μm filter, if no other method is indicated.

1. Spectrophotometer.
2. Shaker with settings 80 °C, 140 rpm (*see* **Note 1**).
3. Longneck flasks.
4. Strain *Sulfolobus acidocaldarius* DSM639.

5. Stock solution A_*Trace mineral solution*: 9.0 g/L $Na_2B_4O_7 \cdot 10H_2O$, 3.6 g/L $MnCl_2 \cdot 4\ H_2O$, 0.44 g/L $ZnSO_4 \cdot 7H_2O$, 0.10 g/L $CuCl_2 \cdot 2H_2O$, 0.06 g/L $Na_2MoO_4 \cdot 2H_2O$, 0.079 g/L $VOSO_4 \cdot 5H_2O$, 0.10 g/L $CoSO_4 \cdot 7H_2O$.
6. Stock solution B_*2× Allen culture medium*: 2.60 g/L $(NH_4)_2SO_4$, 0.56 g/L KH_2PO_4, 0.5 g/L $MgSO_4 \cdot 7H_2O$, 0.14 g/L $CaCl_2 \cdot 2H_2O$, 1 mL/L 148 mM $FeCl_3 \cdot 6H_2O$ solution, 1 mL/L Trace mineral solution (stock solution A), adjust to pH 2.0 using H_2SO_4.
7. Stock solution C_*5 % (w/v) trytone*, sterilized by autoclaving.
8. Working culture medium_1× Allen culture medium: 1/2 volume 2× Allen culture medium (stock solution B), 1/25 volume 5 % tryptone (stock solution C), and complete to one volume with ultrapure water.

2.2 Genomic DNA Extraction

1. *Sulfolobus acidocaldarius* DSM639 culture (*see* **Note 2**).
2. Centrifuge for 15–50 mL tubes, room temperature (22 °C) and 4000 × *g*.
3. Benchtop centrifuge for 15–50 mL tubes, 4 °C, and 10,000 × *g*.
4. Microcentrifuge for 1.5–2.0 mL tubes, 4 °C, and 13,200 rpm.
5. Water bath fixed at 50 °C.
6. Gravity-flow QIAGEN Genomic-Tip 100/G columns (*see* **Note 3**).
7. QIAGEN Genomic DNA Buffer Set.
8. RNase A, DNase, and protease-free (10 mg/mL).
9. Proteinase K (20 mg/mL).
10. Room temperature (22 °C) isopropanol.
11. Room temperature (22 °C) 70 % ethanol solution, sterile filtered through a 0.22 μm filter.
12. Ultrapure water with 18.2 MΩ.cm resistivity at 25 °C, sterilized by autoclaving.

2.3 (Optional Step) Preparation of the Negative Control by Performing a Whole-Genome Amplification (WGA)

This step is optional due to the accuracy of the sequencing technology. However, it is wise to include additional verifications to complement the sequencing results (*see* **Note 4**).

1. Genomic DNA extracted from *Sulfolobus acidocaldarius* (10 ng).
2. QIAGEN REPLI-g Mini Kit (*see* **Note 5**).

3. Buffers prepared the day of the experiment: Buffer D1 (9 μL DLB buffer + 32 μL water nuclease-free), Buffer N1 (12 μL Stop solution + 68 μL nuclease-free water), and Master Mix (29 μL Reaction Buffer + 1 μL DNA polymerase).
4. Thermocycler.

2.4 Genomic DNA Quality Check

1. NanoDrop Spectrophotometer.
2. Qubit Fluorometer: Qubit dsDNA BR Assay Kit.
3. Gel-documentation system (CCD-camera).
4. Agarose gel electrophoresis system.
5. Agarose powder.
6. 1x Tris-acetate-EDTA (TAE) buffer: 40 mM Tris-HCl, pH 7.6, 20 mM acetic acid, 1 mM EDTA.
7. 10 mg/mL Ethidium Bromide stock solution (*see* **Note 6**).
8. DNA Ladder Mix.
9. 6x DNA Loading Dye.
10. Ultrapure water with 18.2 MΩ.cm resistivity at 25 °C, sterilized by autoclaving.

2.5 PacBio Single-Molecule Real-Time (SMRT) Sequencing and Analysis

1. Check for a suitable sequencing provider: https://www.pacb.com/products-and-services/certified-service-providers/
2. If you want to create your own sequencing libraries (*see* **Note 7**), use Megaruptor 3, or similar instrument for shearing, and check for current library construction protocols: https://www.pacb.com/products-and-services/consumables/
3. SMRT Link: https://www.pacb.com/support/software-downloads/

3 Methods

3.1 Sulfolobus acidocaldarius DSM639 Culture

1. Start *Sulfolobus acidocaldarius* DSM639 culture by adding 50 μL from a glycerol stock to a 100 mL longneck flask containing 20 mL 1x Allen culture medium (*see* **Note 8**).
2. Grow the culture at 80 °C with shaking at 140 rpm, and monitor culture growth by measuring the optical density at 600 nm (OD_{600nm}).
3. When OD_{600nm} of the liquid culture 1 reaches 0.4 (mid-exponential phase), transfer 1 mL culture to a new 500 mL longneck flask containing preheated 100 mL 1x Allen culture medium (*see* **Note 8**).
4. Culturing at 80 °C with shaking at 140 rpm is followed, and cells are collected around $OD_{600nm} = 0.1$ (early exponential phase).

3.2 Genomic DNA Extraction from Sulfolobus acidocaldarius *Cells*

1. Collect exponentially growing cells by centrifugation at 4000 × *g* for 5 min, room temperature.
2. Resuspend cell pellet in 3.5 mL of lysis Buffer B1 containing 200 μg/mL RNase A by vortexing at high speed. Buffer B1 containing 200 μg/mL RNase A should be prepared the day of the experiment.
3. Proteinase K is added to the cell lysate to a final concentration of 0.57 mg/mL.
4. Incubate at 37 °C for 1 h, invert after 30 min incubation.
5. Add 1.2 mL of Buffer B2.
6. Incubate at 50 °C for 2 h, invert after 1-h incubation (*see* **Note 9**).
7. Equilibrate a gravity-flow QIAGEN Genomic-Tip 100/G column with 4 mL Buffer QBT.
8. Quickly vortex the sample for a short time at high speed. Then, load the sample onto the equilibrated column.
9. Wash the column with 2 times 7.5 mL (15 mL total) of Buffer QC.
10. Elute with 5 mL of preheated (50 °C) Buffer QF. The eluate is collected in either a 15 or a 50 mL tube.
11. Precipitate the genomic DNA with 3.5 mL (i.e., 0.7 volume) room temperature isopropanol.
12. Mix by gently inverting the tube.
13. Centrifuge directly at 10,000 × *g* for 30 min, 4 °C, and gently discard the supernatant.
14. Wash the genomic DNA pellet with 1 mL room temperature ethanol 70 %.
15. Centrifuge at 13,200 rpm for 30 min, 4 °C, and gently discard the supernatant (*see* **Note 10**).
16. Air-dry the pellet at room temperature for 5–10 min depending on the volume of the pellet (*see* **Note 11**).
17. Resuspend the pellet with ultrapure water with 18.2 MΩ.cm resistivity at 25 °C. The volume varies in function of the volume of the pellet. Store at −20 °C or prepare the negative control and/or proceed directly with SMRT sequencing.

3.3 (Optional Step) Preparation of the Negative Control by Performing a Whole-Genome Amplification (WGA)

1. For one test, prepare 5 μL of DNA sample containing 10 ng of genomic DNA.
2. Mix 5 μL of DNA sample with 5 μL Buffer D1.
3. Vortex and spin down quickly.
4. Incubate the mix for 3 min at room temperature (22 °C).

5. Add 10 μL of Buffer N1.
6. Vortex and spin down quickly.
7. Add 30 μL of Master Mix.
8. Incubate at 30 °C overnight (about 16 h).
9. Inactivate the mix by incubating at 65 °C for 3 min, and store at −20 °C or proceed directly with SMRT sequencing.

3.4 DNA Quality Check (See Note 12)

1. Dissolve 0.8 % (w/v) agarose with 1x TAE buffer by melting the agarose using a microwave oven.
2. Cool down the agarose solution until the mix can be handled with bare hands (approximately 60 °C).
3. Add 1 μL/mL Ethidium Bromide stock solution to reach a final concentration of 0.2 μg/mL.
4. Pour the agarose into the gel tray and place the comb.
5. Let the gel polymerize (20–30 min).
6. Prepare the samples by mixing 5:1 (v/v) DNA sample and 6x DNA loading dye.
7. Load the genomic DNA samples and the DNA ladder mix.
8. Run the electrophoresis in 1x TAE buffer for 60 min at 80 V.
9. Visualize the DNA samples using the gel-documentation system (*see* **Note 13**).
10. Nanodrop: Measure DNA concentration and absorption ratios 260/230 nm and 260/280 nm according to the manufacturer's instructions.
11. Qubit: Measure DNA concentration according to the manufacturer's instructions.

3.5 PacBio SMRT Sequencing and Analysis

3.5.1 Sequencing (See Note 14)

1. Submit your sample to the sequencing provider with specific instructions to preserve the kinetic signals.
2. If you work with your own libraries and machine, start a sequencing run according to the manufacturer's instructions, and make sure to save the kinetic information if you are using a Sequel IIe instrument.

3.5.2 Analysis (See Fig. 1)

1. Import your data into SMRT Link (bam, pbi, and xml file).
2. Import your reference into SMRT Link (fasta-file).
3. Create a "Base Modification Analysis". Select your read-set and your reference. Turn "Find Modified Base Motifs" on. Turn "Consolidate Mapped BAMs for IGV" on if you want to look at the results in IGV (*see* **Note 15**).
4. Check the base modification and motif report to get a quick overview of the modifications, the motifs, and their modification status. To get a more detailed view of the DNA

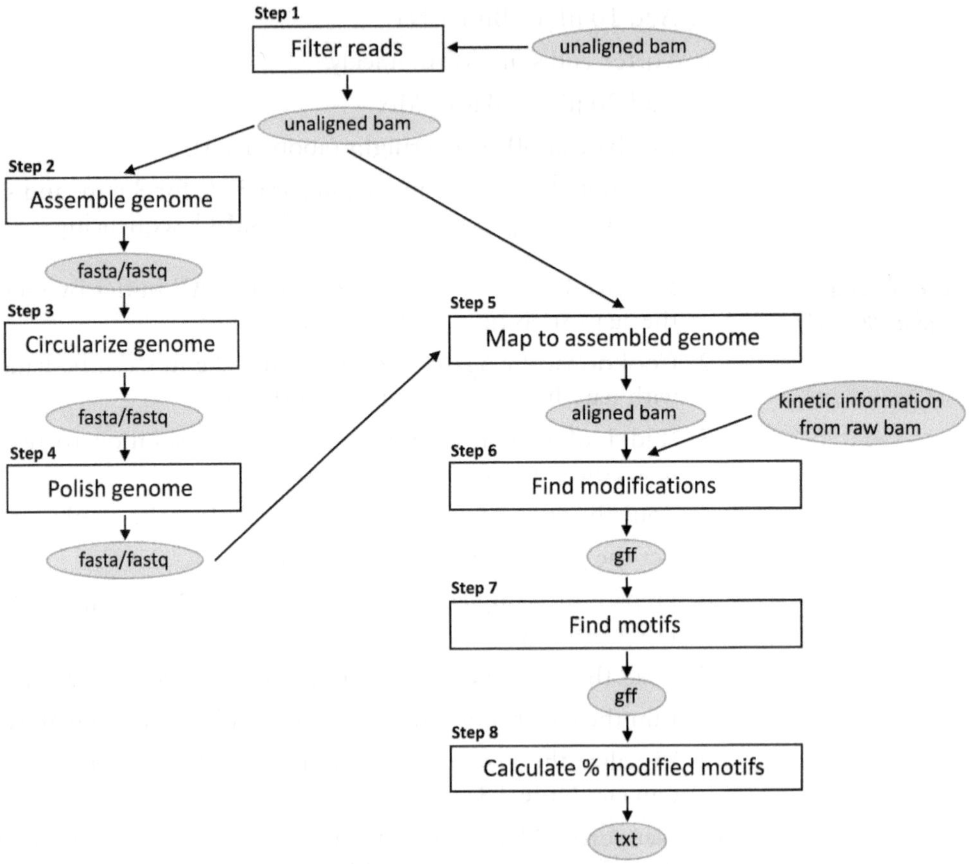

Fig. 1 Flowchart of the methylation analysis. **Steps 1–4** create the reference for the mapping and are combined into one workflow in SMRT Link. If there is a perfect reference available, it can directly be used in **step 5**, and the analysis consists of **step 1** followed by **steps 5–8**. This is also available as a workflow in SMRT Link. White boxes represent tasks performed in the analysis, and grey-shaded boxes indicate input and output file types

methylome, you can visualize the results in IGV by importing the gff-file (*see* **Note 16**).

4 Notes

1. Shaker can be filled with either water, polyethylene glycol (PEG), or low melting oil (mineral oil Rotitherm® K+H, from Roth). The drawback of using water is the fast evaporation at the growth temperature of *Sulfolobus* (70–80 °C), which implies always having a stock of preheated water as a refill. Using PEG, and especially the low melting oil, represents a cost for a lab, but it is a long-term investment.
2. 3.4×10^9 cells were collected in our experiments. This number of *S. acidocaldarius* DSM639 cells has been determined using

the following estimation: $OD_{600nm} = 0.1$ is about 8.4×10^7 cells/mL. This relation has been determined for our growth conditions by plating assays. Thus, it might vary depending on the growth conditions and the *Sulfolobus* strain.

3. Gravity-flow QIAGEN Genomic-Tip 100/G columns have been used to increase the yield of genomic DNA compared to the phenol-chloroform method.
4. The sequencing technology identifies the type of DNA methylation, the sequence motifs, and the methylation state of the motifs. The SMRT sequencing results can be verified by performing a digestion assay to specifically target a sequence motif. The digestion pattern of the native sample should be compared to a negative and positive control.
5. The QIAGEN REPLI-g Mini kit was the only kit available in the laboratory at that time, but any other WGA kits can be used.
6. The agarose gel used in our study was stained with ethidium bromide. However, safer nucleic acid dyes can substitute the use of ethidium bromide.
7. The amount of data produced by an SMRT cell allows for multiplexing of several samples. While pooling of samples is possible for sequencing, libraries have to be constructed individually. Therefore, library preparation costs can add up to more than 50 % of the total sequencing prize, which makes it interesting to perform on-site instead of at the sequencing center.
8. Starting a *Sulfolobus* culture from a glycerol stock can be done in a room temperature medium, while starting a new culture from an ongoing culture requires to preheat the medium to avoid a cold shock.
9. The incubation time to degrade proteins in the genomic DNA extraction protocol is about 2 h. However, this incubation step can be extended up to 4 h if the cell density is higher than 3.4×10^9 cells, and the tube(s) should be inverted every hour.
10. It is really important to remove all traces of ethanol as it affects DNA quality. To continue to remove traces of ethanol, invert the tube to let the last droplets of ethanol on the side of the tube, and remove them using clean absorbant paper.
11. Air-drying the pellet in the genomic DNA extraction protocol is crucial and should not be overdone. Once the white DNA pellet becomes transparent, add the corresponding volume of ultrapure water, sterilized by autoclaving.
12. A good DNA sample should have the following properties:
    (i) The majority of the DNA fragments in the agarose gel picture is longer than 20 kb (*see* Fig. 2).

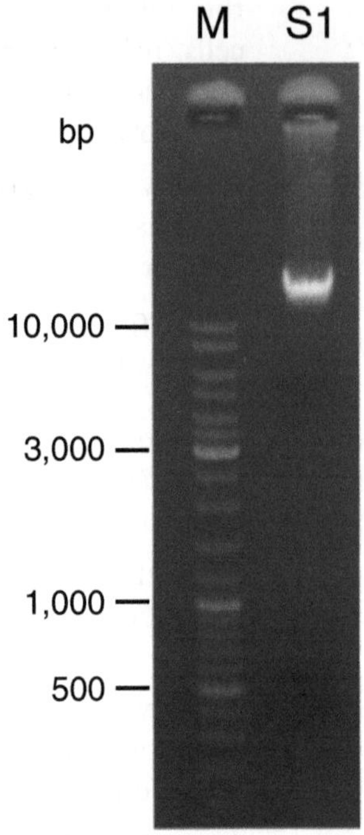

Fig. 2 Genomic DNA purified from *S. acidocaldarius* using QIAGEN Genomic-Tip 100/G column. The same volume (1 μL) of DNA Ladder mix (lane M) and genomic DNA sample (lane S1) was loaded into a 0.8 % agarose gel. Electrophoresis was performed at room temperature in 1x TAE buffer for 60 min at 80 V. Gel was stained with ethidium bromide and digitalized under UV light using a gel-documentation system

(ii) The DNA is RNA-free.

(iii) The 260/230 ratio is within 2.0–2.2.

(iv) The 260/280 ratio is within 1.8–2.0.

(v) The ratio NanoDrop-conc./Qubit-conc. is smaller than 1.5.

13. A rough estimation of the genomic DNA concentration can be obtained by comparing the band intensities of genomic DNA and DNA ladder.

14. Methylation analysis requires a read coverage above 100×. Therefore, one should not pool more than 8 samples, or approximately 20 Mb of genome equivalents, onto one SMRT cell.

15. These instructions are for SMRT Link v10.0 and might look different in other versions of the software.

16. Motifs with a modified fraction lower than 90 % should be analyzed in more detail and verified by a complementary method (*see* **Note 4**).

Acknowledgments

The authors would like to thank the staff at the Uppsala Genome Center, National Genomics Infrastructure, Science for Life Laboratory for helpful comments on the DNA quality check protocol.

Funding

Funding was provided by the Carl R. Woese Institute for Genomic Biology postdoctoral fellowship (to M.C.).

References

1. Jeltsch A (2002) Beyond Watson and Crick: DNA methylation and molecular enzymology of DNA methyltransferases. Chembiochem 3: 274–293. https://doi.org/10.1002/1439-7633(20020402)3:4<274::AID-CBIC274>3.0.CO;2-S
2. Sánchez-Romero MA, Casadesús J (2020) The bacterial epigenome. Nat Rev Microbiol 18: 7–20. https://doi.org/10.1038/s41579-019-0286-2
3. Severin PMD, Zou X, Gaub HE, Schulten K (2011) Cytosine methylation alters DNA mechanical properties. Nucleic Acids Res 39: 8740–8751. https://doi.org/10.1093/nar/gkr578
4. Geahigan KB, Meints GA, Hatcher ME, Orban J, Drobny GP (2000) The dynamic impact of CpG methylation in DNA. Biochemistry 39:4939–4946. https://doi.org/10.1021/bi9917636
5. Derreumaux S, Chaoui M, Tevanian G, Fermandjian S (2001) Impact of CpG methylation on structure, dynamics and solvation of cAMP DNA responsive element. Nucleic Acids Res 29:2314–2326. https://doi.org/10.1093/nar/29.11.2314
6. Diekmann S (1987) DNA methylation can enhance or induce DNA curvature. EMBO J 6:4213–4217. https://doi.org/10.1002/j.1460-2075.1987.tb02769.x
7. Polaczek P, Kwan K, Campbell JL (1998) GATC motifs may alter the conformation of DNA depending on sequence context and N 6 -adenine methylation status: possible implications for DNA-protein recognition. Mol Gen Genet MGG 258:488–493. https://doi.org/10.1007/s004380050759
8. Wion D, Casadesús J (2006) N6-methyl-adenine: an epigenetic signal for DNA–protein interactions. Nat Rev Microbiol 4:183–192. https://doi.org/10.1038/nrmicro1350
9. Camacho EM, Casadesús J (2005) Regulation of *traJ* transcription in the Salmonella virulence plasmid by strand-specific DNA adenine hemimethylation: regulation of *traJ* transcription by Dam methylation. Mol Microbiol 57: 1700–1718. https://doi.org/10.1111/j.1365-2958.2005.04788.x
10. Kumar S, Karmakar BC, Nagarajan D, Mukhopadhyay AK, Morgan RD, Rao DN (2018) N4-cytosine DNA methylation regulates transcription and pathogenesis in Helicobacter pylori. Nucleic Acids Res 46:3429–3445. https://doi.org/10.1093/nar/gky126
11. Clemens AW, Wu DY, Moore JR, Christian DL, Zhao G, Gabel HW (2019) MeCP2 represses enhancers through chromosome topology-associated DNA methylation. Mol Cell 77:1–15. https://doi.org/10.1016/j.molcel.2019.10.033
12. Nye TM, van Gijtenbeek LA, Stevens AG, Schroeder JW, Randall JR, Matthews LA, Simmons LA (2020) Methyltransferase DnmA is responsible for genome-wide N6-methyladenosine modifications at non-palindromic recognition sites in Bacillus subtilis. Nucleic Acids Res 48:5332–5348. https://doi.org/10.1093/nar/gkaa266
13. Fu Y, Luo G-Z, Chen K, Deng X, Yu M, Han D, Hao Z, Liu J, Lu X, Doré LC, Weng X, Ji Q, Mets L, He C (2015) N6-Methyldeoxyadenosine marks active transcription start sites in Chlamydomonas. Cell 161:879–892. https://doi.org/10.1016/j.cell.2015.04.010
14. Beh LY, Debelouchina GT, Clay DM, Thompson RE, Lindblad KA, Hutton ER, Bracht JR, Sebra RP, Muir TW, Landweber LF (2019) Identification of a DNA N6-adenine

methyltransferase complex and its impact on chromatin organization. Cell 177: 1781–1796. https://doi.org/10.1016/j.cell.2019.04.028
15. Lizarraga A, O'Brown ZK, Boulias K, Roach L, Greer EL, Johnson PJ, Strobl-Mazzulla PH, de Miguel N (2020) Adenine DNA methylation, 3D genome organization, and gene expression in the parasite *Trichomonas vaginalis*. Proc Natl Acad Sci U S A 117:13033–13043. https://doi.org/10.1073/pnas.1917286117
16. Roberts RJ, Vincze T, Posfai J, Macelis D (2010) REBASE – a database for DNA restriction and modification: enzymes, genes and genomes. Nucleic Acids Res 38:D234–D236. https://doi.org/10.1093/nar/gkp874
17. Roberts RJ, Vincze T, Posfai J, Macelis D (2015) REBASE – a database for DNA restriction and modification: enzymes, genes and genomes. Nucleic Acids Res 43:D298–D299. https://doi.org/10.1093/nar/gku1046
18. Greer EL, Blanco MA, Gu L, Sendinc E, Liu J, Aristizábal-Corrales D, Hsu C-H, Aravind L, He C, Shi Y (2015) DNA methylation on N6-Adenine in C. elegans. Cell 161:868–878. https://doi.org/10.1016/j.cell.2015.04.005
19. Zhang G, Huang H, Liu D, Cheng Y, Liu X, Zhang W, Yin R, Zhang D, Zhang P, Liu J, Li C, Liu B, Luo Y, Zhu Y, Zhang N, He S, He C, Wang H, Chen D (2015) N6-Methyladenine DNA modification in drosophila. Cell 161:893–906. https://doi.org/10.1016/j.cell.2015.04.018
20. Weber M, Davies JJ, Wittig D, Oakeley EJ, Haase M, Lam WL, Schübeler D (2005) Chromosome-wide and promoter-specific analyses identify sites of differential DNA methylation in normal and transformed human cells. Nat Genet 37:853–862. https://doi.org/10.1038/ng1598
21. Down TA, Rakyan VK, Turner DJ, Flicek P, Li H, Kulesha E, Gräf S, Johnson N, Herrero J, Tomazou EM, Thorne NP, Bäckdahl L, Herberth M, Howe KL, Jackson DK, Miretti MM, Marioni JC, Birney E, Hubbard TJP, Durbin R, Tavaré S, Beck S (2008) A Bayesian deconvolution strategy for immunoprecipitation-based DNA methylome analysis. Nat Biotechnol 26:779–785. https://doi.org/10.1038/nbt1414
22. Flusberg BA, Webster DR, Lee JH, Travers KJ, Olivares EC, Clark TA, Korlach J, Turner SW (2010) Direct detection of DNA methylation during single-molecule, real-time sequencing. Nat Methods 7:461–465. https://doi.org/10.1038/nmeth.1459
23. Clark TA, Murray IA, Morgan RD, Kislyuk AO, Spittle KE, Boitano M, Fomenkov A, Roberts RJ, Korlach J (2012) Characterization of DNA methyltransferase specificities using single-molecule, real-time DNA sequencing. Nucleic Acids Res 40:e29–e29. https://doi.org/10.1093/nar/gkr1146
24. Laszlo AH, Derrington IM, Brinkerhoff H, Langford KW, Nova IC, Samson JM, Bartlett JJ, Pavlenok M, Gundlach JH (2013) Detection and mapping of 5-methylcytosine and 5-hydroxymethylcytosine with nanopore Msp A. Proc Natl Acad Sci 110:18904–18909. https://doi.org/10.1073/pnas.1310240110
25. Rand AC, Jain M, Eizenga JM, Musselman-Brown A, Olsen HE, Akeson M, Paten B (2017) Mapping DNA methylation with high-throughput nanopore sequencing. Nat Methods 14:411–413. https://doi.org/10.1038/nmeth.4189
26. Prangishvili DA, Vashakidze RP, Chelidze MG, Gabriadze IY (1985) A restriction endonuclease *Sua* I from the thermoacidophilic archaebacterium *Sulfolobus acidocaldarius*. FEBS Lett 192:57–60. https://doi.org/10.1016/0014-5793(85)80042-9
27. Grogan DW (2003) Cytosine methylation by the suaI restriction-modification system: implications for genetic fidelity in a hyperthermophilic archaeon. J Bacteriol 185:4657–4661. https://doi.org/10.1128/JB.185.15.4657-4661.2003
28. Ouellette M, Jackson L, Chimileski S, Papke RT (2015) Genome-wide DNA methylation analysis of Haloferax volcanii H26 and identification of DNA methyltransferase related PD-(D/E)XK nuclease family protein HVO_A0006. Front Microbiol 6. https://doi.org/10.3389/fmicb.2015.00251
29. Blow MJ, Clark TA, Daum CG, Deutschbauer AM, Fomenkov A, Fries R, Froula J, Kang DD, Malmstrom RR, Morgan RD, Posfai J, Singh K, Visel A, Wetmore K, Zhao Z, Rubin EM, Korlach J, Pennacchio LA, Roberts RJ (2016) The epigenomic landscape of prokaryotes. PLoS Genet 12:e1005854. https://doi.org/10.1371/journal.pgen.1005854
30. Couturier M, Lindås A-C (2018) The DNA methylome of the hyperthermoacidophilic crenarchaeon Sulfolobus acidocaldarius. Front Microbiol 9. https://doi.org/10.3389/fmicb.2018.00137
31. Chimileski S, Dolas K, Naor A, Gophna U, Papke RT (2014) Extracellular DNA metabolism in Haloferax volcanii. Front Microbiol 5. https://doi.org/10.3389/fmicb.2014.00057

Chapter 4

ProD: A Tool for Predictive Design of Tailored Promoters in *Escherichia coli*

Friederike Mey, Jim Clauwaert, Maarten Van Brempt, Michiel Stock, Jo Maertens, Willem Waegeman, and Marjan De Mey

Abstract

A major goal in synthetic biology is the engineering of synthetic gene circuits with a predictable, controlled and designed outcome. This creates a need for building blocks that can modulate gene expression without interference with the native cell system. A tool allowing forward engineering of promoters with predictable transcription initiation frequency is still lacking. Promoter libraries specific for σ^{70} to ensure the orthogonality of gene expression were built in *Escherichia coli* and labeled using fluorescence-activated cell sorting to obtain high-throughput DNA sequencing data to train a convolutional neural network. We were able to confirm in vivo that the model is able to predict the promoter transcription initiation frequency (TIF) of new promoter sequences. Here, we provide an online tool for promoter design (ProD) in *E. coli*, which can be used to tailor output sequences of desired promoter TIF or predict the TIF of a custom sequence.

Key words Synthetic biology, Machine learning, Promoter design, Metabolic engineering, Biotechnology, Toolbox

1 Introduction

As metabolic engineering advances in both fundamental knowledge on the functioning of microorganisms as well as in providing tools to achieve precise engineering goals, the required level of precision increases simultaneously [1–3]. To improve current production titers by engineering a desired expression level, it is often not sufficient to find and turn the right screw. Combinatorial pathway optimization has been one strategy to reduce the time invested in searching for the optimal pathway which maximizes production [4, 5]. In general, control can be executed on three levels: transcriptional, translational and post-translational. While libraries exist

Authors Friederike Mey and Jim Clauwaert have equally contributed to this chapter.

These authors jointly supervised this work: Willem Waegeman, Marjan De Mey

Eveline Peeters and Indra Bervoets (eds.), *Prokaryotic Gene Regulation: Methods and Protocols*,
Methods in Molecular Biology, vol. 2516, https://doi.org/10.1007/978-1-0716-2413-5_4,

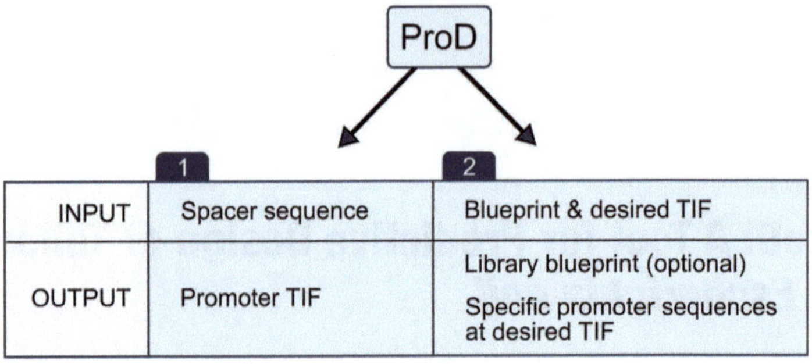

Fig. 1 The two-way usage of ProD, a promoter transcription initiation frequency (TIF) prediction tool supporting de novo design of promoter sequences. ProD can perform 2 types of predictions: In (1), the tool takes a custom spacer sequence and predicts the TIF. In (2), an input blueprint and desired TIF are used to output the blueprint for a library at the desired TIF, as well as specific spacer sequences for each requested TIF, or a single spacer sequence for each requested TIF

with promoters and ribosome binding site (RBS) sequences at predefined strengths [6], such libraries lack the flexibility that might be required in new setups. The success of the RBS (Library) Calculator from the Salis lab [7, 8] has shown the high demand of the synthetic biology community for such predictive tools that output a de novo library. In this work, we present ProD, a promoter transcription initiation frequency (TIF) prediction tool supporting de novo design of promoter sequences. This tool was built using promoters with randomized spacer regions in *Escherichia coli* that remained orthogonal by being specific to sigma factor σ^{70}. Orthogonality was ensured by leaving the -35 and -10 boxes adjacent to the fully randomized spacer region intact [9]. Data was collected by cloning the promoters upstream a fluorescent reporter gene, fluorescent-activated cell sorting (FACS) and subsequent DNA sequencing, allowing for the collection of considerably large datasets (250,000–400,000 unique sequences per setting) that mapped each sequence to fluorescence output. A custom-built convolutional neural network for ordinal regression was then trained on this dataset which is able to evaluate a custom spacer sequence, returning an estimated promoter TIF (0–10) based on the DNA sequence. Alternatively, ProD is able to output a promoter spacer sequence library at desired TIF or a library with specific promoter sequences in the range of the desired TIF (0–10, or very low to very high expression) [10]. An overview of the two workflows is depicted in Fig. 1.

2 Materials

ProD is available at https://github.com/MEMO-group/ProD and written in Python 3.7. The tool can either be run through Binder, which does not require a local installation, or installed on a local machine (*see* **Note 1**). To ensure compatibility in the latter case, the required package versions are listed in the repository.

2.1 Relevant Repository Files

1. ProD.py: the actual python script for the predictive promoter tool (*see* **Note 2**).
2. ProD_Notebook.ipynb: a Jupyter Notebook used for running code in an interactive environment. Contains detailed instructions on the functionality of the tool.
3. README.md: markdown readme file containing instructions on how to use the tool.
4. environment.yml: contains the package versions with which the tool was built. This custom environment ensures smooth running of the program. For ProD, Conda is used for environment management and therefore provided as a YAML file.
5. ex_seqs.fa: example input file used in the README.MD file. The fasta file is read as input and results in an output prediction csv file containing the predicted TIF for the input sequences (in this repository, it is called my_predictions.csv).

2.2 Prerequisites If ProD Is Run from the Terminal

1. Clone the ProD GitHub repository in your working directory (can also be downloaded as ZIP file from the website).
2. For Linux users, an environment can be created using the environment.yml file provided in the repository. For Mac and Windows users, verify that required Python libraries are available in your active environment:
   (a) Pytorch 1.4.0
   (b) Numpy 1.19.1
   (c) Pandas 1.1.0
   (d) Scipy 1.5.2

2.3 Prerequisites If ProD Is Utilized Using Jupyter

1. Install Jupyter Notebook or Jupyter Lab (https://jupyter.org).
2. Open ProD_Notebook.ipynb with Jupyter
3. A brief explanation on how to use the file is described in the file. In short, a code cell can be run by clicking on it and pressing Ctrl + Enter.

3 Methods

3.1 The ProD Tool

The ProD tool's motor is a convolutional neural network trained on 284,421 unique promoter sequences mapped to a fluorescent output value, which is representative for promoter TIF (*see* **Note 3**). This vast library of fluorescent output was achieved by introducing a spacer library. The spacer is framed by the −10 and −35 boxes to ensure an orthogonal promoter-σ^{70} pair. Davis et al. [11] have shown that by using an insulator sequence up- and downstream of the promoter element, the latter's TIF can be reliably predicted with minimal influence of the genetic context, both on plasmid and genome level. Hence, the promoter library was created using an insulator sequence as described by Davis et al. [11], containing a fixed transcription start site (UP sequence) upstream of the promoter, which facilitates polymerase binding, and a fixed untranslated region (UTR) downstream, facilitating promoter escape. The design of the insulator sequence is shown below:

GGTCTATGAGTGGTTGCTGGATAAC**TTTACG**
(NNNNNNNNNNNNNNNNN)TATAATATATT
CA*GGGAGAGCACAACGGTTTCCCTCTACAAATAA*
TTTTGTTTAACTTT

The promoter region (−60 to +1) is shown in bold; the conserved −35 and −10 box are bold underlined and the UP region bold italic. The UTR is shown in italics. N_{17} represents the spacer region, whose sequence is predicted by the ProD tool (*see* **Note 4**). The TIF of an input spacer sequence, given the usage of the same conserved -35 and -10 boxes, can be predicted. The tool will be most accurate when used in the same insulator context, as Davis et al. [11] have shown the influence of surrounding regions on minimal promoters. When a promoter library is constructed (*see* **Note 5**), the tool outputs the entire insulator sequence as described above, as well as the spacer sequence, which can subsequently be assembled and cloned upstream an RBS and ORF of choice (*see* **Note 6**). The tool accepts blueprints to accommodate constraints in spacer design and takes 17 base pairs as input (*see* **Note** 7). The TIF is represented in classes from 0 to 10, with 0–2 = zero to low expression, 3–5 = low to medium expression, 6–8 = medium to high expression and 9–10 = high to very high expression. Figure 2 represents the workflow in the library setting.

3.2 Construction of Promoters with Desired TIF

The outcome of the tool is a mixed-base spacer sequence called library blueprint, as well as discrete spacer sequences for each requested TIF (0–10) and the resulting sequence of the entire insulator. Alternatively, a single spacer sequence at a specific TIF can be calculated. The generated csv file contains, in addition, the probability values of the spacer for each TIF class. A specific spacer

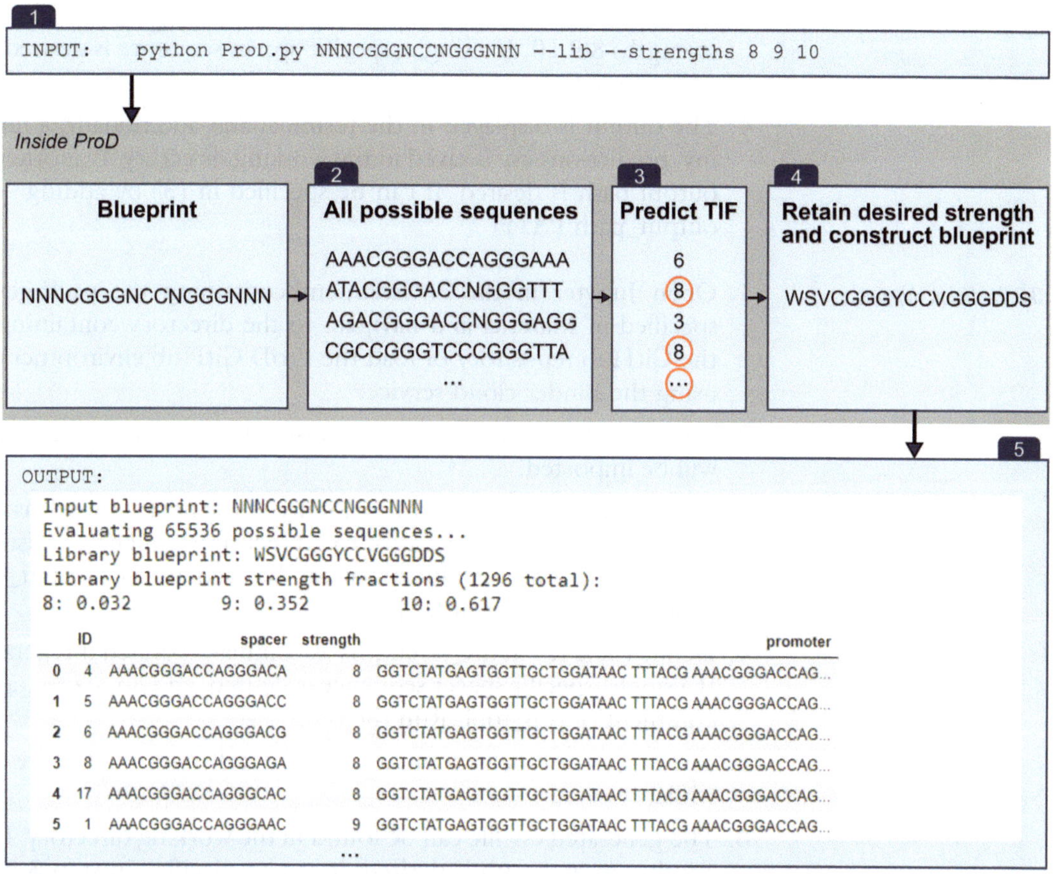

Fig. 2 The different steps in the process through which the ProD model obtains sequences with desired transcription initiation frequency (TIF). When constructing a library, the input (1) consists of a blueprint and desired TIF. The tool evaluates all possible sequences (2) and predicts their respective TIF (3). Only sequences of the desired TIF are retained from the previous step and a blueprint is constructed (4). The output (5) consists of the library blueprint, the fraction of sequences classified to each category of strength and a table with the spacer, TIF and insulator sequence. The CSV file my_predictions.csv further contains the probability for each TIF class for each spacer sequence

sequence or the library blueprint may be selected for subsequent cloning. The process through which the model obtains the final outputs is depicted in Fig. 2.

3.2.1 Using the Terminal

1. Open the terminal, activate the environment containing the packages described in Material.
2. Navigate to the directory containing the cloned GitHub repository.
3. Call the ProD tool inserting the desired library backbone and TIF(s), for example: python ProD.py NNNCGGG NCCNGGGNNN --lib --strengths 8 9 10. Alternative to a blueprint sequence, a fasta file containing a single blueprint

can be used as input: python ProD.py my_blueprint.fa --lib --strengths 8 9 10. If a single, specific spacer sequence is desired, add --lib_size 1.

4. The output is displayed in the terminal and additionally, a file my_predictions.csv is saved in the working directory. If another output path is desired, it can be specified in (3) by adding --output_path PATH

3.2.2 Using Jupyter

1. Open Jupyter in the environment containing the packages specified in Material and navigate to the directory containing the GitHub repository or load the ProD GitHub environment using the Binder cloud service.
2. Run the cell by selecting it and pressing Ctrl + Enter. The tool will be imported.
3. Enter your custom blueprint sequence in the next code cell as a string of 17 bases. Alternative to a blueprint sequence, a fasta file containing a single blueprint can be used as input: input_-data = ["my_blueprint.fa"].
4. Define the TIFs in my_strengths as comma-separated integers. If a single, specific spacer sequence is desired, add lib_size=1 as argument of run_tool. Run the cell.
5. The output will be generated below the cell. This might take a while.
6. The generated csv file can be found in the working directory. If another output path is desired, it can be specified in **step 3** of this section by adding output_path=PATH as argument for run_tool.

3.2.3 Cloning Design

1. Select the library blueprint or a discrete spacer sequence.
2. Insert your selection in the insulator sequence (see Subheading 3.1 for reference of the insulator sequence).
3. Assemble and clone the resulting promoter sequence upstream of RBS + ORF of choice.

3.3 Predicting TIF by Evaluation of Custom Spacers

As discussed in Subheading 3.1, TIF can be dependent on the local context, which is why the prediction task is only expected to work reliably if the insulator sequence (except spacer) is kept intact.

3.3.1 Using the Terminal

1. Open the terminal, activate the environment containing the packages described in Material.
2. Navigate to the directory containing the cloned GitHub repository.
3. Evaluate one or multiple spacer sequences by calling for example python ProD.py NNNCGGGNCCNGGGAAA AAACCCGGGTTTTAAAC. Alternatively, the sequence (s) can be provided as fasta file: python ProD.py ex_seqs.fa

4. The output is displayed in the terminal and additionally, a file my_predictions.csv is saved in the working directory. If another output path is desired, it can be specified in the previous step by adding -output_path PATH.

3.3.2 Using Jupyter

1. Open Jupyter in the environment containing the packages specified in Material and navigate to the directory containing the GitHub repository or open Binder via the ProD GitHub website.
2. Locate the 'Evaluate Custom Spacers' section at the bottom of the notebook.
3. Provide input data either as list input_data = ["AAAACCCCGGGGTTTTT", "TTTTCCCCGGGGTTTTT", "TTTTGGGGCCCCAAAAA"] or as file input_file = ["ex_seqs.fa"]
4. The output is displayed below the cell and additionally, a file my_predictions.csv is saved in the working directory. If another output path is desired, it can be specified in the previous step by adding output_path=PATH as argument for run_tool.

4 Notes

1. Binder is a service that can execute git repositories on the cloud for free. This can be achieved by visiting https://mybinder.org, or follow the link provided in the README. As such, no prerequisites are necessary. When using Binder, changes in the file are not saved.
2. The trained model is saved as model_RPOD.pt in the model folder and is used by the ProD tool to make predictions on novel input.
3. The 17 nucleotides sequence(s) are processed by a convolutional neural network of four 1×1, 16 1×4, and 32 1×2 convolutions and two fully connected layers of 128 and 64 nodes.
4. When evaluating the model, no specific nucleotide was found to occur with high frequency that could not be explained by the input blueprint. Therefore, we cannot confirm the findings of other studies that a single nucleotide has an effect on promoter TIF [12–14], but rather suggest the influence of more complex structural features.
5. When creating a library, the tool first creates all possible sequences from the blueprint input, second, it predicts the TIFs of each sequence and retains those matching the requested TIF. Spacers are then sampled to construct a

degenerate sequence for the library blueprint (unless the input allows for more than 500,000 possible sequences). If the input sequence for generating libraries allows for more than 500,000 possible sequences, the tool will randomly sample 100,000 possible sequences to evaluate. The processing time might be longer, and no library blueprint will be generated.

6. While the original insulator was successfully tested on plasmid and genome level, the in vivo validation of our tool was performed on plasmids.
7. The input DNA sequence may contain any letter of the mixed-base code (R Y M K S W H B V D N).

Acknowledgements

F.M. holds a PhD grant (61478) from FWO (Fonds Wetenschappelijk Onderzoek – Vlaanderen). This research was also supported by the BOF-IOP project "MLSB" (BOF16/ IOP/040) of the Bijzonder Onderzoeksfonds, the FWO research project "SynSysBio4COS" (G0B8118) of the Research Foundation – Flanders (FWO) and by the H2020 project "BioRoboost" of the European Union (820699). W.W. also received funding from the Flemish Government under the "Onderzoeksprogramma Artificielë Intelligentie (AI) Vlaanderen" Programme.

References

1. Nielsen AA, Segall-Shapiro TH, Voigt CA (2013) Advances in genetic circuit design: novel biochemistries, deep part mining, and precision gene expression. Curr Opin Chem Biol 17:878–892. https://doi.org/10.1016/j.cbpa.2013.10.003
2. Bradley RW, Buck M, Wang B (2016) Tools and principles for microbial gene circuit engineering. J Mol Biol 428:862–888. https://doi.org/10.1016/j.jmb.2015.10.004
3. Brophy JAN, Voigt CA (2014) Principles of genetic circuit design. Nat Methods 11:508–520. https://doi.org/10.1038/nmeth.2926
4. Jeschek M, Gerngross D, Panke S (2017) Combinatorial pathway optimization for streamlined metabolic engineering. Curr Opin Biotechnol 47:142–151. https://doi.org/10.1016/j.copbio.2017.06.014
5. Coussement P, Bauwens D, Maertens J, De Mey M (2017) Direct combinatorial pathway optimization. ACS Synth Biol 6:224–232. https://doi.org/10.1021/acssynbio.6b00122
6. De Mey M, Maertens J, Lequeux GJ et al (2007) Construction and model-based analysis of a promoter library for E. coli: an indispensable tool for metabolic engineering. BMC Biotechnol 7(34). https://doi.org/10.1186/1472-6750-7-34
7. Salis HM (2011) Chapter two – the ribosome binding site calculator. In: Voigt C (ed) Methods in enzymology. Academic Press, pp 19–42
8. Salis HM, Mirsky EA, Voigt CA (2009) Automated design of synthetic ribosome binding sites to control protein expression. Nat Biotechnol 27:946–950. https://doi.org/10.1038/nbt.1568
9. Bervoets I, Van Brempt M, Van Nerom K et al (2018) A sigma factor toolbox for orthogonal gene expression in Escherichia coli. Nucleic Acids Res 46:2133–2144. https://doi.org/10.1093/nar/gky010

10. Van Brempt M, Clauwaert J, Mey F et al (2020) Predictive design of sigma factor-specific promoters. Nat Commun 11:5822. https://doi.org/10.1038/s41467-020-19446-w
11. Davis JH, Rubin AJ, Sauer RT (2011) Design, construction and characterization of a set of insulated bacterial promoters. Nucleic Acids Res 39:1131–1141. https://doi.org/10.1093/nar/gkq810
12. Meng H, Wang J, Xiong Z et al (2013) Quantitative design of regulatory elements based on high-precision strength prediction using artificial neural network. PLoS One 8:e60288. https://doi.org/10.1371/journal.pone.0060288
13. Kiryu H, Oshima T, Asai K (2005) Extracting relations between promoter sequences and their strengths from microarray data. Bioinformatics 21:1062–1068. https://doi.org/10.1093/bioinformatics/bti094
14. Rhodius VA, Mutalik VK, Gross CA (2012) Predicting the strength of UP-elements and full-length E. coli σE promoters. Nucleic Acids Res 40:2907–2924. https://doi.org/10.1093/nar/gkr1190

Chapter 5

Computational Study on the Dynamics of Mycobacterium Tuberculosis RNA Polymerase Assembly

Sneha Bheemireddy and Narayanaswamy Srinivasan

Abstract

Gene regulation is an intricate phenomenon involving precise function of many macromolecular complexes. Molecular basis of this phenomenon is highly complex and cannot be fully understood using a single technique. Computational approaches can play a crucial role in overall understanding of functional and mechanistic features of a protein or an assembly. Large amounts of structural data pertaining to these complexes are publicly available. In this project, we took advantage of the availability of the structural information to unravel functional intricacies of *Mycobacterium tuberculosis* RNA polymerase upon interaction with RbpA. In this article, we discuss how the knowledge on protein structure and dynamics can be exploited to study function using various computational tools and resources. Overall, this article provides an overview of various computational methods which can be efficiently used to understand the role of any protein. We hope especially the nonexperts in the field could benefit from our article.

Key words Protein complex, Computational methods, Protein dynamics, Allostery, Structural features

1 Introduction

Prokaryotes are single-celled organisms and lack distinct nucleus or organelles, yet they manage to perform all essential functions using intracellular organization [1, 2]. They depend on large macromolecular complexes for all critical processes ranging from cell division to differentiation [3]. Eukaryotes also use this compartmentalization and organization strategies for their functions like nucleolus formation [4], lipid metabolism, and many more. Prokaryotes are perfect model organisms to study fundamental cellular processes and are far less complex to understand than eukaryotes. They can synthesize and regulate genes based on environmental conditions [5–7]. The potential of the cells to express genetic information is

Dedication This book chapter is dedicated to Prof. N. Srinivasan, who passed away on September 3rd, 2021.

Eveline Peeters and Indra Bervoets (eds.), *Prokaryotic Gene Regulation: Methods and Protocols*,
Methods in Molecular Biology, vol. 2516, https://doi.org/10.1007/978-1-0716-2413-5_5,

known as gene regulation. It involves various mechanisms which include DNA replication, transcription, and translation. An understanding of gene regulation will provide us insights into evolution [8], help us with advancements in biotechnology [9], aid in drug discovery [10], and solve many complex questions.

In general, macromolecules do not function in isolation, but form functional complexes in association with other macromolecules. The disruption of these macromolecular complex interaction networks forms the molecular basis for many diseases [10]. Hence it is imperative to study their properties, such as kinetics, and dynamics of complex formation [11]. Three-dimensional (3D) structure is required for a detailed characterization of its biochemical and biophysical behavior. Biophysical techniques like NMR, X-ray crystallography, and cryo-EM [12] can unearth the structural intricacies of the proteins, but every method has own its strengths and weaknesses. Structure determination techniques have undergone many advancements during the past few decades [13]. Even after all these developments, biophysical experiments have few limitations. Many membrane protein complexes are difficult to crystallize [14], and solution NMR studies are limited by the molecular weight of complex, and cryo-EM is challenging when protein size is small (<150 kDa).

In an attempt to further our understanding of macromolecular complexes, structural biology is supported with development of computational tools which are based on strong principles from physics, chemistry, and data analysis techniques. There has been a surge of data in biology hence demanding the need for computation. From visualization to data analysis, sequencing to structure prediction, many *in silico* methods and techniques have been developed. Various computational tools are available for protein modeling [15], analyzing high-throughput genome sequencing [16], drug development [17], structural analysis, and many more. All these computational tools also have limitations and the results obtained using computational tools seek experimental validation.

The regulation of gene expression in prokaryotic cells occurs at the transcriptional level. A multi-step process called transcription initiation is the major regulatory determinant for gene expression [18]. RNA polymerase (RNAP) holoenzyme carries out the transcription, which consists of the five-subunit core ($\alpha1$, $\alpha2$, β, β' and ω) and a σ subunit involved in specific promoter recognition [19]. Transcription initiation begins when RNAP holoenzyme (R) binds with a promoter sequences (P) to form the closed complex (RPc). The free energy associated with binding then drives multiple conformational changes in both the RNAP and promoter resulting in the isomerization of RPc to RPo (open complex). After binding, RNA polymerization begins as nucleotides complementary to the template DNA are incorporated into a growing RNA transcript. Actinobacteria have unique transcription factors, including CarD and RbpA, which cooperatively stabilize the RPo [20, 21], and, as a result, have been characterized as transcriptional activators. RbpA is upregulated when subjected to bacterial stresses and is essential for growth under nutrient-rich conditions [22]. RbpA is vital for maintaining cellular homeostasis and modulating *Mycobacterium tuberculosis* (*Mtb*) gene expression to drive the pathogenesis of tuberculosis. As RNAP is a direct target for first-line antibiotics used to treat *Mtb* infection [23], further mechanistic studies of the mycobacterial transcription may lead to novel therapeutic strategies.

In this article, we discuss a few *in silico* techniques which are used in our group to analyze macromolecular complexes exemplifying RNA polymerase as a case study. Interaction of a protein with another protein might result in a change in the 3D structure or dynamics or both. Most of the protein structural changes occur at the site of interaction (protein-protein interface). Sometimes, they occur at a location far from the interface suggesting allostery. Here we discuss techniques to study structural similarity (RMSD), properties of the interface (interfacial and hot spot residues), normalized square fluctuations at the residue level, the correlation between residue movements, and allostery in proteins.

2 Materials

This section describes the databases and tools used in our methodology to understand the structural and dynamical features of a protein complex.

2.1 Protein Data Bank (PDB)

Experimentally determined macromolecular structures (proteins, nucleic acids, etc.) and their complexes obtained using X-ray, NMR, and cryo-EM are archived in a worldwide repository called Protein Data Bank (PDB). PDB contains 3D coordination data, information about polymer sequence, experimental method, and other structure descriptions. As of August 2021, PDB hosts 181,163 structures. The structures available through PDB are subjected to various validation measures before making them available online. PDB is updated weekly and is managed by Research Collaboratory for Structural Bioinformatics (RCSB) since 1998. We used PDB to retrieve the data of experimentally determined structures of protein complexes. RCSB PDB [24] is available at https://www.rcsb.org. PDB entries are mainly available from three data centers: the RCSB PDB, PDB Europe (PDBe) [25] and PDB Japan (PDBj) [26], and are centralized by worldwide PDB (wwPDB) [27].

2.2 PyMOL

Molecular visualization can provide structural, biological, and pharmacological insights into a macromolecular complex. PyMOL [28] is a widely used open-source graphical visualization program that can be employed for the visualization and analysis of protein structures. This software runs on a variety of platforms operating with command line and visual interfaces. Python scripts can be incorporated into PyMOL. It is used to produce high-quality 3D images of macromolecules and small molecules. PyMOL's latest version can be downloaded from https://pymol.org/2/.

There are other molecular visualization programs, namely VMD (Visual Molecular Dynamics) [29], Chimera [30], 3Dmol.js [31] and NGL viewer [32]. VMD has been designed with the ability to animate MD simulation trajectories. Chimera has the capability to visualize large-scale molecular assemblies like viral coats. 3Dmol.js is a JavaScript library that can be used to provide an interactive 3D representation of molecular data. The NGL viewer is a web application used for visualization and can be embedded in other websites.

2.3 Modeller

The modeller [15] program performs homology and comparative protein structure modeling where a three-dimensional (3D) model is obtained by optimally satisfying spatial restraints derived from the alignment of the query sequence with the structure of its homologues. Modeller can also be used to perform other tasks like *de novo* modeling of loops in protein structures, optimization of protein models, comparison of protein structures, etc. Modeller is available both as a stand-alone package and can be imported as a python package. Modloop is a webserver based on modeller and is available to automate loop modeling. Modbase is found at https://modbase.compbio.ucsf.edu/modloop/.

2.4 TM-Align

TM-align [33] is an algorithm for sequence-independent protein structure comparisons. It aligns topologically equivalent residues based on structural similarity using heuristic dynamic programming iterations. TM-align recognizes local structural matches between protein pairs and assigns TM-score, a measure of structural similarity. An optimal superposition of two structures is obtained based on TM-score, which is used to scale structural similarity. TM-score varies from 0 to 1, where 1 indicates a perfect match between two structures. TM-score less than 0.2 corresponds to randomly chosen unrelated proteins. TM-align as a webserver and downloadable suite can be found at https://zhanglab.ccmb.med.umich.edu/TM-align/.

Dali Lite [34] and CE [35] are structural alignment methods that directly search for optimal alignment by joining small aligned substructures together. Deep align [36] is similar to TM-align where structural alignment is found after iteratively optimizing a rough initial alignment using cycles of geometric superposition.

2.5 The FoldX Suite

FoldX [37] is an empirical force field that can be used to determine the interaction energy of protein complexes, evaluation of protein stability, and the effect of point mutations. FoldX can be found at http://foldxsuite.crg.eu/. It contains terms for de-solvation energies, coulombic interactions, van der Waal's forces, hydrogen bonding, entropic changes, and electrostatics. The FoldX suite includes many sub-modules that can perform various functions. Here we use two modules of FoldX, namely AlaScan and AnalyseComplex.

AlaScan module of the FoldX mutates every residue to alanine one by one and estimates the difference in energy between "mutant" and "wild-type" structure. FoldX can help mutate protein and DNA side chains using a probability-based rotamer library while exploring alternative conformations of the surrounding side chains. AnalyseComplex module determines the interaction energy between two molecules. The interaction energy is obtained by the difference between global energy and the sum of individual energies while target molecule stability is calculated through unfolding.

2.6 NACCESS

NACCESS is a stand-alone program and is freely accessible. It calculates the accessible area of a 3D structure of proteins and nucleic acids using atomic and residue accessibilities. The program uses the Lee & Richards method [38], whereby a probe of a given radius (1.4 Å) is rolled around the surface of the molecule and the path traced out by its center is the accessible surface. NACCESS is available at http://wolf.bms.umist.ac.uk/naccess/.

2.7 Protein Interactions Calculator (PIC)

PIC [39] is a webserver that recognizes intra- and inter-protein interactions for a given PDB file. PIC recognizes various kinds of interactions such as disulphide bonds, hydrophobic interactions, ionic interactions, hydrogen bonds, and aromatic interactions. It considers both solvent accessibility and distance criteria to identify interfacial interactions. PIC is accessible as a webserver from http://pic.mbu.iisc.ernet.in/.

2.8 Programming Tools

Programming knowledge is often necessary while dealing with extensive data. Python is a rapidly developing programming language that is widely used in the latest bioinformatic tools. Here are some python packages which help to obtain information about protein dynamics.

2.8.1 ProDy

ProDy [40] is an open-source python package for protein structure and dynamics analysis. It allows for quantitative characterization of structural variations in heterogeneous datasets of structures experimentally resolved for a given biomolecular system and compares these variations with the theoretically predicted equilibrium dynamics. ProDy allows systematic comparison of proteins and helps with the intrinsic dynamics and helps to gain insight into the relationship between structure, dynamics, and function. ProDy tools and documentation are available at http://prody.csb.pitt.edu/.

We make use of Perturbation Response Scanning (PRS) Package to study allostery in proteins. Other web servers are available for the prediction of allosteric sites in a protein. These include SPACER (based on binding leverage approach), the PARS method (difference in the normal modes of ligand-bound and unbound structures), AlloPred (make the allosteric site rigid and calculate the effects at the active site), DynOmics (finds hinge residues that control two slowest normal modes). PRS can also be used using results of molecular dynamics simulation. Machine learning methods to predict allostery have been implemented in AlloSite (based on support vector machines). Allosteric sites are shown to be conserved in evolution. Few studies developed methodologies for the prediction of these allosteric pockets using evolutionary principles.

2.8.2 Biopython

Biopython [41] is a set of freely available computational tools written in python. Biopython contains various features, which include parsers for different file formats (BLAST, Clustalw, FASTA, GenBank), various clustering modules, access to online servers (NCBI), interfaces to multiple programs (DSSP), etc. One can find Biopython tools and documentation at https://biopython.org/docs/1.74/api/index.html.

2.8.3 Bio3d

Bio3d [42] is a set of tools written in the R language developed in the interest of structural bioinformatics. Bio3d contains various features, which include clustering, distance matrix analysis, principal component analysis, etc. It takes advantage of the extensive graphical and statistical capabilities of R programming. Bio3d-cna package performs protein structure and correlation network analysis where a protein structure network is constructed based on contact map and network edges are weighted by values derived from residue dynamic-cross correlations. Bio3d documentation and details are available at http://thegrantlab.org/bio3d/.

3 Methods

This section describes a pipeline to study structural features and dynamics of RNA polymerase using *in silico* methods. A diagram of the described pipeline is shown in Fig. 1. First, it is essential to look up the availability of structural information of RNA polymerase. Then we study their structural features and elucidate their dynamic behavior using the methods described below. We explain all these methods with the help of RNA polymerase as a case study at every step.

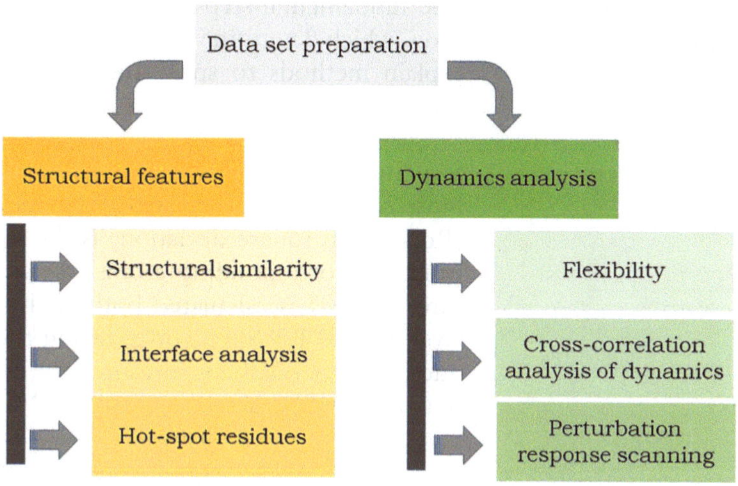

Fig. 1 A simplified flow chart of the methods used to examine structural and dynamics features of protein complexes

3.1 Dataset Preparation

Preparing a curated dataset is a significant step. Here are the steps that were followed in dataset preparation.

1. We begin by identifying 3-D structures of *Mtb* RNA polymerase from PDB. It is advisable to prepare a table of available structures with details of resolution of the structures, method of structure determination, length of the proteins in the structure, details of missing residues, and information on mutations.
2. All the properties mentioned above will help us filter the structures. Better resolution of an X-ray structure increases the confidence in the positions of atoms (*see* **Note 1**).
3. Proteins with high length coverage should be selected. Sometimes deposited structures may lack coordinates of a few residues in the protein chain. It is better to consider proteins with a lesser number of missing residues.
4. We should not select a structure with mutations as we often aim to understand the protein dynamics in its native structure.

We have identified two structures RbpA bound (PDB code: 6C04) and RbpA unbound (PDB code: 5UH5). We could not find any crystal structure of *Mtb* RNA polymerase bound with RbpA. So, we chose the cryo-EM structure. Both structures had atomic coordinate entries of a few residues (1011–1021) of the β' chain missing. We used loop modeling. We generated 100 models and a model with a high DOPE score was selected for further analysis (*see* **Note 2**).

3.2 Structural Features

The fundamental step would be to study the available 3D structures, which determine the functional properties. In this section, we explain methods to study the structural properties of the RNA polymerase complex.

3.2.1 Structural Similarity

Structural similarity is measured using various parameters. The most common measure is RMSD. As the term suggests, RMSD (Root mean square deviation) is the square root of the deviation from the corresponding atoms' mean position from both structures. RMSD is calculated both at the global and individual Cα-level. Global RMSD involves taking the mean among all the residues, but Cα-RMSD is measured for every residue. Here we demonstrate RMSD calculations using TM-align.

The steps involved are:

1. PDB structures are submitted to TM-align, and the input options include normalizing TM-score by the length of the structures, specified alignment, algorithm for alignment. Here we have used default options. The command used for locally executable TM-align is given below:

   TMalign PDB1.pdb PDB2.pdb -o PDB_sup.pdb

2. We wrote a python code using the RMSD formula [43] to calculate RMSD per residue.

Global RMSD was ranging from 0.7 Å to 2 Å where 2 Å corresponds to σ-factor. RbpA is involved in direct interaction with the σ-factor and from our observation, it is possible that RbpA binding causes structural changes in RNA polymerase. From the local Cα-RMSD (*see* Fig. 2), we could identify residues with considerable local structural change. Notably, we have also observed structural changes at a site far from the interface. These observations give us clues about allostery (*see* **Note 3**).

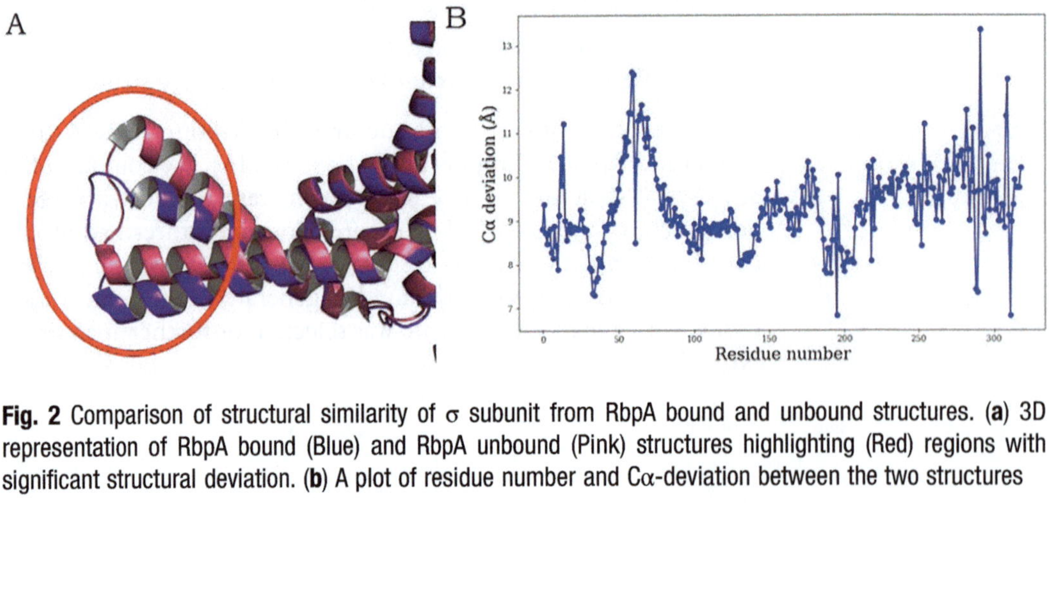

Fig. 2 Comparison of structural similarity of σ subunit from RbpA bound and unbound structures. (**a**) 3D representation of RbpA bound (Blue) and RbpA unbound (Pink) structures highlighting (Red) regions with significant structural deviation. (**b**) A plot of residue number and Cα-deviation between the two structures

3.2.2 Analysis of Protein-Protein Interface

A protein-protein interface usually shows high degrees of geometric and chemical complementarities [44]. Electrostatic and hydrophobic forces are believed to play an essential role in protein-protein interactions [45]. It is vital to see the structural changes that happen at interfaces upon interaction with another protein. In this section, we discuss various methods to study the properties of protein interfaces.

Interfacial Residues

One can either use distance criteria or solvent accessibility criteria for the identification of interfacial residues. If any two atoms (from two different proteins) fall within a distance cut-off (4.5 Å or 5 Å), residues corresponding to these atoms are regarded as interfacial residues. In solvent accessibility criteria, if relative solvent accessibility (RSA) of two residues is above a threshold, then we consider them as interfacial residues. There is no single number used as a cut-off, but it ranges from 5–16%, with higher thresholds leading to a lower number of solvent-exposed residues.

Here we explain the steps to study protein interface. We have submitted the PDB structures to the PIC server, and interfacial residues obtained were mapped onto the protein structure. Figure 3 highlights the interfacial residues between α1 and α2 proteins upon association with RbpA. It is observed that the interactions between them were majorly due to hydrophobic interactions supported by ionic and hydrogen bonding. We have also observed that the intra-protein interactions change upon interaction with RbpA.

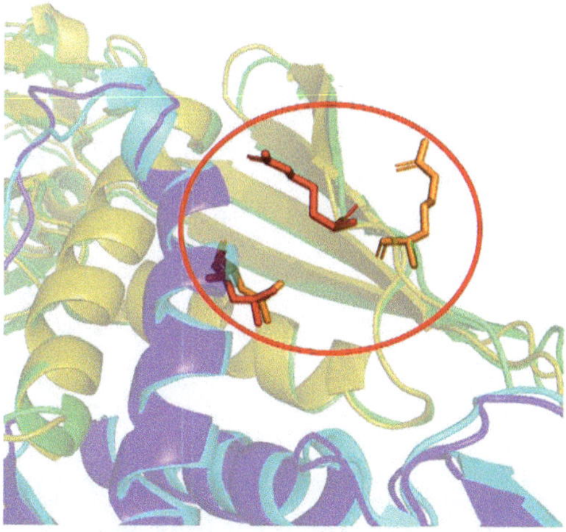

Fig. 3 Comparing interfacial residues of two proteins. 3D representation of RbpA bound (Cyan and Mustard) and RbpA unbound (Purple and Green) structures highlighting interfacial residues in bound (Red) and unbound (Orange) form

Calculation of Interaction Energy

Protein-protein interactions involve various kinds of interactions, including van der Waals, electrostatic, hydrogen bond, and hydrophobic interactions. Experiments can give us the equilibrium dissociation constant (Kd) for protein complexes which will help us calculate Gibbs free energy (interaction energy) of binding. However, calculating protein-protein binding free energy theoretically is still a challenge due to the complexity of various interactions at play [46]. Four significant factors limit theoretical calculations: the rotamer approximation [47], the inaccuracy of electrostatics and solvation [48], the denatured state models [49], and the errors in side-chain entropy calculations [50]. Attempts are being made to solve each one of these problems. Here we describe one of the widely used methods for interaction energy calculations. FoldX mainly relies on empirical data derived from experimental work on proteins and is the fast and accurate estimation of mutational free energy changes on the stability of a protein or a protein complex.

We have executed the following command to obtain interaction energy using FoldX. This input also has other options to specify chain names and complex with nucleic acids:

FoldX –command=AnalyseComplex –pdb=PDB1.pdb

Figure 4 is an example output of the above command. Upon analyzing results from RNA polymerase complex, it was observed

```
interaction between A and B

BackHbond         =            -1.59
SideHbond         =            -3.14
Energy_VdW        =            -20.09
Electro           =            0.85
Energy_SolvP      =            18.62
Energy_SolvH      =            -32.95
Energy_vdwclash =              4.10
energy_torsion  =              0.50
backbone_vdwclash=             0.77
Entropy_sidec     =            14.35
Entropy_mainc     =            2.99
water bonds       =            0.00
helix dipole      =            0.00
loop_entropy      =            0.00
cis_bond          =            0.00
disulfide         =            0.00
kn electrostatic=              0.57
partial covalent interactions = 0.00
Energy_Ionisation =            0.04
Entropy Complex =              2.38
------------------------------------------------------------
Total             =                                   -15.74
```

Fig. 4 Sample output obtained from FoldX (module: AnalyseComplex)

that binding of RbpA improves the interaction between the subunits making stronger associations.

Hot Spot Residues

With improvement in the understanding of protein-protein interactions, it was realized that there are regions at the interfaces which contribute richly to binding free energy. These regions are termed "Hot spots" and the residues present in these regions are called "Hot spot residues" [51]. Hot spot residues are those residues that bring in energy change greater than 2kcal/mol when mutated to alanine. These residues are essential for protein function and might also play a significant role in allosteric regulation.

We have executed the following command to obtain interaction energy using FoldX.

FoldX --command=AlaScan --pdb=PDB1.pdb

The output file contains $\Delta\Delta G$ of mutating a residue to alanine. We can use this file to identify those interfacial residues based on the criteria mentioned above for hot spot residues (*see* **Note 4**). The hot spot residues of RbpA bound and unbound structures were identified, and we learned that association of RbpA brings changes in the hot spot residue network. Figure 5 highlights the distribution of the hot spot residue network in RbpA bound and unbound forms.

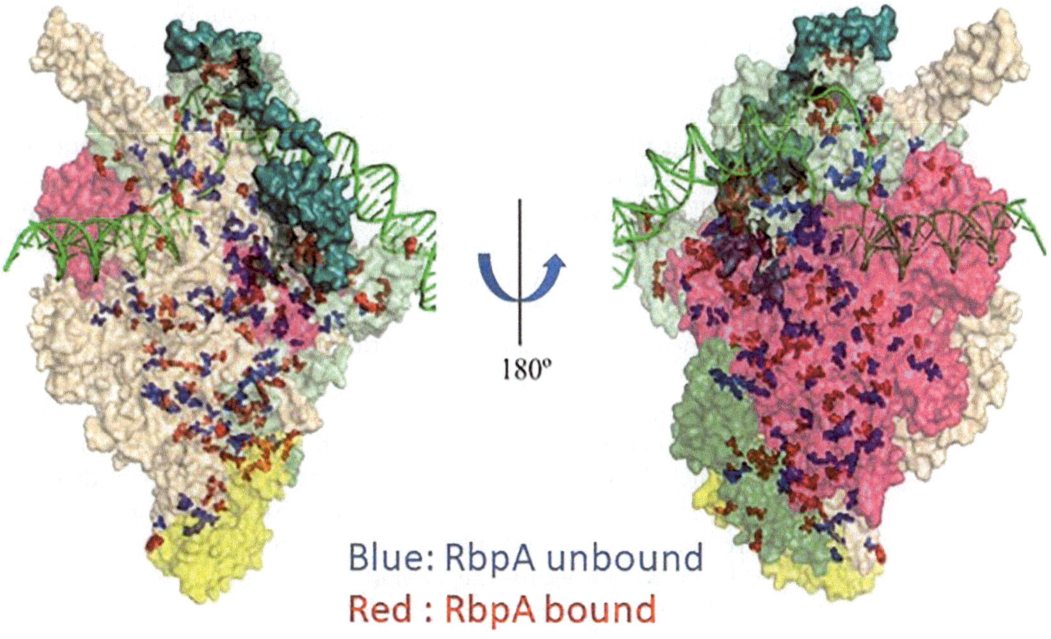

Fig. 5 Hot spot residues of RbpA bound (Red) and unbound (Blue) polymerase complexes were mapped onto its 3D structure

3.3 Features of Dynamics

Although proteins adopt unique 3D structures, they are not static entities. Proteins can sample numerous conformations in the neighborhood of their native conformation. They also transition between various conformational states' ensembles [52–55]. This dynamic nature enables the proteins to perform various functions, including allostery [56]. The study of protein dynamics is concerned with this transition of protein over multiple states. One can look at protein dynamics from kinetic (concerned with the movement of protein transition between states) and dynamics (nature and equilibrium populations). In this article, we discuss methods to understand this kinetic perspective of protein dynamics.

One can use techniques like NMR spectroscopy to study the protein dynamics [56], but it has its limitations in terms of time scale, protein size, etc. Despite recent advances in biophysical techniques, protein dynamics is primarily a task for theoretical methods. One of the most common and powerful of them is atomistic molecular dynamics (MD) [57, 58]. It is a rigorous way that enables the accurate representation of protein dynamics under a physiological-like environment. In MD simulation of a protein, dynamics is studied using the classical Newtonian dynamic equations. Unfortunately, MD is complex and computationally expensive. Coarse-grained methods coupled to simple potentials are alternatives to MD [59]. By using coarse-grained methods, we lose atomic detail but gain computational simplicity.

Normal Mode Analysis (NMA) is a method for studying long-range motions associated with proteins [60]. Low-frequency collective motions termed as global motions of proteins have been shown to signify a biologically relevant function. NMA can be used to obtain these details of molecular motions of proteins. NMA takes a set of cartesian coordinates of a protein structure and a forcefield that defines the interaction between atoms. A "Hessian" matrix is then generated from the second derivate of the potential energy and is diagonalized to yield eigen vectors and eigen values. An all-atom NMA poses a significant computational problem due to massive calculations to solve complete spectra of motions. Hence generally, Cα-NMA is used.

3.3.1 Square Fluctuations

The most used dynamical similarity measure is the correlation of aligned residues. Concerted motions are conditioned by the lowest energy modes of fluctuations around their equilibrium conformation. Comparing the magnitude of this correlated movement of residues is one of the properties that characterize dynamics [61]. These atomic fluctuations are comparable to X-ray B-factors (Temperature factors). Commonly, C-alpha fluctuations are considered to represent the overall magnitude of fluctuations of residues.

Here are the steps followed to obtain square fluctuations:

1. Parse the PDB file and specify atom type (if required).

   *from prody import **

   structure = parsePDB('PDB_ID')

   calpha = structure.select('name CA')

2. We perform normal mode analysis using the ANM (Anisotropic Network Model) (NMA) function and build hessian matrix.

   anm = ANM('PDB_code anm')

   anm.buildhessian(calpha)

3. We calculate modes from the Hessian obtained and write it to a file and visually inspect the modes to select normal modes manually.

   anm.calcModes()

   writeNMD ('PDB_ code.nmd',anm,calpha)

4. Square fluctuations are normalized and plotted with their residue number.

   showSqFlucts(anm)

Figure 6 shows square fluctuations obtained using the NMA, and it can be observed that there are regions in the proteins that show higher fluctuations (*see* **Note 5**). From our analysis, we could identify changes in the fluctuations in the proteins that are not involved in RbpA interaction.

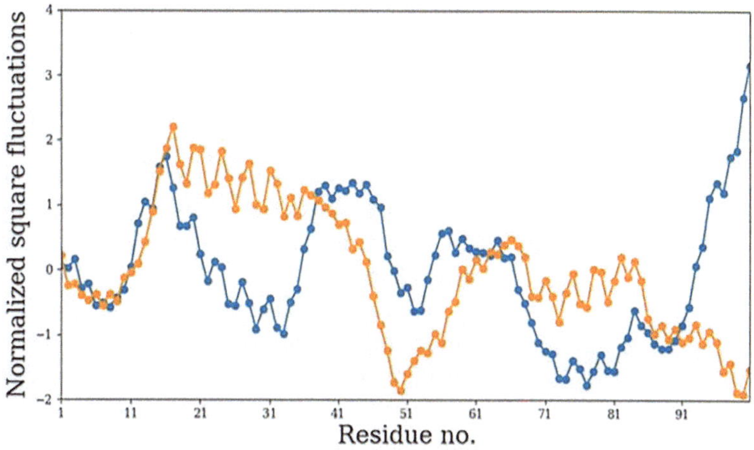

Fig. 6 A sample plot comparing normalized square fluctuations of two proteins

3.3.2 Dynamic Cross-Correlation Matrix

The correlation of the internal motions suggests that the structure of the molecule controls collective effects present in a protein [62]. The degree of correlation of the local fluctuation would be difficult to detect from cross-correlations if the region itself is involved in a large-amplitude correlated motion. Dynamic cross-correlation matrix informs us about the extent of correlation in the motions among the protein complex residues and their effect upon interaction with another protein.

Steps used to calculate dynamic cross-correlation were described as follows:

1. Parse the PDB file.

   pdb <- read.pdb (system.file("PDB1.pdb", package= "bio3d"))

2. We perform normal mode analysis using anm function.

   modes <- nma(pdb)

3. We calculate the correlation matrix using output obtained from NMA.

   cm <- dccm.nma(modes)

4. The correlation matrix obtained was plotted, and the values were normalized to an interval of -1 and 1.

   plot (cm, sse = pdb, contour = FALSE, col.regions = bwr.colors (20), at = seq(-1, 1, 0.1))

We could not find any differences in terms of cross-correlation matrix. It can be inferred that RbpA does not change correlations of residue motions of RNA polymerase.

3.3.3 Perturbation Response Scanning (PRS)

Allosteric regulation of protein function is critical in regulating cellular processes [63], so its underlying mechanisms are of primary concern for protein engineering [64] and drug design [65]. Allostery can happen in many ways; generally, it involves an effector modifying the protein's behavior at the active site. Effectors can either enhance or decrease the activity of a protein. The general mechanism of allostery is based predominantly on mechanical operations at the molecular and atomic scales. The long-range nature of allosteric communication is often well described by low-frequency modes that involve the motion of the majority of atoms [56].

We can measure the response of effectors through the determination of kinetic parameters. This can be achieved by applying linear response theory (LRT) to proteins under selected external perturbations [66]. PRS is based on applying this LRT to understand the basis for structural changes in proteins. A protein is represented as an elastic network model using C-alpha atoms. A force of magnitude ΔF unit is applied on the Cα atom of each residue, and the direction of the force applied is chosen randomly (isotropic). Then the whole protein is scanned for the response of a single perturbation. The result of perturbation is recorded both in terms of magnitude and direction of the response. These perturbations are repeated at least 1000 times on each residue, and the average of all responses is considered for further analysis.

Steps to study allostery in proteins is described below:

1. PRS is performed using ProDy package. We begin by performing Cα-NMA.

   calpha = parsePDB('PDB_code', subset='ca')

   anm = ANM('Protein PRS')

   anm.buildHessian(calpha)

   anm.calcModes()

2. Perturbation response is obtained using the "calcPerturbResponse" function and is represented as a heatmap.

   show = showPerturbResponse(anm_ampar, atoms=ampar_ca)

Using PRS, we were able to obtain effector and sensor residues of RNA polymerase complex. Figure 7a represents the effectivity of all the residues in RNA polymerase where red represents the highest effectiveness and blue represents the least effectiveness. Figure 7b represents the average sensitivity of the residues, where red implies the most sensitive residues and blue indicates the least sensitive residues. From our analysis, we could identify regions at the site of DNA binding as sensors upon perturbation of residues interacting with RbpA.

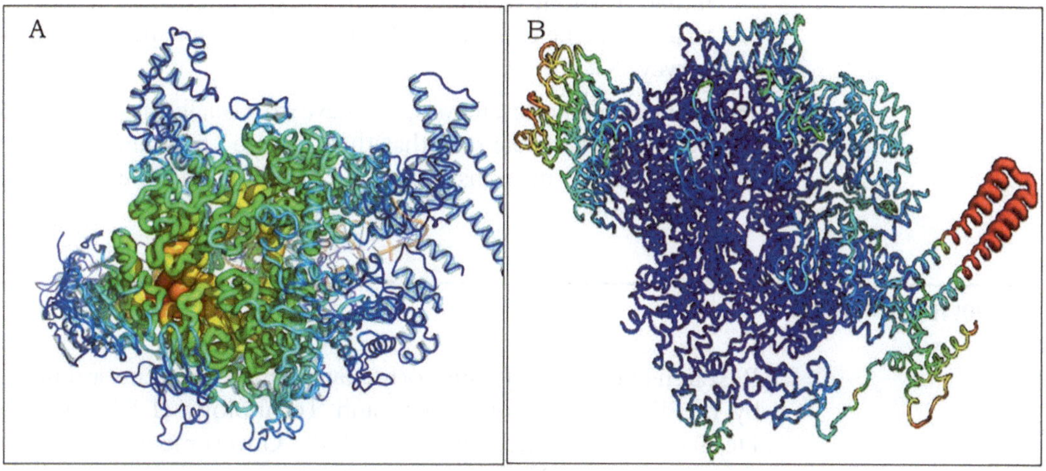

Fig. 7 (**a**) Average effectiveness profile of RNA polymerase where red represents the highest effectiveness and blue represents the least effectiveness. (**b**) Average sensitivity of the RNA polymerase, where red implies most sensitive residues and blue indicates least sensitive residues

4 Conclusion

Using our pipeline, we could study the structural and dynamic changes that occur in RNA polymerase complex upon interacting with RbpA. RbpA brings significant structural changes in the polymerase complex and improves the inter-protein interactions. Binding of RbpA changes the hot spot residue network and magnitude of fluctuations in all the proteins. RbpA regulates the transcription by acting upon DNA binding residues.

5 Notes

1. If two structures are of same resolution and one of them is determined using X-ray crystallography and another using cryo-EM, one selects the X-ray structure. It is because resolution of a cryo-EM structure is an average number. Hence it should not be considered equivalent to the X-ray structure resolution.
2. User has to specify number of models to be generated by modeller. It is advised to generate at least 100 models and select the best among them using DOPE score.
3. Visualization plays an important role in understanding the underlying mechanism. We should inspect the regions with higher RMSD visually for a better understanding about macromolecular complex.
4. Although we use in silico methods for detection of hot spot residues, many of the protein-protein interactions have accumulated alanine scanning data from experiments. So, it is better to look for the experimental data if any available from literature survey.
5. In normal mode analysis, we have to visually inspect all the 20 modes to make sure that there are no knots and physically inaccessible motions of the proteins included in our calculations.

Acknowledgments

Research in our group is supported by the FIST program sponsored by the Department of Science and Technology (DST) and the Department of Biotechnology (DBT), Government of India in the form of IISc-DBT partnership programme and Bioinformatics and Computational Biology Centre. SB acknowledges CSIR, Government of India toward the research fellowship and NS is a J. C. Bose National Fellow.

References

1. Rudner DZ, Losick R (2010) Protein subcellular localization in bacteria. Cold Spring Harb Perspect Biol 2:a000307
2. Govindarajan S, Amster-Choder O (2016) Where are things inside a bacterial cell? Curr Opin Microbiol 33:83–90
3. Allison C, Hughes C (1933) Bacterial swarming: an example of prokaryotic differentiation and multicellular behaviour. Sci Progr 75(3–4): 403–422
4. Brangwynne CP (2013) Phase transitions and size scaling of membrane-less organelles. J Cell Biol 203:875–881. https://doi.org/10.1083/jcb.201308087
5. Dekel E, Alon U (2005) Optimality and evolutionary tuning of the expression level of a protein. Nature 436:588–592. https://doi.org/10.1038/nature03842
6. Keren L, Hausser J, Lotan-Pompan M et al (2016) Massively parallel interrogation of the effects of gene expression levels on fitness. Cell 166:1282–1294.e18. https://doi.org/10.1016/j.cell.2016.07.024
7. Bauer CR, Li S, Siegal ML (2015) Essential gene disruptions reveal complex relationships between phenotypic robustness, pleiotropy, and fitness. Mol Syst Biol 11:773. https://doi.org/10.15252/msb.20145264
8. Dayhoff MO, Schwartz RM (2020) Prokaryote evolution and the symbiotic origin of eukaryotes. In: Endosymbiosis and cell biology. De Gruyter, Berlin, pp 63–84
9. Bertram R, Hillen W (2008) The application of Tet repressor in prokaryotic gene regulation and expression. Microb Biotechnol 1:2–16. https://doi.org/10.1111/j.1751-7915.2007.00001.x
10. Brötz-Oesterhelt H, Bandow JE, Labischinski H (2005) Bacterial proteomics and its role in antibacterial drug discovery. Mass Spectrom Rev 24:549–565. https://doi.org/10.1002/mas.20030
11. Berendsen HJ, Hayward S (2000) Collective protein dynamics in relation to function. Curr Opin Struct Biol 10:165–169
12. Dobson CM (2019) Biophysical techniques in structural biology. Annu Rev Biochem 88: 25–33. https://doi.org/10.1146/annurev-biochem-013118-111947
13. Papageorgiou AC, Poudel N, Mattsson J (2021) Protein structure analysis and validation with X-ray crystallography. In: Methods in molecular biology. Humana Press Inc., New York, pp 377–404
14. Levy D, Chami M, Rigaud JL (2001) Two-dimensional crystallization of membrane proteins: The lipid layer strategy. FEBS Lett. No longer published by Elsevier, pp 187–193
15. Šali A, Blundell TL (1993) Comparative protein modelling by satisfaction of spatial restraints. J Mol Biol 234:779–815
16. Pareek CS, Smoczynski R, Tretyn A (2011) Sequencing technologies and genome sequencing. J Appl Genet 52:413–435
17. Shekhar C (2008) In silico pharmacology: computer-aided methods could transform drug development. Chem Biol 15:413–414
18. Browning DF, Busby SJW (2016) Local and global regulation of transcription initiation in bacteria. Nat Rev Microbiol 14:638–650
19. Feklístov A, Sharon BD, Darst SA, Gross CA (2014) Bacterial sigma factors: a historical, structural, and genomic perspective. Annu Rev Microbiol 68:357–376
20. Rammohan J, Ruiz Manzano A, Garner AL et al (2016) Cooperative stabilization of Mycobacterium tuberculosis rrnAP3 promoter open complexes by RbpA and CarD. Nucleic Acids Res 44:7304–7313. https://doi.org/10.1093/nar/gkw577
21. Hubin EA, Fay A, Xu C et al (2017) Structure and function of the mycobacterial transcription initiation complex with the essential regulator RbpA. elife 6. https://doi.org/10.7554/eLife.22520
22. Prusa J, Jensen D, Santiago-Collazo G et al (2018) Domains within RbpA serve specific functional roles that regulate the expression of distinct mycobacterial gene subsets. J Bacteriol 200. https://doi.org/10.1128/JB.00690-17
23. Ma C, Yang X, Lewis PJ (2016) Bacterial transcription as a target for antibacterial drug development. Microbiol Mol Biol Rev 80: 139–160. https://doi.org/10.1128/mmbr.00055-15
24. Berman HM, Westbrook J, Feng Z et al (2000) The protein data bank. Nucleic Acids Res 28: 235–242
25. Gutmanas A, Alhroub Y, Battle GM et al (2014) PDBe: protein data bank in Europe. Nucleic Acids Res 42:D285–D291. https://doi.org/10.1093/nar/gkt1180
26. Kinjo AR, Suzuki H, Yamashita R et al (2012) Protein Data Bank Japan (PDBj): maintaining a structural data archive and resource description framework format. Nucleic Acids Res 40: D453–D460. https://doi.org/10.1093/nar/gkr811

27. Berman H, Henrick K, Nakamura H, Markley JL (2007) The worldwide Protein Data Bank (wwPDB): ensuring a single, uniform archive of PDB data. Nucleic Acids Res 35: D301–D303. https://doi.org/10.1093/nar/gkl971
28. DeLano WL (2002) Pymol: an open-source molecular graphics tool. CCP4 Newsl protein Crystallogr 40:82–92
29. Humphery W, Dalke A, Schulten K (1996) Visual molecular dynamics. Pdf. J Mol Graph 14:33–38
30. Pettersen EF, Goddard TD, Huang CC et al (2004) UCSF Chimera – a visualization system for exploratory research and analysis. J Comput Chem 25:1605–1612. https://doi.org/10.1002/jcc.20084
31. Rego N, Koes D (2015) 3Dmol.js: molecular visualization with WebGL. Bioinformatics 31: 1322–1324. https://doi.org/10.1093/bioinformatics/btu829
32. Rose AS, Hildebrand PW (2015) NGL viewer: a web application for molecular visualization. Nucleic Acids Res 43:W576–W579. https://doi.org/10.1093/nar/gkv402
33. Zhang Y, Skolnick J (2005) TM-align: a protein structure alignment algorithm based on the TM-score. Nucleic Acids Res 33: 2302–2309. https://doi.org/10.1093/nar/gki524
34. Holm L, Kääriäinen S, Rosenström P, Schenkel A (2008) Searching protein structure databases with DaliLite v.3. Bioinformatics 24: 2780–2781. https://doi.org/10.1093/bioinformatics/btn507
35. Levitt M, Gerstein M (1998) A unified statistical framework for sequence comparison and structure comparison. Proc Natl Acad Sci U S A 95:5913–5920. https://doi.org/10.1073/pnas.95.11.5913
36. Wang S, Ma J, Peng J, Xu J (2013) Protein structure alignment beyond spatial proximity. Sci Rep 3. https://doi.org/10.1038/srep01448
37. Schymkowitz J, Borg J, Stricher F et al (2005) The FoldX web server: an online force field. Nucleic Acids Res 33:382–388. https://doi.org/10.1093/nar/gki387
38. Lee B, Richards FM (1971) The interpretation of protein structures: estimation of static accessibility. J Mol Biol 55. https://doi.org/10.1016/0022-2836(71)90324-X
39. Tina KG, Bhadra R, Srinivasan N (2007) PIC: protein interactions calculator. Nucleic Acids Res 35:473–476. https://doi.org/10.1093/nar/gkm423
40. Bakan A, Meireles LM, Bahar I (2011) ProDy: protein dynamics inferred from theory and experiments. Bioinformatics 27:1575–1577. https://doi.org/10.1093/bioinformatics/btr168
41. Cock PJA, Antao T, Chang JT et al (2009) Biopython: freely available Python tools for computational molecular biology and bioinformatics. Bioinformatics 25:1422–1423. https://doi.org/10.1093/bioinformatics/btp163
42. Grant BJ, Rodrigues APCC, ElSawy KM et al (2006) Bio3d: an R package for the comparative analysis of protein structures. Bioinformatics 22:2695–2696. https://doi.org/10.1093/bioinformatics/btl461
43. Carugo O, Pongor S (2008) A normalized root-mean-square distance for comparing protein three-dimensional structures. Protein Sci 10:1470–1473. https://doi.org/10.1110/ps.690101
44. Larsen TA, Olson AJ, Goodsell DS (1998) Morphology of protein-protein interfaces. Structure 6:421–427. https://doi.org/10.1016/S0969-2126(98)00044-6
45. Tsai C-J, Lin SL, Wolfson HJ, Nussinov R (1997) Studies of protein-protein interfaces: a statistical analysis of the hydrophobic effect. Protein Sci 6:53–64. https://doi.org/10.1002/pro.5560060106
46. Mendes J, Guerois R, Serrano L (2002) Energy estimation in protein design. Curr Opin Struct Biol 12:441–446
47. Mendes J, Baptista AM, Carrondo MA, Soares CM (1999) Improved modeling of side-chains in proteins with rotamer-based methods: a flexible rotamer model. Proteins Struct Funct Genet 37:530–543. https://doi.org/10.1002/(SICI)1097-0134(19991201)37:4<530::AID-PROT4>3.0.CO;2-H
48. Baptista AM, Scares CM (2001) Some theoretical and computational aspects of the inclusion of proton isomerism in the protonation equilibrium of proteins. J Phys Chem B 105: 293–309. https://doi.org/10.1021/jp002763e
49. Pande VS, Grosberg AY, Tanaka T (1997) Statistical mechanics of simple models of protein folding and design. Biophys J 73:3192–3210. https://doi.org/10.1016/S0006-3495(97)78345-0
50. Leach AR, Lemon AP (1998) Exploring the conformational space of protein side chains using dead-end elimination and the A* algorithm. Proteins Struct Funct Genet 33: 227–239. https://doi.org/10.1002/(SICI)

1097-0134(19981101)33:2<227::AID-PROT7>3.0.CO;2-F

51. Keskin O, Ma B, Nussinov R (2005) Hot regions in protein-protein interactions: the organization and contribution of structurally conserved hot spot residues. J Mol Biol 345: 1281–1294. https://doi.org/10.1016/j.jmb.2004.10.077
52. Ishima R, Torchia DA (2000) Protein dynamics from NMR. Nat Struct Biol 7:740–743
53. Lippincott-Schwartz J, Snapp E, Kemvorthy A (2001) Studying protein dynamics in living cells. Nat Rev Mol Cell Biol 2:444–456
54. Frauenfelder H, Chen G, Berendzen J et al (2009) A unified model of protein dynamics. Proc Natl Acad Sci U S A 106:5129–5134. https://doi.org/10.1073/pnas.0900336106
55. McCammon JA (1984) Protein dynamics. Rep Prog Phys 47:1–46. https://doi.org/10.1088/0034-4885/47/1/001
56. Tzeng SR, Kalodimos CG (2011) Protein dynamics and allostery: an NMR view. Curr Opin Struct Biol 21:62–67
57. Karplus M, Kuriyan J (2005) Molecular dynamics and protein function. Proc Natl Acad Sci U S A 102:6679–6685
58. Levitt M (1983) Molecular dynamics of native protein. J Mol Biol 168:595–617. https://doi.org/10.1016/S0022-2836(83)80304-0
59. Derreumaux P, Mousseau N (2007) Coarse-grained protein molecular dynamics simulations. J Chem Phys 126:025101. https://doi.org/10.1063/1.2408414
60. Case DA (1994) Normal mode analysis of protein dynamics. Curr Opin Struct Biol 4: 285–290. https://doi.org/10.1016/S0959-440X(94)90321-2
61. Cooper A (1976) Thermodynamic fluctuations in protein molecules. Proc Natl Acad Sci U S A 73:2740–2741. https://doi.org/10.1073/pnas.73.8.2740
62. Abseher R, Nilges M (1998) Are there non-trivial dynamic cross-correlations in proteins? J Mol Biol 279:911–920. https://doi.org/10.1006/jmbi.1998.1807
63. Wodak SJ, Paci E, Dokholyan NV et al (2019) Allostery in its many disguises: from theory to applications. Structure 27:566–578
64. Raman S, Taylor N, Genuth N et al (2014) Engineering allostery. Trends Genet 30: 521–528
65. Gaczynska M, Osmulski PA (2014) Harnessing proteasome dynamics and allostery in drug design. Antioxid Redox Signal 21: 2286–2301. https://doi.org/10.1089/ars.2013.5816
66. Ikeguchi M, Ueno J, Sato M, Kidera A (2005) Protein structural change upon ligand binding: linear response theory. Phys Rev Lett 94: 078102. https://doi.org/10.1103/PhysRevLett.94.078102

Chapter 6

In Vitro Transcription Assay for Archaea Belonging to *Sulfolobales*

David Sybers, Daniel Charlier, and Eveline Peeters

Abstract

Archaeal transcription and its regulation are characterized by a mosaic of eukaryotic and bacterial features. Molecular analysis of the functioning of the archaeal RNA polymerase, basal transcription factors, and specific promoter-containing DNA templates allows to unravel the mechanisms of transcription regulation in archaea. In vitro transcription is a technique that allows the study of this process in a simplified and controlled environment less complex than the archaeal cell. In this chapter, we present an in vitro transcription methodology for the study of transcription in *Sulfolobales*. It is described how to purify the RNA polymerase and the basal transcription factors TATA-binding protein and transcription factor B of *Saccharolobus solfataricus* and how to perform in vitro transcription reactions and transcript detection. Application of this protocol for other archaeal species could require minor modifications to protein overexpression and purification conditions.

Key words In vitro transcription, Archaea, *Sulfolobales*, *Saccharolobus*, TATA-binding protein, Transcription factor B, RNA polymerase

1 Introduction

While the archaeal transcription machinery is very similar to the eukaryotic RNA polymerase II (RNAPII) system, albeit less complex, archaeal transcriptional regulators are generally bacterial-type regulators [1, 2]. This raises many questions with regards to the underlying molecular mechanisms. The study of the archaeal transcription machinery is therefore an opportunity to better understand transcription and its regulation and how this evolved, not only in Archaea but also in the other domains of life. A range of techniques and approaches to study (archaeal) transcription are available, including techniques to map transcription on a system level, such as RNA sequencing; targeted transcriptional analysis, such as quantitative RT-PCR or reporter gene analysis; and in vitro techniques, such as in vitro transcription as described here. In vitro transcription allows us to study the mechanisms of

Eveline Peeters and Indra Bervoets (eds.), *Prokaryotic Gene Regulation: Methods and Protocols*, Methods in Molecular Biology, vol. 2516, https://doi.org/10.1007/978-1-0716-2413-5_6,

RNA synthesis, the control of gene expression on transcriptional level, functional studies of recombinant DNA molecules, and the preparation of RNA molecules for further studies.

In vitro transcription constitutes the following steps for an archaeal system: first, purified RNAP and the TATA-binding protein (TBP) and transcription factor B (TFB) are combined with a DNA template minimally containing an archaeal promoter with all *cis*-acting transcription initiation sequences and the first segment of an open reading frame. This DNA template can either be linear, resulting in a run-off transcription [3], or circular with a supercoiled [4] or relaxed topology [5]. However, circular DNA topology mimics more closely the native topology of the genomic DNA than linear DNA, leading to better predictions of in vivo behavior of the pre-initiation complex that is formed by the template, RNAP, TBP, and TFB and determines transcriptional strength. Different methods for detecting the resulting RNA strand can be employed. When using radioisotopes, α-^{32}P-UTP can be used, incorporating it into the RNA product. This will lead to strong signals but has the disadvantage of visualizing partial transcripts at different positions in a denaturing polyacrylamide gel, complicating quantitative analysis. An alternative is to use unlabeled NTPs for the in vitro transcription reaction with detection being performed using a 5′ radioactively labeled primer in a reverse transcription reaction. The resulting cDNA can then be detected with gel electrophoresis in primer extension. Even though this will lead to a lower signal compared to using α-^{32}P-UTP, it will increase the specificity of the transcript detection, facilitating interpretation.

In vitro transcription procedures have been optimized for the archaeal model organisms *Methanocaldococcus jannaschii* [3, 5–7], *Pyrococcus furiosus* [4], and *Sulfolobales* (*Saccharolobus solfataricus*) [8, 9]. Here, we describe the procedure for the latter.

2 Materials

2.1 Safety Instructions

Several chemicals used in the procedure, including acrylamide/*bis*-acrylamide, ammonium persulfate, *N*,*N*,*N*′,*N*′-tetramethylethylenediamine (TEMED), ethylenediaminetetraacetic acid (EDTA), ethanol, and acetic acid are toxic, neurotoxic, corrosive, potentially carcinogenic, and/or flammable. Therefore, it is advised to take protective measures, including wearing gloves, and to work in a fume hood when manipulating these chemicals. Avoid inhalation and skin contact. Consult the accompanying safety sheets prior to handling one of these chemicals.

All work with the [^{32}P] radioisotope should be performed in a dedicated room that is specifically equipped for the manipulation of radioisotopes. Authorization to work with radioisotopes must be granted by the institute's health and safety office and all safety instructions should be strictly followed. This includes the use of a personal dosimeter to monitor exposure to radiation, performing manipulation with [^{32}P]-labeled material behind plexiglass shields, covering working surfaces with bench coat paper, and collecting contaminated waste in separate authorized waste containers. [^{32}P] is a high energy β-ray emitter; it is therefore advised to make use of a Geiger counter to regularly verify the working area, equipment, and gloves for absence of radioisotope spills. In case of spills, a clean-up and/or decontamination should be performed as soon as possible.

2.2 Purification of Transcription Factors

2.2.1 Solutions, Enzymes, Chemicals

1. Chemically competent *Escherichia coli* Rosetta (DE3) strain for heterologous overexpression containing an IPTG-inducible plasmid construct for the expression of either His-tagged *Saccharolobus solfataricus* TBP or TFB.
2. Lysogeny Broth (LB) medium.
3. Antibiotic for selection of overexpression vector.
4. Isopropyl-ß-thiogalactopyranisode (IPTG).
5. 0.9% NaCl.
6. Lysis buffer: 20 mM Tris-HCl (pH 8.0), 500 mM NaCl, 0.1% Triton-X-100, 1 mM $MgCl_2$, and cOmplete™ EDTA-free Protease Inhibitor Cocktail.
7. Buffer A: 20 mM Tris-HCl (pH 8.0), 500 mM NaCl, 30 mM NaH_2PO_4, and 30 mM imidazole.
8. Buffer B: 20 mM Tris-HCl (pH 8.0), 500 mM NaCl, 30 mM NaH_2PO_4, and 500 mM imidazole.
9. NuPAGE® LDS Sample Buffer.
10. PageRuler™ Prestained Protein Ladder.
11. Freezing buffer: 50 mM Tris–HCl (pH 7.5), 200 mM KCl, 1 mM DTT, 1 mM EDTA, 0.01% Triton X-100, and 50% glycerol.

2.2.2 Other Materials and Equipment

1. Shaking incubator able to reach 37 °C.
2. Culture flasks.
3. Spectrophotometer.
4. Ice or equivalent cooling.
5. Centrifuge with cooling.
6. Sonicator.
7. Heating block able to reach up to 70 °C.
8. Fast Protein Liquid Chromatography (FPLC) installation for protein purification.
9. Ni-NTA-Agarose column (HisTrap™ 5 mL Chelating HM).

10. Acrylamide gel for sodium dodecyl sulfate (SDS) polyacrylamide gel electrophoresis (PAGE) and SDS buffer.
11. Dialysis membranes with a molecular weight cutoff of 10 kDa or lower.

2.3 Purification of RNA Polymerase

2.3.1 Solutions, Enzymes, Chemicals

1. Brock Basal salt medium [10].
2. 0.1% NZAmine solution.
3. 0.2% sucrose solution.
4. *Saccharolobus solfataricus* P2 strain (*see* **Note 1**).
5. RNAP lysis buffer: 50 mM MES (pH 6.0), 100 mM NaCl, 1 mM EDTA, 1 mM DTT, and cOmplete™ Protease Inhibitor Cocktail.
6. Buffer 1: 20 mM MES (pH 6.0), 1 mM EDTA, 0.5 mM DTT, and 10 mM NaCl.
7. Buffer 2: 20 mM MES (pH 6.0), 1 mM EDTA, 0.5 mM DTT, and 1 M NaCl.
8. *S. solfataricus* RpoB1-specific antibodies.
9. Gel filtration buffer: 20 mM MES (pH 6.0), 200 mM NaCl, 1 mM EDTA, and 0.5 mM DTT.
10. Freezing buffer: 50 mM Tris–HCl (pH 7.5), 200 mM KCl, 1 mM DTT, 1 mM EDTA, 0.01% Triton X-100, and 50% glycerol.

2.3.2 Other Materials and Equipment

1. Shaking incubator or water/oil bath able to reach 75 °C.
2. Centrifuge.
3. Sonicator.
4. FPLC installation for protein purification.
5. HiTrap™ 5 mL Heparin column.
6. HiLoad™ 26/60 Superdex 200 pg size exclusion chromatography column.
7. Acrylamide gel for SDS-PAGE and SDS buffer.
8. Dialysis membrane with a molecular weight cutoff of 100 kDa or lower.

2.4 Preparation of [5′-^{32}P] Single-End Labeled Oligonucleotides and DNA Fragments

2.4.1 Solutions, Enzymes, Chemicals

1. Fresh [γ-^{32}P]-ATP (3000 Ci/mmol).
2. T4 polynucleotide kinase (10 U/μL).
3. 10× forward kinase buffer.
4. Nuclease-free water.
5. Probe-specific oligonucleotides (dissolved in nuclease-free water), which can be used as primers for in vitro transcription and for PCR amplification, with a working solution concentration of 10 pmol/μL.

6. Plasmid template DNA harboring the target sequence.
7. Commercial PCR amplification 2× master mix (e.g., GoTaq™) or separate PCR components, including a DNA polymerase and dNTPs.

2.4.2 Other Materials and Equipment

1. Sterile 1.5 mL tubes, 0.25 mL tubes, and pipette tips.
2. Heating thermostatic water baths or heating blocks at 37 °C and 65 °C.
3. Microcentrifuge.
4. Thermocycler.

2.5 Native Gel Electrophoresis for Purification of Oligonucleotides

2.5.1 Solutions and Chemicals

1. A 5× concentrated stock solution for the preparation of TBE electrophoresis buffer (1× buffer: 89 mM Tris-base, 89 mM boric acid, and 2.5 mM EDTA).
2. A 0.1 g/mL stock solution of ammonium persulfate (dissolved in double-distilled water (ddH_2O)). This solution can be stored up to a few weeks at 4 °C.
3. TEMED solution. This should be stored in the dark at 4 °C.
4. Acrylamide/*N*, *N′*-methylene *bis*-acrylamide solution (30%, ratio 29:1). This should be stored in the dark at 4 °C.
5. ddH_2O.
6. A repellent solution to siliconize the glass plates (e.g., Blue Slick™ spray).
7. 10× retardation loading dye mix (consisting of 25% ficoll, 0.1% xylene cyanol, and 0.1% bromophenol blue).

2.5.2 Other Materials and Equipment

1. Pair of glass plates with dimensions that are compatible with the dimensions of the electrophoresis tank.
2. Pair of spacers (0.1 cm thick) and cognate comb to create wells.
3. Commercial rubber sealing system for glass plate assembly or, alternatively, tape and clamps, to create a sandwich construction of glass plates and spacers and comb.
4. 50 mL syringe.
5. Büchner funnel and rubber tubing to connect to a vacuum pump.
6. Filter paper to use with the Büchner funnel.
7. Vacuum pump.
8. Power supply and electric cabling.
9. Electrophoresis tank for vertical gel electrophoresis.

2.6 Autoradiography

2.6.1 Solutions

1. A developer solution for photography, which is fivefold diluted in water. This should be stored at room temperature but needs to be protected from light. It can be repeatedly used but should be refreshed when it obtains a brownish color.

2. A fixer solution for photography, which is also fivefold diluted in water and should also be stored at room temperature, protected from light. This solution should be refreshed when becoming slightly pink-colored.

2.6.2 Other Materials and Equipment

1. Radioisotope and chemiluminescent imaging film.
2. Plastic foil.
3. Black plastic films or bags that are large enough to pack the gel and film and protect it from light.
4. A dark room equipped with red light and a sink with running tap water and demineralized water.
5. Two trays for develope and fixer solutions, which are slightly larger than the autoradiograph film.

2.7 Recovery of DNA from Polyacrylamide Gel

2.7.1 Solutions, Chemicals, Materials, and Equipment

1. RNase-free water treated with diethylpyrocarbonate (DEPC): 1 mL DEPC/L H_2O (shake vigorously), incubated at 37 °C overnight, followed by autoclaving for 15 min to remove remaining DEPC.
2. Sterile glass tubes.
3. Sterile 1 mL pipette tips.
4. Pasteur pipettes (150 mm).
5. Drawn-out 150 mm Pasteur pipettes with thin tip, to allow removal of a maximum of supernatant without touching the DNA pellet upon DNA precipitation in 1.5 mL tubes.
6. Sterile glass wool (sterilized by autoclaving).
7. Self-sealing tape (e.g., parafilm).
8. Nuclease-free scalpel.
9. Water bath at 45 °C.
10. Microcentrifuge.
11. Vacuumcentrifuge (e.g., speedvac).
12. Vortex.

2.8 In Vitro Transcription Reactions

2.8.1 Solutions, Chemicals, Materials, and Equipment

1. TBE electrophoresis buffer: 89 mM Tris-base, 89 mM boric acid, and 2.5 mM EDTA. Make a 10× concentrated stock solution.
2. 2 mM rNTP solution.
3. Purified TBP and TFB (*see* Subheading 3.1).
4. Purified RNAP (*see* Subheading 3.2).
5. Nuclease-free water.
6. NEW buffer: 10 mM Tris (pH 8.0), 750 mM NaCl, 10 mM EDTA, and 0.5% SDS.

7. Formamide dye mix: 100 mL deionized formamide, 0.03 g xylene cyanol, 0.03 g bromophenol blue, and 20 mM EDTA.
8. Heating block that heats up to 70 °C.
9. Microcentrifuge.

2.9 Reverse Transcription Reactions

2.9.1 Solutions, Chemicals, Materials, and Equipment

1. 5× reverse transcriptase buffer.
2. AMV reverse transcriptase (20 U/μL).
3. dNTP mix (2 mM of each).
4. Protector RNase inhibitor (40 U/μL).
5. Nuclease-free H_2O.
6. Formamide dye mix: 100 mL deionized formamide, 0.03 g xylene cyanol, 0.03 g bromophenol blue, 20 mM EDTA.
7. Heating block.

2.10 Denaturing Gel Electrophoresis

2.10.1 Solutions and Chemicals

1. A 5× concentrated stock solution for the preparation of TBE electrophoresis buffer (1× buffer: 89 mM Tris-base, 89 mM boric acid, and 2.5 mM EDTA).
2. A 0.1 g/mL stock solution of ammonium persulfate (dissolved in ddH_2O). This solution can be stored at 4 °C for up to a few weeks.
3. TEMED solution. This should be stored in the dark at 4 °C.
4. Acrylamide/*N*, *N*′-methylene *bis*-acrylamide solution (40%, ratio 19:1). This should be stored in the dark at 4 °C.
5. Urea.
6. ddH_2O.
7. A repellent solution to siliconize the glass plates (e.g., Blue Slick™ spray).
8. Sticky solution: 48.5 mL ethanol, 1.5 mL 10% acetic acid, and 150 μL γ-methacryloxypropyltrimethoxysilane.
9. 2 L of 10% acetic acid. This solution can be reused for the treatment of 5 gels.

2.10.2 Other Materials and Equipment

1. A pair of heat-resistant, sequencing-type glass plates that are compatible with the dimensions of the electrophoresis tank.
2. Pair of spacers (0.25 mm thick) and cognate 20-well comb to create wells.
3. Commercial rubber sealing system for plate assembly or, alternatively, tape and clamps, to create a sandwich construction of glass plates and spacers and comb.
4. 50 mL syringe.
5. Büchner funnel and rubber tubing to connect to vacuum pump.

6. Filter paper to use with the Büchner funnel.
7. Vacuum pump.
8. Power supply and electric cabling allowing electrophoresis at a constant power of at least 50 W (±1500 V).
9. Electrophoresis tank for vertical sequencing-type gel electrophoresis.
10. Plastic tray slightly larger than the dimensions of the glass plates.
11. Small electric heating system to blow hot air.
12. Heating magnetic stirrer.

3 Methods

3.1 Purification of Transcription Factors

3.1.1 Transformation of Plasmids into Overexpression Strain

1. Add 20 ng of plasmid (pET-based expression plasmid encoding His-tagged TBP or TFB) to 50 μL of competent *E. coli* Rosetta (DE3) cells (*see* **Note 2**).
2. Incubate at 42 °C for 40 s.
3. Let cool down at room temperature for several minutes.
4. Add 1 mL LB medium and incubate at 37 °C for 1 h (when using ampicillin resistance marker, incubation time varies for other antibiotics).
5. Spin down the cells, discard supernatans, resuspend in 50–100 μL LB, and a plate on agar plates that contain LB medium with the specific antibiotic for selection of the TBP- or TFB-overexpressing plasmid.
6. Incubate plates overnight at 37 °C.
7. Pick a single colony and inoculate it into LB medium containing the specific antibiotic (a few mL should suffice).
8. Incubate in a shaking incubator at 37 °C until the culture reaches mid-exponential phase (optical density at 600 nm (OD_{600}) between 0.3 and 0.6).
9. Make a glycerol stock from the culture for long-term storage at −80 °C.

3.1.2 Heterologous Expression (in E. coli) and Purification of S. solfataricus TBP and TFB

The procedure below is executed for each of the two basal transcription factors.

1. Inoculate 50 mL LB medium containing the appropriate antibiotic with a colony of the *E. coli* Rosetta (DE3) containing the overexpression vector.
2. Incubate overnight in a shaking incubator at 37 °C.
3. Inoculate 300 mL LB medium containing the appropriate antibiotic with part of the preculture (until reaching a starting culture OD_{600} of between 0.05 and 0.1) (*see* **Note 3**).

4. Incubate at 37 °C while shaking until an OD_{600} of 0.6 is reached.
5. Induce by adding 1 mM IPTG to the culture and incubate for another 3 h. An IPTG stock solution can be prepared in sterile water using aseptic techniques.
6. Cool the culture down on ice for 10 min.
7. Centrifuge at 3500 × *g* for 10 min at 4 °C.
8. Resuspend the pellet in 10 mL ice-cold 0.9% NaCl solution.
9. Centrifuge at 6500 × *g* for 10 min at 4 °C (*see* **Note 4**).
10. Resuspend the pellet in 6 mL lysis buffer by pipetting up and down.
11. Sonicate at 20% of maximal amplitude (Vibracell 75,043, Bio-block Scientific) for 15 min while continuously cooling the sample.
12. Centrifuge at 12000 × *g* for 10 min.
13. Transfer the supernatants to a new tube and incubate at 70 °C for 10 min.
14. Centrifuge at 12000 × *g* for 10 min.
15. Transfer the supernatants to a new tube and add an equal volume of buffer A.
16. Apply the sample to a column containing Ni-NTA-Agarose that has been pre-equilibrated with buffer A.
17. Elute the target protein using a linearly increasing gradient of buffer B.
18. Take a 7.5 μL aliquot of elution fractions suspected to contain the target protein, mixed with 2.5 μL NuPAGE® LDS Sample Buffer.
19. Incubate at 70 °C for 10 min.
20. Load the samples (10 μL) and a protein ladder with appropriate size range (10 μL) onto the SDS-gel.
21. Run at 200 V for 35 min.
22. Pool all positive fractions and dialyse against 1 L freezing buffer at 4 °C for 1 h.
23. Dialyse against a fresh batch of 1 L freezing buffer and incubate at 4 °C overnight.
24. Measure the protein concentration. If needed, concentrate the protein (*see* Subheading 3.5.2 for appropriate protein concentration).
25. Store the purified protein in small-volume aliquots (e.g., 10 μL) at −80 °C until needed.

3.2 Expression and Purification of S. solfataricus RNA Polymerase

1. Prepare 1 L Basal Brock medium with 0.1% NZAmine and 0.2% sucrose.
2. Inoculate with *S. solfataricus* P2 (*see* **Note 1**).
3. Incubate at 75 °C and 180 rpm until the culture has grown to mid-exponential phase (OD_{600} between 0.3 and 0.6).
4. Centrifuge at 6500 × *g*.
5. Resuspend in 15 mL RNAP lysis buffer.
6. Sonicate at 20% of maximal amplitude (Vibracell 75,043, Bio-block Scientific) for 15 min while continuously cooling the sample.
7. Centrifuge at 13000 × *g* for 10 min.
8. Transfer the supernatants to a new tube and add an equal volume of buffer 1.
9. Load the sample onto a HiTrap 5 mL Heparin column that has been pre-equilibrated with buffer 1.
10. Elute the target protein using a linearly increasing gradient of buffer 2 (increasing NaCl concentration).
11. Determine which fractions contain RNAP by dot-blot immunodetection using RpoB1-specific antibodies (*see* **Note 1**) and pool together positive fractions.
12. Load the pooled samples onto a HiLoad 26/60 Superdex 200 pg (GE Healthcare) size exclusion column equilibrated with gel filtration buffer.
13. Analyze elution fractions that are suspected to contain RNAP for the presence of the protein by performing SDS-PAGE.
14. Pool all positive fractions and dialyse against 1 L freezing buffer at 4 °C for 1 h.
15. Dialyse against a fresh batch of 1 L freezing buffer and incubate at 4 °C overnight.
16. Measure the protein concentration. If needed, concentrate the protein (*see* Subheading 3.5.2 for appropriate protein concentration).
17. Store the purified protein in small-volume aliquots (e.g., 10 μL) at −80 °C until needed.

3.3 Preparation of [5′-^{32}P] Single-End Labeled Primer

3.3.1 Primer Characteristics

An important element of the in vitro transcription experiment is a well-thought design of the primer used for the reverse transcription reaction to detect the transcript. This primer is designed as a reverse complementary sequence to the open reading frame of the target gene. Design a primer with a length between 20 and 25 nucleotides, a lack of secondary structures, and that ends with a G or C nucleotide. Ideally, this primer generates a reverse transcriptase product with a length between 30 and 60 nucleotides. The transcription start site of the gene of interest could be inferred from a high-resolution transcriptomic database for *S. solfataricus* and *Sulfolobus acidocaldarius* [11]. If data are unavailable for the gene of interest, one could start counting from the translational start of the gene, given that a large fraction of genes in *Sulfolobales* is leaderless.

3.3.2 Oligonucleotide Labeling

1. Mix in a 1.5 mL sterile tube 60 pmol of oligonucleotide with 3.0 μL of commercial 10× forward kinase buffer (buffer A), 1.0 μL T4 polynucleotide kinase (10 U/μL), 2.5 μL [γ-^{32}P]-ATP (3000 Ci/mmol) and adjust the volume to 30 μL with nuclease-free water.
2. Mix gently by pipetting a couple of times up and down, spin briefly to collect the sample at the bottom of the tube.
3. Incubate at 37 °C for 40 min.
4. Transfer the tube to 65 °C and incubate for 10 min to inactivate the enzyme and stop the reaction.
5. Proceed to the purification of the amplicon of interest by gel electrophoresis and recovery from gel (see below) or store the reaction mixes overnight at −20 °C until proceeding with the experiment.

3.4 Purification of [^{32}P]-Labeled DNA by Non-denaturing Polyacrylamide Gel Electrophoresis

3.4.1 Aim

In order to separate the labeled primer from non-incorporated labeled [γ-^{32}P]-ATP and other reaction products from the labeling reaction, it will be migrated on a polyacrylamide gel followed by excision from the gel and elution (*see* **Note 5**).

3.4.2 Treatment of Glass Plates

1. Clean the two glass plates of the gel casting system with detergent, rinse first with tap water, then with demineralized water, and wipe dry with absorbing paper.
2. Siliconize the inside face of the large front plate by spraying repellent solution on the plate, spread equally all over the plate with absorbing paper, and let evaporate for a few minutes.
3. Use a commercial rubber gel sandwich system to assemble both glass plates or alternatively, seal the plates using tape and clamps. Place the assembly slightly tilted on a working surface that is easily cleanable (Teflon plate or disposable bench coat paper) in case of acrylamide spills.

3.4.3 Preparation of a 10% Non-denaturing Polyacrylamide Gel

1. Mix in a 50 mL cylinder 10 mL of 5× TBE buffer with 16.6 mL of 30% acrylamide/*bis*-acrylamide (ratio 29:1) stock solution, 0.5 mL ammonium persulfate solution (0.1 g/mL) and adjust to 50 mL with demineralized water.

2. Filter the solution using a Büchner funnel connected to a vacuum pump and let it degas for at least 10 min.
3. Take a 10 mL aliquot from the solution, add 10 μL TEMED, mix and use a syringe to let the solution flow down in the space between the plates along the spacers, 5 mL on either side of the assembly.
4. Allow the gel to polymerize and verify by tilting the assembly. Casting only a part of the gel reduces the risk of further leakage while pouring the bulk of the gel.
5. Add 20 μL TEMED to the remaining 40 mL of polyacrylamide solution and use a syringe to carefully fill up the space in between the plates. Stop when the acrylamide solution has filled the entire space until the top of the smaller glass plate. Avoid trapping air bubbles in the gel solution by pouring slowly.
6. Insert the comb in the solution on top of the sandwiched plates, while avoiding trapping air bubbles beneath the teeth of the comb.
7. Let the gel polymerize for at least 30–45 min.
8. Remove the sealing, comb, and potentially spilled polyacrylamide on the outer surface of the glass plates and mount the gel in the electrophoresis tank (*see* **Note 6**).
9. Add 1× TBE running buffer to the upper and lower chamber of the electrophoresis tank.
10. Use a syringe to rinse the wells, thereby ensuring that no polyacrylamide remains in the wells.
11. Switch on the current and prerun the gel for 30 min at 180 V (11 V/cm) in order to remove the excess non-polymerized reagents.

3.4.4 Sample Preparation and Gel Electrophoresis

1. Mix 20 μL aliquots of the labeled primer reaction mix (*see* Subheading 3.3.2) with 2 μL of 10× retardation loading dye mix. Centrifuge briefly.
2. Switch off the current and use long, thin pipette tips to load the samples in the wells. Avoid blowing air bubbles in the samples while pipetting.
3. Switch on the current and let migrate at 180 V for 15 min to enable the primer to quickly penetrate the gel matrix.
4. Reduce the current to 150 V and continue the migration to obtain a good separation. As a guide, the fastest migrating dye (bromophenol blue) should be in the lower part of the gel while not yet reaching the bottom, as it runs corresponding to a double-stranded DNA fragment of 33 bp on a 10% polyacrylamide gel.

3.4.5 Imaging by Autoradiography

1. Stop the electrophoresis.
2. Remove the assembly from the electrophoresis tank and wipe the glass plates dry. Make sure to remove all running buffer from the wells before disassembling the glass plates by shaking heavily above the sink with the wells turned downwards.
3. Put the assembly on a flat surface with the large siliconized front plate on top. Work behind a plexiglass shield.
4. Use a thin spatula as a lever to carefully pry the plates apart and remove the top plate. Remove the spacers while avoiding to distort the gel.
5. Cover the gel still laying on the small backplate with plastic foil (fold back on the edges to avoid disturbance upon later manipulations).
6. Close the glass plate assembly again but place a sheet of paper in between the foil-covered gel and glass plate to avoid that they would stick to each other.
7. Move the assembly to the darkroom (work with red lights).
8. Remove the front plate and the paper or cardboard.
9. Cut a piece of X-ray-sensitive film that is sufficiently large to cover the area of interest and place it as such that the film perfectly aligns with the bottom of the backplate and slightly sticks out on at least one side.
10. Place the front plate on top of the X-ray-sensitive film and fix the whole assembly with clamps.
11. Use the tip of scissors to scratch the contour of the small backplate on the film and indicate the orientation by making a small incision at one of the top corners of the film. This will allow a perfect realignment of the film and gel after development.
12. Expose the film for 10 min, then remove it from the assembly.
13. Pour a sufficient volume of 5× diluted developer in a tray to cover the entire surface of the film and develop the film for 2–3 min while gently shaking the tray regularly.
14. Abundantly rinse the film with tap water.
15. Then move it to a second tray with 5× diluted fixer. Fix the film during 2–3 min while gently shaking the tray regularly.
16. Rinse abundantly with tap water, then rinse with demineralized water. From this point on, working with red light is no longer necessary.
17. Suspend the film with a clamp and let air-dry until the film is completely dry.

18. Remove the running buffer from the electrophoresis tank but be aware that the bottom chamber may be heavily contaminated with [γ-^{32}P]-ATP that may have migrated out of the gel, depending on the electrophoresis conditions. Verify with the Geiger counter. Rinse the electrophoresis apparatus abundantly with tap water and verify with the Geiger counter that there is no remaining contamination.

3.4.6 Recovery of Labeled Primer from the Polyacrylamide Gel Matrix

1. Prepare an assembled elution set-up for every primer that has to be purified. Start by inserting a small amount of sterile glass wool in a sterile 1 mL pipette tip and very gently push the glass wool towards the bottom of the tip with a Pasteur pipette without pressing too hard. In this way, a plug will be formed that retains the polyacrylamide particles while allowing buffer to flow through. Seal the tip of the 1 mL pipette tip with Parafilm (or any other self-sealing tape) and insert the lower part in a sterile glass tube.
2. Work behind a plexiglass shield and deposit the autoradiography on a piece of bench coat. Take the orientation into account and put the gel with the front plate on top of the autoradiography. Align the small backplate with the marked contour on the autoradiography and remove the front plate.
3. Remove the plastic foil.
4. Use a scalpel to cut out a slice of acrylamide that corresponds to the band of labeled primer on the autoradiography film. Keep the size of the gel slice as small as possible while encompassing the entire band.
5. Use the scalpel to transfer the gel slice into the 1 mL pipette tip construction and use a Pasteur pipette to push it further down without touching the glass wool plug.
6. Crush the polyacrylamide between the wall of the 1 mL pipette tip and the Pasteur pipette by making a rotating movement (be careful not to touch or displace the plug and do it gently to avoid breaking the tip of the Pasteur pipette). Repeat this movement till the slice has been transformed into small bead-like particles (this will improve the recovery by facilitating diffusion of the DNA) that stick to the Pasteur pipette.
7. Let 200 μL of RNase-free water flow down along the tip of the Pasteur pipette to collect the beads at the bottom of the 1 mL pipette tip, on top of the plug.
8. Repeat this operation a second time to collect all acrylamide beads. Make sure that all the acrylamide is submerged in the elution buffer and that air bubbles are removed from the solution.

9. Use parafilm to seal the top of the 1 mL pipette tip and to create a connection between the pipette tip and the glass tube.
10. Put the assembly in a floater and incubate the whole assembly overnight in a water bath at 45 °C.
11. Remove the tube from the water bath after overnight incubation and remove the parafilm sealing from the connection between the pipette tip and the glass tube.
12. Carefully remove the Parafilm sealing at the tip of the 1 mL pipette tip and immediately insert it in the glass tube (*see* **Note 7**).
13. Remove the parafilm sealing at the top of the 1 mL pipette tip and collect the liquid containing labeled primer in the glass tube.
14. When no more liquid comes out of the tip (*see* **Note 8**), transfer the collected liquid into a clean 1.5 mL tube.
15. Estimate the radioactive counts (cpm/μL) for an aliquot of the purified labeled primer using the Geiger counter.

3.5 In Vitro Transcription (IVT)

3.5.1 Design and Preparation of Template DNA

Any cloning vector can be used that harbors the promoter and a first part of the open reading frame of the gene of interest (minimally encompassing the region to which the reverse complementary primer will hybridize). However, it is advised not to clone the entire open reading frame of the gene thereby avoiding the generation of long transcript products. A preparation of the template DNA can be obtained through plasmid miniprep extraction from *E. coli*, thereby resulting in supercoiled plasmid DNA.

3.5.2 In Vitro Transcription Reactions

1. Prepare the following mixtures: (i) mix 1 containing per reaction 2.5 μL 10× TBE buffer, 2.5 μL 2 mM rNTPs, 20 ng TBP, 20 ng TFB, add water up to 25 μL; (ii) mix 2 containing per reaction 2.5 μL 10× TBE buffer, 100 ng DNA template, 75 ng RNAP, add water up to 25 μL. These mixtures can be prepared for several IVT reactions, which can be designed to have differences in the concentrations of TBP, TFB, or RNAP, or to which different concentrations of regulatory transcription factors are added.
2. Mix 25 μL of each mixture together (from now on referred to as IVT mixture), spin down to collect the sample at the bottom of the tube.
3. Incubate the mixture at 70 °C for 30 min.
4. Spin tubes for 20–30 s and keep them at room temperature.
5. Prepare the following primer mixture (indicated amounts for one reaction, multiply for the total number of reactions): 1 μL glycogen (20 μg/μL), about 60,000 cpm labeled primer, NEW buffer up to a total volume of 250 μL (*see* **Note 9**). Although

this is not compulsory, it is useful to add an aliquot of gel-purified end-labeled double-stranded DNA fragment of a length significantly different from the length of the expected primer extension product and with a sequence unrelated to the gene under study. For example, an aliquot with an activity between 5000 and 10,000 cpm can be added to each reaction. This so-called recovery marker can be used for the normalization after densitometry of IVT bands.

6. For each reaction: add 250 μL phenol/chloroform (1:1) to an empty 1.5 mL tube.
7. For each reaction: add 250 μL primer mixture to the IVT mixture to each sample.
8. Add the reaction mixture (from **step** 7) to the tube from **step 6**.
9. Vortex for 20 s.
10. Spin for 5 min at 8000 rpm.
11. Take 250 μL of the upper phase and transfer it to a new tube.
12. Add 900 μL of 100% ice-cold ethanol.
13. Mix well by inverting the tube a couple of times and allow to precipitate at −20 °C for at least 1 h.
14. Spin down for 15 min at maximum speed in a microcentrifuge to pellet the precipitated DNA.
15. Immediately remove the supernatant with a drawn-out Pasteur pipette. Use a different one for every reaction.
16. Add 1 mL of 80% ice-cold ethanol to the pellet and centrifuge 3 min at maximum speed.
17. Remove the supernatant and dry the pellet for a few minutes in the vacuum centrifuge.
18. Proceed to the reverse transcription reaction or, alternatively, store the dry pellet at −20 °C.

3.6 Reverse Transcription

1. Prepare the RT-mix on ice by assembling per reaction: 4 μL 5× reverse transcriptase buffer, 2 μL dNTP mix (2 mM of each), 0.5 μL protector RNase inhibitor (40 U/μL), 1 μL AMV reverse transcriptase (20 U/μL), 12.5 μL nuclease-free water.
2. Resuspend each pellet in 20 μL RT-mix by adding the mix and pipetting up and down about 30 times (*see* **Note 10**).
3. Incubate at 42 °C for 60 min.
4. Add 20 μL formamide dye.

3.7 Analysis of the Reaction Products by Gel Electrophoresis Under Denaturing Conditions

3.7.1 Preparation of an 8% Denaturing Acrylamide/bis-acrylamide Stock Solution

1. For 500 mL of an 8% polyacrylamide gel: weigh 210 g urea in a 1 L glass beaker.
2. Add 100 mL 5× TBE electrophoresis buffer, 100 mL demineralized water and 100 mL of 40% acylamide/*bis*-acrylamide solution (ratio 19/1).
3. Dissolve the urea by stirring while heating the solution.
4. When all urea is dissolved, transfer to a 500 mL cylinder and adjust the volume to 500 mL with demineralized water.
5. Filter the solution using a Büchner funnel connected to a vacuum pump and let it degas for at least 10 min.
6. Transfer the solution to a glass bottle and store at 4 °C in the dark. This solution can be kept for a few months.

3.7.2 Treatment and Assembly of the Glass Plates

1. Use a set of glass plates to prepare thin large sequencing-type gels. Thoroughly clean the front and back glass plates with detergent, rinse abundantly with tap water, then with demineralized water, wipe dry with absorbing paper, and degrease the inside face of both plates with technical grade ethanol.
2. Treat the inside face of the large front plate with 'sticky' solution to ensure tight adhesion of the gel by pouring a few mL of the solution on the plate, spreading all over the plate with absorbing paper and letting it dry for at least a couple of hours (or overnight) (*see* **Note 11**).
3. Siliconize the inside face of the small backplate by spraying repellent solution on the plate, spreading equally all over the plate with absorbing paper, and letting it evaporate a few minutes.
4. Assemble the small and large plates with 0.25 mm thick spacers. Use a commercial rubber gel sandwich system to assemble both glass plates or alternatively, seal the plates using tape and clamps. Place the assembly slightly tilted on a working surface that is easily cleanable (Teflon plate or disposable bench coat paper) in case of acrylamide spills.

3.7.3 Gel Preparation

1. Filter 60 mL of cold denaturing polyacrylamide stock solution (*see* Subheading 3.7.1) using a Büchner funnel connected to a vacuum pump and let it degas for at least 10 min. Transfer the solution to a beaker.
2. Add 600 μL ammonium persulfate solution (0.1 g/mL) and 30 μL TEMED to the acrylamide solution.
3. Mix by gentle swirling and use a 50 mL syringe to carefully fill up the space in between the plates. Stop when the acrylamide solution has filled the entire space until the top of the smaller glass plate. Avoid trapping air bubbles in the gel solution by pouring slowly but consistently (without interruption).

4. Insert the comb in the solution on top of the sandwiched plates, while avoiding to trap air bubbles beneath the teeth of the comb.
5. Let the gel polymerize for at least 45 min.

3.7.4 Sample Preparation, Gel Electrophoresis Under Denaturing Conditions, and Visualization by Autoradiography

1. Start by measuring the radioactivity of every dry pellet from the primer extension reaction mixtures with a Geiger counter (*see* **Note 12**).
2. Transfer aliquots with an equal number of counts (approximately 10,000 cpm/lane) of all reactions to 1.5 mL tubes and adjust all samples to the same number of counts in the same volume with formamide dye mix (*see* **Note 13**).
3. Remove the sealing and comb from the gel assembly. Remove all remaining acrylamide above the wells on the front plate and use a capillary (made from the tip of a Pasteur pipette) to clean the wells, if required. Clean the outer surface of the glass plates.
4. Install the assembly in an appropriate electrophoresis tank. Fill the upper and lower chamber with TBE buffer and use a syringe to rinse and remove air bubbles from the wells.
5. Connect to the power supply and perform a prerun of 30 min at constant power of 60 W.
6. Just before loading the samples, heat the samples for 3 min at 90 °C in a heating block to completely denature the DNA strands and to remove any secondary structures in single-stranded DNA.
7. Use a syringe to rinse the wells and load the samples.
8. Perform the electrophoresis at a constant power of 60 W. Take into account that bromophenol blue and xylene cyanol co-migrate with approximately 19 nt and 75 nt long fragments, respectively, on an 8% polyacrylamide gel to estimate the required migration distance for the expected primer extension products and recovery marker.
9. Stop the electrophoresis, remove the gel assembly from the electrophoresis tank and deposit it with the small backplate on top on a flat surface.
10. Use a small wedge to pry open the plates and remove the backplate and spacers.
11. Incubate the front plate with the gel sticking to it for 10 min in a plastic tray that contains 2 L of 10% acetic acid and shake regularly. Work in a chemical fume hood.
12. Remove the plate from the tray, allow the liquid to drain off, and wipe the backside of the plate dry with absorbing paper.
13. Place the plate on a flat surface and remove liquid from the gel by rolling an absorbing paper roll over the gel.

14. Use a small electric heating device blowing hot air to completely dry the gel. Blow from different sides to ensure homogenous drying.
15. Allow the plate to cool down.
16. Cover the gel with plastic foil, cover the backplate to protect from radiation, and move to the darkroom.
17. Proceed to imaging by autoradiography as described in Subheading 3.4.5 but use a longer, appropriate exposure time. If the signals are very low, an intensifying screen placed on top of the X-ray-sensitive film and exposure at −80 °C can be used. This will increase the signal intensity by approximately a factor two.

3.8 Densitometry and Interpretation

Once the result is obtained, a scan can be made of the autoradiography film, which can be subjected to densitometric analysis using ImageJ software [12]. Typically, two bands are observed in each lane: one for the primer extension product and one for the recovery marker. By measuring integrated band densities and normalizing this using the recovery marker band densities, a quantitative analysis of transcriptional activity can be performed (*see* Fig. 1b, c).

4 Notes

1. An alternative to purification of untagged RNAP from *Sulfolobales spp.* is the genomic insertion of a His-tag to one of the subunits in the genome (RpoB1, for example), followed by the use of a purification protocol similar to the one used for His-tagged TBP/TFB purification (using Ni beads and imidazole for elution). Such an approach would be advised for *Sulfolobus acidocaldarius* or *S. islandicus*, for which genetic tools are available [13].
2. For optimization of expression and solubility of TBP and TFB, the following conditions can be varied: overexpression strain, incubation temperature/rpm, time of induction, inducer concentration, etc. In case of a suspected difficult protein folding, the overexpression strain could be induced after a small heat shock, leading to presence of chaperones in the overexpression strain during induction.
3. Use culture flasks with a sufficiently large volume (e.g., 1-L flask for a culture of 300 mL), to ensure sufficient aeration during shaking.
4. After this step it is best to store the pellet at −20 °C until the rest of the protocol can be fully finished. Once imidazole has been added to the sample, it is advised to finalize the purification protocol until the end. After purification the sample will be

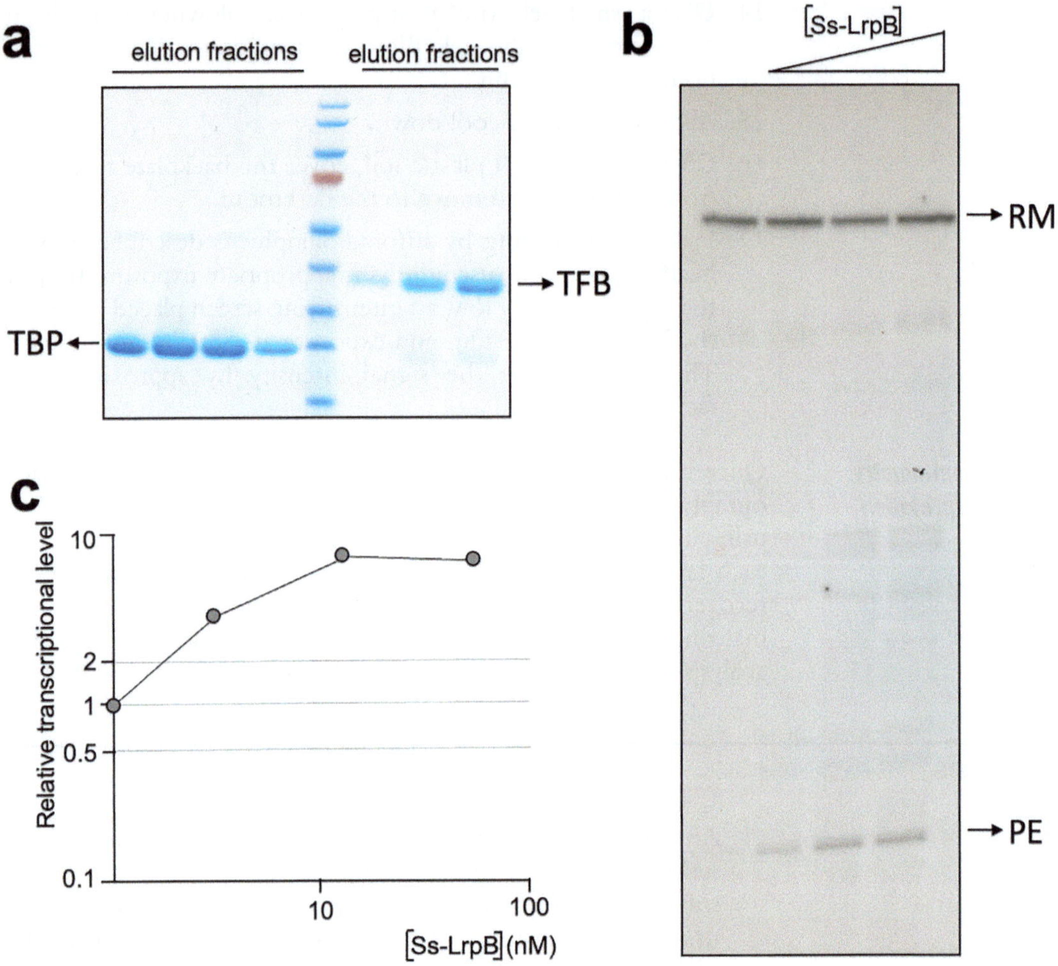

Fig. 1 Analysis of the effect of Ss-LrpB concentrations on transcription. Example of an in vitro transcription experiment performed by the procedure presented in this chapter. The addition of different concentrations of the regulatory transcription factor Ss-LrpB of *S. solfataricus* is analyzed for in vitro transcription mediated by the *S. solfataricus leuA* promoter. (**a**) SDS-PAGE analysis of the purification of the basal transcription factors TBP and TFB, with a molecular weight of 25 and 36 kDa, respectively. The MW ladder represents following MWs from bottom to top: 15, 25, 35, 40, 55, 70, 100, 130 and 170 kDa. (**b**) Autoradiograph of an in vitro transcription experiment for a plasmid template encompassing the *leuA* promoter and initial part of the open reading frame, performed in the presence of different Ss-LrpB concentrations. PE represents the primer extension product and RM the recovery marker. (**c**) Densitometric analysis of the result, calculated relative with respect to the signal measured in absence of Ss-LrpB and normalized to the RM signal

in a buffer with a high concentration of imidazole, this is known to cause protein precipitation, so getting rid of it as fast as possible by dialyzing the sample is necessary.

5. Alternatively, as the sample only contains the oligonucleotide of interest, purification of the labeled oligonucleotide could be performed using commercially available DNA purification columns.

6. If not proceeding immediately with the experiment, the polyacrylamide gel could alternatively be stored at 4 °C until further use. In this case, leave the rubber assembly system and comb in place, put absorbing paper with some water on top of the gel, and wrap the entire gel in plastic foil.
7. Be careful because the liquid might come out quickly. Wear a double pair of gloves when inserting the pipette tip in the glass tube and perform this manipulation rapidly to avoid contamination of gloves and working surface. After performing this step, verify with the Geiger counter that no contamination has occurred before proceeding further.
8. Verify that most of the radioactivity is present in the collected liquid. If this is not the case, an additional 100–200 μL RNase-free water could be added to the acrylamide for further elution. However, be advised that this will dilute the radioactively labeled primer.
9. Often, precipitation is visible in the NEW buffer. In this case, warm up NEW buffer by placing the tube in warm water.
10. Check if the pellet is properly resuspended with the Geiger counter. This can be done by taking up the resuspension into the pipet tip and checking both the tip and the empty tube for radioactivity.
11. Always treat the same plate with 'sticky' solution and the same plate with repellent solution because once this has been done it is difficult to remove it completely. Switching between treatments of plates with both solutions upon repeated use of the plates increases the risk that the gel gets disrupted when prying the plates apart at the end of the electrophoresis. Once the autoradiograph has been made and the gel no longer required, the plate can be reutilized. To this end, the plate has to be completely submerged for at least 1 h in a plastic tray with tap water. Use a scraper to remove the gel from the glass plate and collect this for the radioactive waste. Thoroughly clean the plate with detergent by using a brush, abundantly rinse with tap water and demineralized water, wipe dry, and finally degrease with ethanol.
12. It is also possible to run sequencing ladders along the primer extension products, which are prepared with an end-labeled DNA fragment that is PCR-amplified using the labeled primer in combination with a non-labeled primer and using the in vitro transcription template as template DNA. In this case, the in vitro transcription experiment will also reveal information about the transcription start site location(s).
13. It is important to adjust all samples to the same number of counts in approximately the same volume. Loading identical volumes on gel can homogenize this loss over the different samples.

Acknowledgments

This work was supported by the Vrije Universiteit Brussel and funded by Research Foundation Flanders (FWO-Vlaanderen) (FWO Research Project G021118 and FWO-SB PhD fellowship 1S19717N to David Sybers).

References

1. Blombach F, Matelska D, Fouqueau T, Cackett G, Werner F (2019) Key concepts and challenges in archaeal transcription. J Mol Biol [Internet] 431:4184–4201. Available from: https://linkinghub.elsevier.com/retrieve/pii/S0022283619304097
2. Kyrpides NC, Ouzounis CA (1999) Transcription in archaea. Proc Natl Acad Sci [Internet] 96:8545–8550. Available from: http://www.pnas.org/cgi/doi/10.1073/pnas.96.15.8545
3. Ouhammouch M, Dewhurst RE, Hausner W, Thomm M, Geiduschek EP (2003) Activation of archaeal transcription by recruitment of the TATA-binding protein. Proc Natl Acad Sci [Internet] 100:5097–5102. Available from: http://www.pnas.org/cgi/doi/10.1073/pnas.0837150100
4. Hethke C, Bergerat A, Hausner W, Forterre P, Thomm M (1999) Cell-free transcription at 95 degrees: thermostability of transcriptional components and DNA topology requirements of *Pyrococcus* transcription. Genetics [Internet] 152:1325–1333. Available from: http://www.ncbi.nlm.nih.gov/pubmed/10430563
5. Wilkinson SP, Ouhammouch M, Geiduschek EP (2010) Transcriptional activation in the context of repression mediated by archaeal histones. Proc Natl Acad Sci [Internet] 107:6777–6781. Available from: http://www.pnas.org/cgi/doi/10.1073/pnas.1002360107
6. Smollett K, Blombach F, Werner F (2015) Transcription in archaea: preparation of *Methanocaldococcus jannaschii* transcription machinery. In: Bacterial transcriptional control, pp 291–303. Available from: http://link.springer.com/10.1007/978-1-4939-2392-2_17
7. Smollett K, Blombach F, Werner F (2015) Transcription in archaea: in vitro transcription assays for mjRNAP. Methods Mol Biol [Internet] 1276:305–314. Available from: http://www.ncbi.nlm.nih.gov/pubmed/25665572
8. Guillière F, Peixeiro N, Kessler A, Raynal B, Desnoues N, Keller J et al (2009) Structure, function, and targets of the transcriptional regulator SvtR from the hyperthermophilic archaeal virus SIRV1. J Biol Chem [Internet] 284:22222–22237. Available from: https://linkinghub.elsevier.com/retrieve/pii/S002192581849429X
9. Qureshi SA (1997) Factor requirements for transcription in the Archaeon *Sulfolobus shibatae*. EMBO J [Internet] 16:2927–2936. Available from: http://emboj.embopress.org/cgi/doi/10.1093/emboj/16.10.2927
10. Brock TD, Brock KM, Belly RT, Weiss RL (1972) Sulfolobus: a new genus of sulfur-oxidizing bacteria living at low pH and high temperature. Arch for Mikrobiol [Internet] 84:54–68. Available from: http://link.springer.com/10.1007/BF00408082
11. Cohen O, Doron S, Wurtzel O, Dar D, Edelheit S, Karunker I et al (2016) Comparative transcriptomics across the prokaryotic tree of life. Nucleic Acids Res [Internet] 44:W46–W53. Available from: https://academic.oup.com/nar/article-lookup/doi/10.1093/nar/gkw394
12. Schneider CA, Rasband WS, Eliceiri KW (2012) NIH image to ImageJ: 25 years of image analysis. Nat Methods [Internet] 9:671–675. Available from: http://www.nature.com/articles/nmeth.2089
13. Peng N, Han W, Li Y, Liang Y, She Q (2017) Genetic technologies for extremely thermophilic microorganisms of *Sulfolobus*, the only genetically tractable genus of crenarchaea. Sci China Life Sci [Internet] 60:370–385. Available from: http://link.springer.com/10.1007/s11427-016-0355-8

Chapter 7

Prediction of DNA-Binding Transcription Factors in Bacteria and Archaea Genomes

Leonardo Ledesma, Rafael Hernandez-Guerrero, and Ernesto Perez-Rueda

Abstract

DNA-binding transcription factors (TFs) play a central role in the gene expression of all organisms, from viruses to humans, including bacteria and archaea. The role of these proteins is the fate of gene expression in the context of environmental challenges. Because thousands of genomes have been sequenced to date, predictions of the encoded proteins are validated through the use of bioinformatics tools to obtain the necessary experimental, posterior knowledge. In this chapter, we describe three approaches to identify TFs in protein sequences. The first approach integrates the results of sequence comparisons and PFAM assignments, using as reference a manually curated collection of TFs. The second approach considers the prediction of DNA-binding structures, such as the classical helix-turn-helix (HTH); and the third approach considers a deep learning model. We suggest that all approaches must be considered together to increase the possibility of identifying new TFs in bacterial and archaeal genomes.

Key words Transcription factors, Bacteria, Archaea, Orthologs, Sequence comparison

1 Introduction

Transcriptional regulation is a central mechanism for controlling gene expression in all organisms. To regulate the synthesis of RNA from DNA, a class of proteins specifically binding to DNA has been described. These proteins are called DNA-binding transcription factors (TFs), and they respond to changes in the cellular environment by altering gene expression. In this context, TFs promote (as activators) or inhibit (as repressors) the transcription initiation of a gene, as a function of the DNA-binding site, around the promoter and the physiological status of the cell. In general, activators are able to promote the binding of the RNA polymerase

Supplementary Information The online version contains supplementary material available at [https://doi.org/10.1007/978-1-0716-2413-5_7].

Eveline Peeters and Indra Bervoets (eds.), *Prokaryotic Gene Regulation: Methods and Protocols*,
Methods in Molecular Biology, vol. 2516, https://doi.org/10.1007/978-1-0716-2413-5_7,

(RNAP), by a cooperative mechanism. In contrast, repressors block the access of the RNAP to the promoter, by steric hindrance [1–3].

Due to the crucial role of TFs in coordinating the gene expression kinetics of a genome, they have been studied in many ways, including mutation analysis and elucidation of three-dimensional structures [4]. In particular, sequence comparisons have been an excellent tool to classify and predict additional TFs in Bacterial organisms, such as *Escherichia coli* K-12 or *Bacillus subtilis* 168. In this regard, TFs in bacteria have been classified into families [5], where the most outstanding groups correspond to AraC/XylS, LysR, and TetR/AcrR, among others. In archaea, some seminal works have described the existence of TFs that are similar (in sequence) to bacteria; however, archaeal TFs with experimental evidence are limited [6, 7].

In the last decade, the advent of next-generation sequencing, the increase of available sequenced genomes, and accessibility to more sophisticated informatics tools have led to an increase in the number of TFs, basically considering the sequence comparisons. We must remember that the identification of the repertoire of TFs in a genome sequence is central to understanding the regulation of gene expression and, on a global scale, for the elucidation of regulatory networks.

In this chapter, we describe three main approaches to identify TFs based on sequence comparisons.

2 Materials

2.1 Equipment

A laptop or desktop computer with Linux operating system with a minimum of 8GB RAM and with high-speed internet connection is recommended.

2.2 Retrieving Genome Sequences

Prokaryotic genomes must be downloaded from the NCBI web server, in the section "refseq". Diverse programs can be used, such as the "ncbi-genome-download-master" program, with the following command:

% ncbi-genome-download-runner.py -s refseq -F all -l all bacteria (or archaea)

With this command, the genbank, fasta, features, gff, protein-fasta, genpept, cds-fasta, rna-fasta, assembly-report, and assembly-stats can be retrieved. To achieve a comparative analysis, open reading frames (ORFs) that encode predicted proteins in all organisms must be considered, i.e., the protein-fasta file.

2.3 Retrieving a Repertoire of DNA-Binding TFs with Experimental Evidence

A dataset of TFs with experimental evidence can be downloaded from the Encyclopedia of Transcription Factors website (ENTRAF; web.pcyt.unam.mx/EntrafDB). This dataset was obtained from scanning the Uniprot database (www.uniprot.org, version October 10, 2018) with the following keywords (and their combinations): Transcription; Gene regulation; DNA-binding; Repression; Activation; Repressor; Activator; Regulator; Bacteria; Archaea. These records were manually verified, by literature look-up and BLAST searches against a nonredundant (NR) database at NCBI, with E-values less than 10^{-3}. Proteins with no known regulatory roles, such as peptidases, lipolytic enzymes, integrases, and transporters, were excluded. Therefore, for each collected TF, its protein ID, protein sequence, and functional description were considered, along with the resources for its corresponding IDs in available databases.

In addition, TFs from specialized databases on gene regulation, including RegulonDB [8], DBTBS [9], and DBD [10] were retrieved and integrated into the previous dataset. Finally, a total of 668 proteins were considered TFs, and experimental evidences for these proteins included some of the following: Mutagenesis, Footprint, Crystallographic information, Transcriptional fusions, Protein structural information (from PDB), Binding of purified proteins (BPP), and Binding of cellular extracts (BCE), among others [4].

2.4 PFAM Assignments to Predict TFs

TFs can be scrutinized to assess their domain organization by using the PFAM database assignments [11], with an E-value at $\leq 10^{-3}$ to be considered significant. From this collection of TFs, 111 PFAMs associated with DNA binding were identified. These assignments are manually curated to identify proteins devoted to gene regulation. (The complete list of PFAM IDs is included as Supplementary Material S1).

2.5 Identification of Orthologous Proteins

To identify proteins with non-conserved DBDs, the program Proteinortho [12] can be used. This program implements an extended version of the reciprocal best heuristic alignment, and can be used as follows:

```
% perl proteinortho5.pl Entraf.fas infile -project=infile -cov=70
```

with an E-value of 10^{-5} and a coverage of $\geq 70\%$ considered significant to identify orthologous proteins against the collection of well-known proteins (ENTRAF proteins).

2.6 BLAST Searches

BLAST (basic local alignment search tool) is a suite of programs for comparing sequence information, such as the amino acid sequences of proteins or the nucleotides of DNA and/or RNA sequences. Protein-protein BLAST (blastp) uses a protein query and returns the most similar protein sequences from the protein database that the user specifies [13].

2.7 HMMER Searches

HMMER (http://hmmer.org/) is a suite of programs designed for searching sequence databases for sequence homologs and for making sequence alignments. It implements probabilistic models called profile hidden Markov models (profile HMMs). These programs are usually used together with a profile database, such as PFAM; however, HMMER can also work with profiles constructed by the user or with a protein query sequence against a database with phmmer, or an iterative search with jackhmmer [14].

3 Methods

3.1 PFAM Assignments in Sequenced Genomes

Because the description of TFs with new structural domains is a dynamic process, to evaluate the protein architecture of TFs we suggest scanning sequence genomes (protein-fasta files) with the complete 16,712 hidden Markov models (HMMs) information from the PFAM database.

The program pfam_scan.pl is useful for searching one or more protein sequences against the library of HMMs from PFAM (obtained from ftp.sanger.ac.uk/pub/database/Pfam) with the HMMER3 binaries installed, with an E-value of 10^{-3} (this E-value is considered a conservative default) and with the option of clan_overlap "activated", to show overlapping hits within clan member families. The script can be used as follows:

% pfam_scan.pl -fasta <fasta_file> -dir <directory location of Pfam files> -outfile <file>: output file, otherwise send to "stdout"

PFAM groups families with a common evolutionary ancestor together into clans. If there are overlapping matches within a clan, *pfam_scan.pl* only shows the most significant (with lowest E-value) match within the clan.

3.2 Identification of Protein Orthologs

In addition, the program Proteinortho can be applied to identify the orthologous proteins in the complete collection of all 668 well-known TFs against bacterial and archaeal genomes, with an E-value of 10^{-5} and a coverage of $\geq$70%. Finally, a set of TFs considers the union of those proteins identified by PFAM and Proteinortho searches.

3.3 Integration of Outputs to TF Assignments

In order to identify the repertoire of DNA-binding TFs, two approaches can be implemented and the output integrated. The first approach considers an exhaustive scanning with the PFAM assignments on the prokaryotic genomes; from there, 111 PFAM domains associated with DNA-binding TFs can be retrieved. The second approach considers an orthologous analysis with the Proteinortho program, taking as reference the collection of 668 well-known TFs (*see* Fig. 1).

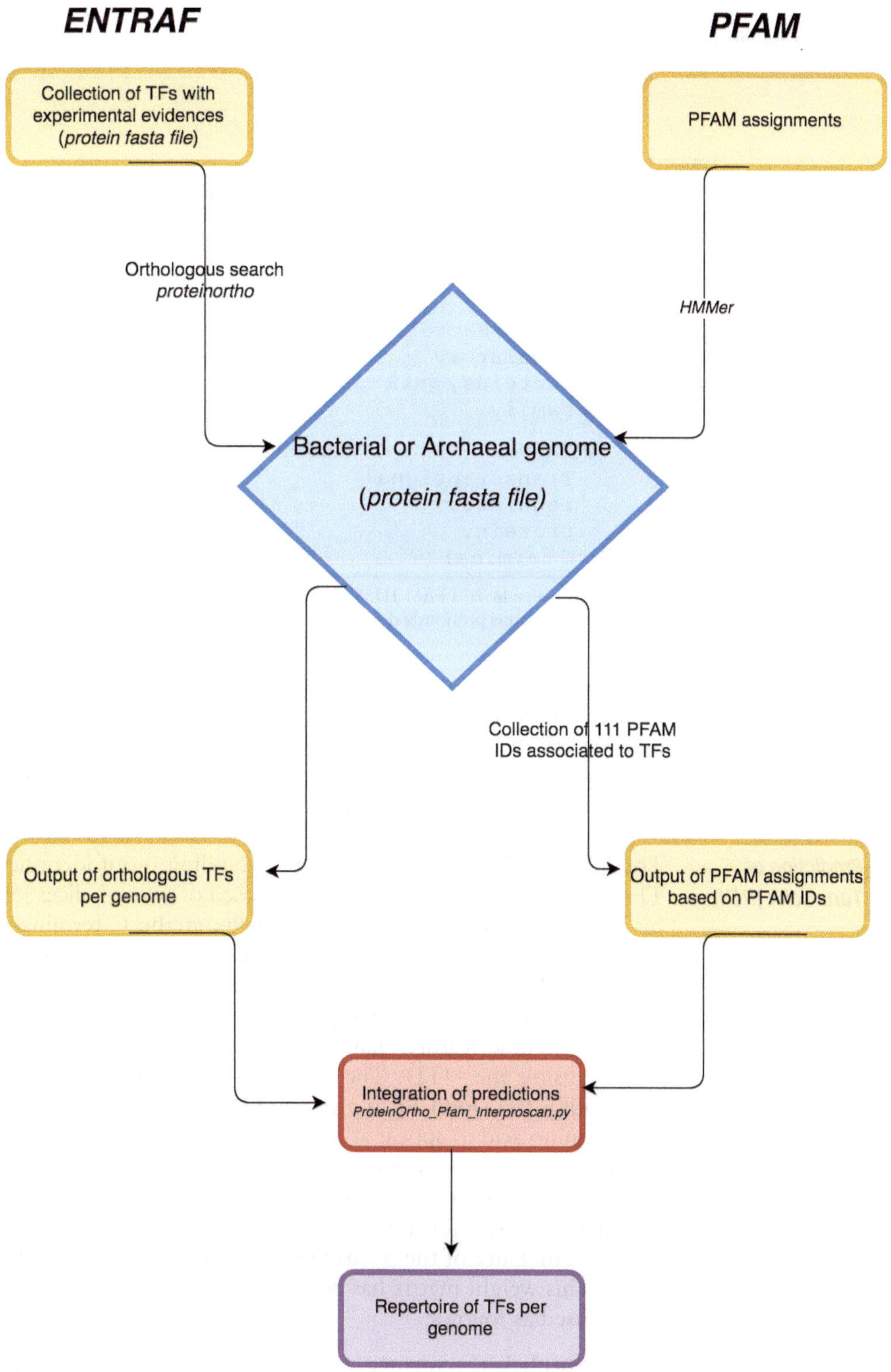

Fig. 1 Flux diagram of TF prediction

Table 1
Output of the integration of PFAM IDs and Proteinortho

SEQ ID	Length (aa)	PFAM ID	PFAM position (start, end)	Evalue	Ortho ID
WP_003592463.1	309	PF13936 Helix-turn-helix domain	4-42	9.4E-9	-
WP_004265054.1	243	PF00392 Bacterial regulatory proteins, gntR family	6-67	3.3E-11	-
WP_004265073.1	228	PF00486 Transcriptional regulatory protein, C terminal	151-225	1.1E-23	YRKP_BACSU

Columns are as follows: Sequence ID, length size in aa, PFAM ID, Start-end position, e-value, and ortholog protein when they were identified considering the parameters previously described

The following is a script to integrate the information that can be used and its corresponding output (*see* Table 1 and Supplementary Material S2):

% proteinOrthoLastVersion.py genome.pfam. output genome.ortho.output genome_tf_prediction.out

3.4 Prediction of Helix Turn Helix (HTH)

The HTH is the most common DNA-binding motif identified in TFs of Prokaryotes. The motif is composed of two α-helices joined by a short strand [15]. The second helix at the C-terminal end makes specific contacts with the DNA major groove. The positively charged and hydrophobic amino acids residing in the recognition helix increase the bonding capacity in DNA-protein complexes [16, 17]. These features have been used to classify and predict TFs by using the HTH. The most classical program to identify this motif is the Dodd-Egan program [18], which assigns a weight for an amino acid to appear at a certain location in the motif, i.e., a weight matrix. These weights are then totaled to compute a score for every 20-amino-acid subsequences of the original sequence. The highest score for any subsequence of the given sequence is used as a measure of the probability that the sequence is an HTH motif. This weight matrix has been implemented in EMBOSS and can be used as follows:

% helixturnhelix -sequence <protein-fasta> -outfile <output> (default -rformat motif)

Here is sample output for the repressor LacI of the bacterium *Escherichia coli* K-12:

```
# Sequence: LACI_ECOLI from: 1 to: 360
# HitCount: 1
# Hits above +2.50 SD (972.73)

Maximum_score_at at "*"

(1) Score 2160.000 length 22 at residues 4->25
 *
 Sequence: VTLYDVAEYAGVSYQTVSRVVN
 | |
 4 25
 Maximum_score_at: 4
 Standard_deviations: 6.54
```

Additional (optional) qualifiers can be used, such as -mean or -sdvalue, to increase the sensitivity of the search. Posterior to this search, BLAST searches against the Uniprot database are recommended (E-value of 10^{-3}) to exclude those proteins associated with functions beyond gene regulation.

3.5 Prediction of TFs Based on Deep Learning Methods

In order to solve the main tasks related to classification, prediction, and detection, methods based on artificial intelligence have been developed in different areas, such as computer vision, natural language processing, and recently in the analysis of biological data. In this regard, a recent approach has been described that implements deep learning models for the identification of TFs by using convolutional neural networks (CNNs) as a feature extractor, DeepTFactor [19]. In brief, the protein sequence is recorded to a binary matrix using a technique called one-hot encoding. Having the binary matrix as input tensor, three different CNNs are trained with several arrays of kernels filters to obtain the more relevant features on each convolution layer. All the features from each CNN are concatenated to be introduced to a new CNN and multilayer perceptron in order to predict if the initial protein sequence is a TF or should be excluded.

The application of a CNN such as a deep learning model places it on the foundations of bio-inspired learning, as a powerful tool to generalize the distribution of a data domain in order to predict and classify proteins based on features and embedded patterns (*see* Fig. 2).

The source code can be downloaded from

https://bitbucket.org/kaistsystemsbiology/deeptfactor/src/master

and can be used with the following command:

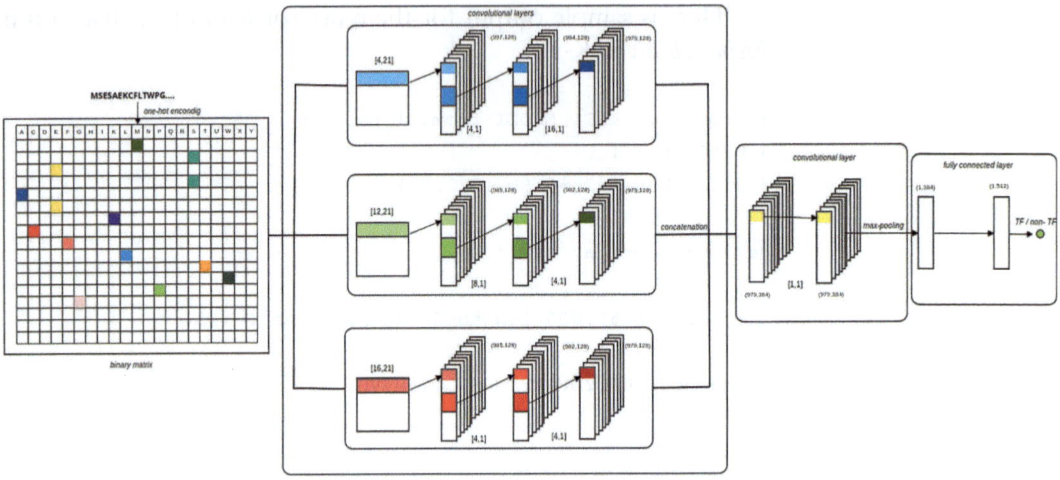

Fig. 2 DeepTFactor architecture. In the first stage, a binary image from an amino acid sequence is created using one-hot encoding. This matrix is then used through three CNNs with arrays of kernel size in their respective convolutional layers. In the third stage, the resulting feature arrays are concatenated to use in another CNN, extracting the most relevant features by using max-pooling. Finally, a fully connected layer is used to predict a TF. (Modified from Ref. [19])

% python tf_running.py -i escherichia_coli_K-12.fasta -o ./result -g cuda:1

With the following output:

```
sequence_ID description prediction score
AAC73448.1 LacI E. coli str. K-12 MG1655 True 0.9817
AAC73449.2 MhpR E. coli str. K-12 MG1655 True 0.9992
AAC73433.1 PrpR E. coli str. K-12 MG1655 True 0.8720
AAC73441.3 CynR E. coli str. K-12 MG1655 True 0.9991
AAC73460.2 FrmR E. coli str. K-12 MG1655 True 0.9949
```

4 Discussion

The identification of DNA-binding TFs is central in the understanding of gene regulatory networks in Bacteria and Archaea, and it opens the possibility to explore new regulatory interactions based on the resulting predictions. In this chapter, we describe three approaches to identify TFs in protein sequences. The first approach considers the integration of sequence comparisons and PFAM assignment using as reference a manually curated collection of TFs. The second approach considers the prediction of DNA-binding structures, such as the classical HTH. Finally, the third approach considers a deep learning model. All these approaches can be used in an integrative fashion to achieve more robust predictions, to better define the role of TFs, and for their posterior experimental validation.

Acknowledgments

We thank Joaquin Morales, Sandra Sauza, and Israel Sanchez for their technical support. Leonardo Ledesma is a doctoral student from Programa de Doctorado en Ingeniería y Ciencias de la Computación at UNAM and received a fellowship from Consejo Nacional de Ciencia y Tecnología (CONACYT CVU 857463). This work was supported by Dirección General de Asuntos del Personal Académico-Universidad Nacional Autónoma de México (IN-209620), Programa Iberoamericano de Ciencia y Tecnología para el Desarrollo (P918PTE0261), and CONACYT (320012). There was no additional external funding received for this study. The funders had no role in study design, data collection and analysis, decision to publish, or preparation of the manuscript.

References

1. Martinez-Antonio A, Janga SC, Salgado H, Collado-Vides J (2006) Internal-sensing machinery directs the activity of the regulatory network in Escherichia coli. Trends Microbiol 14(1):22–27
2. Browning DF, Busby SJ (2016) Local and global regulation of transcription initiation in bacteria. Nat Rev Microbiol 14(10):638–650
3. Browning DF, Busby SJ (2004) The regulation of bacterial transcription initiation. Nat Rev Microbiol 2(1):57–65
4. Flores-Bautista E, Cronick CL, Fersaca AR, Martinez-Nuñez MA, Perez-Rueda E (2018) Functional prediction of hypothetical transcription factors of Escherichia coli K-12 based on expression data. Comput Struct Biotechnol J 16:157–166
5. Perez-Rueda E, Hernandez-Guerrero R, Martinez-Nuñez MA, Armenta-Medina D, Sanchez I, Ibarra JA (2018) Abundance, diversity and domain architecture variability in prokaryotic DNA-binding transcription factors. PLoS One 13(4):e0195332
6. Cortés-Avalos D, Martínez-Pérez N, Ortiz-Moncada MA, Juárez-González A, Baños-Vargas AA, Estrada-de Los Santos P et al (2021) An update of the unceasingly growing and diverse AraC/XylS family of transcriptional activators. FEMS Microbiol Rev
7. Pérez-Rueda E, Janga SC (2010) Identification and genomic analysis of transcription factors in archaeal genomes exemplifies their functional architecture and evolutionary origin. Mol Biol Evol 27(6):1449–1459
8. Gama-Castro S, Salgado H, Santos-Zavaleta A, Ledezma-Tejeida D, Muniz-Rascado L, Garcia-Sotelo JS et al (2016) RegulonDB version 9.0: high-level integration of gene regulation, coexpression, motif clustering and beyond. Nucleic Acids Res 44(D1):D133–D143
9. Sierro N, Makita Y, de Hoon M, Nakai K (2008) DBTBS: a database of transcriptional regulation in Bacillus subtilis containing upstream intergenic conservation information. Nucleic Acids Res 36(Database issue):D93–D96
10. Kummerfeld SK, Teichmann SA (2006) DBD: a transcription factor prediction database. Nucleic Acids Res 34(Database issue):D74–D81
11. Mistry J, Chuguransky S, Williams L, Qureshi M, Salazar GA, Sonnhammer ELL et al (2021) Pfam: the protein families database in 2021. Nucleic Acids Res 49(D1):D412–D4d9
12. Lechner M, Findeiss S, Steiner L, Marz M, Stadler PF, Prohaska SJ (2011) Proteinortho: detection of (co-)orthologs in large-scale analysis. BMC Bioinf 12:124
13. Altschul SF, Gish W, Miller W, Myers EW, Lipman DJ (1990) Basic local alignment search tool. J Mol Biol 215(3):403–410
14. Eddy SR (2004) What is a hidden Markov model? Nat Biotechnol 22(10):1315–1316
15. Brennan RG, Matthews BW (1989) The helix-turn-helix DNA binding motif. J Biol Chem 264(4):1903–1906

16. Nishikawa T, Okamura H, Nagadoi A, König P, Rhodes D, Nishimura Y (2001) Solution structure of a telomeric DNA complex of human TRF1. Structure 9(12):1237–1251
17. Wintjens R, Rooman M (1996) Structural classification of HTH DNA-binding domains and protein-DNA interaction modes. J Mol Biol 262(2):294–313
18. Dodd IB, Egan JB (1990) Improved detection of helix-turn-helix DNA-binding motifs in protein sequences. Nucleic Acids Res 18(17): 5019–5026
19. Kim GB, Gao Y, Palsson BO, Lee SY (2021) DeepTFactor: A deep learning-based tool for the prediction of transcription factors. Proc Natl Acad Sci U S A 118(2)

Chapter 8

In Vivo Screening Method for the Identification and Characterization of Prokaryotic, Metabolite-Responsive Transcription Factors

Amber Joka Bernauw, Veerke De Kock, and Indra Bervoets

Abstract

In prokaryotes, transcription factors (TFs) are of uttermost importance for the regulation of gene expression. However, the majority of TFs are not characterized today, which hampers both the understanding of fundamental processes and the development of TF-based applications, such as biosensors, used in metabolic engineering, synthetic biology, diagnostics, etc. One way of analyzing TFs is through in vivo screening, enabling the study of TF-promoter interactions, ligand inducibility, and ligand specificity in a high-throughput fashion. Here, an approach is described for the selection and cloning of TF-promoter pairs, the development of a reporter system, and the measurement and analysis of fluorescent reporter assays. Furthermore, the importance of a suitable inducible plasmid system is illustrated together with prospective adaptations to modify a reporter system's output signal. The given approach can be used for the investigation of native, heterologous, or even artificially created TFs in *Escherichia coli*, and can be extended toward use in other microorganisms.

Key words In vivo screening, Transcription factor, Gene regulation, Ligand-induced response, *Escherichia coli*, High-throughput, Biosensor, Heterologous, Metabolite-responsive

1 Introduction

Prokaryotes can adapt themselves to fluctuating environments by tightly regulating their gene expression. Important key elements that enable this regulation on a transcriptional level are transcription factors (TFs), composed of a DNA binding and—in most cases—a ligand-binding domain. TFs can recognize and bind certain regions of the DNA, thereby activating or repressing the expression of neighboring genes [1]. When interacting with a ligand, the TF's regulatory activity is modulated, affecting the previously exercised activation or repression. This ability to sense the presence of metabolites and to alter gene expression levels in response enables microorganisms to switch on/off full metabolic

Eveline Peeters and Indra Bervoets (eds.), *Prokaryotic Gene Regulation: Methods and Protocols*,
Methods in Molecular Biology, vol. 2516, https://doi.org/10.1007/978-1-0716-2413-5_8,

pathways when coping with sudden changes in their environment, such as the absence or presence of certain nutrients [2].

Although thousands of TFs have already been predicted by genome-wide computational studies, only a fraction has been studied in detail so far [3]. Moreover, most experiments to characterize TFs in depth are being performed in vitro (e.g., structure determination studies, DNA- and ligand-binding assays), which can provide results that are not completely in accordance with the TF's behavior in vivo. TFs can however also be used to build in vivo inducible systems that sense a certain input—located intra- or extracellularly—and convert this proportionally into a measurable output. The input depends on the TF's properties and can vary from a certain metabolite to changes in light intensity [4], oxygen level [5], temperature [6], or even the presence of neighboring bacterial cells [7, 8]. The output, on the other hand, is independent of the TF and can for example be a fluorescent or luminescent signal, resistance to antibiotics, or the expression of a specific gene of interest. This technique allows to study the regulatory response, inducibility, and specificity of TFs in vivo in detail, thereby contributing to both the discovery and further characterization of TFs with interesting properties. Furthermore, due to the more high-throughput nature of this method, not only native, but also heterologous and artificially created TFs can be easily studied under different conditions. An illustration of this is the extensive screen that has been performed in *Cupriavidus necator*, in which the regulatory response and ligand specificity of 16 TFs were characterized in vivo in *C. necator*, in *Escherichia coli* and in *Pseudomonas putida* [9].

The construction of TF-based inducible systems is not only interesting for the study of gene regulation but also to obtain knowledge about a TF in its natural context. When optimized for specific applications, for example in the fields of synthetic biology, metabolic engineering, diagnostics or environmental studies, TF-based inducible systems can be applied as biosensors [10]. A very useful practice is the high-throughput screening of samples or strains for the presence of a certain molecule. An example of this is the TF-based biosensor utilized to screen a mutant library of *Corynebacterium glutamicum* strains for L-methionine and branched-chain amino acid production [11]. Other examples show the application of TF-based biosensors to measure lactate concentrations during biopharmaceutical production by mammalian cell cultures [12] or for the detection of the micronutrient zinc in human serum [13]. On the other hand, TF-based biosensors can also be employed to mimic the ingeniously established gene regulatory systems that exist in nature, aiming to improve bio-based production of industrially relevant compounds [14]. An illustration of this is the biosensor based on FadR from *E. coli* that was used to regulate genes in the fatty acid ethyl ester (FAEE) production pathway, aiming to balance concentrations of the intermediate metabolites, ethanol, and acyl-CoA, while increasing FAEE titers [15]. Layered regulatory circuits, combining multiple biosensors or a biosensor together with a CRISPRi or RNAi system, have also been built, allowing a tighter regulation that mimics regulation occurring in nature even better [16–19].

The protocol described here entails an approach to construct and test TF-based metabolite-inducible systems in vivo in *E. coli*. Topics that are covered are the design and construction of a proper plasmid chassis, the selection and introduction of TFs and their corresponding promoters, fluorescent reporter measurements, and illustrative results to show both potential and pitfalls of the system. This method is both appropriate for high-throughput screens of large libraries of unknown, mutated, or re-engineered TFs as well as for small-scale experiments, such as the investigation of a specific TF-promoter interaction or the development and optimization of a TF-based biosensor for a specific purpose. In addition, the reported approach can be adapted for use in other microorganisms.

2 Materials

2.1 Bacterial Strains, Media, and Solutions

1. *E. coli* MG1655 (*see* **Note 1**).
2. *E. coli* DH10B carrying the thermo-sensitive pKD46 plasmid [20] (*see* **Note 2**).
3. Chemically competent *E. coli* MG1655 cells (*see* **Notes 1** and **3**). Aliquoted per 100 μL and stored at −80 °C.
4. 2YT medium: 16 g/L tryptone, 10 g/L yeast extract, and 5 g/L NaCl. Autoclaved.
5. LB medium: 25 g/L LB broth high salt. Supplemented with 15 g/L microagar in case of solid media. Autoclaved.
6. Base medium: 2.50 g/L NH_4Cl, 6.25 g/L $(NH_4)_2SO_4$, 3.74 g/L KH_2PO_4, 9.15 g/L K_2HPO_4, 10.46 g/L MOPS, 0.63 g/L NaCl, and pH 7.0. Autoclaved.
7. Glucose and SO_4 solution: 2.5 g/L $MgSO_4{\cdot}7H_2O$ and 82.5 g/L glucose·H_2O. Autoclaved.
8. Trace element solution: 3.6 g/L $FeCl_2{\cdot}4H_2O$, 5 g/L $CaCl_2{\cdot}2H_2O$, 1.3 g/L $MnCl_2{\cdot}2H_2O$, 0.38 g/L $CuCl_2{\cdot}2H_2O$, 0.5 g/L $CoCl_2{\cdot}6H_2O$, 0.94 g/L $ZnCl_2$, 0.0311 g/L H_3BO_4, 0.4 g/L $Na_2EDTA{\cdot}2H_2O$, and 1.01 g/L thiamine·HCl. Filter sterilized (0.22 μm).
9. Molybdate solution: 0.967 g/L $Na_2MoO_4{\cdot}2H_2O$. Filter sterilized (0.22 μm).
10. Defined supplemented medium 1 L (*see* **Note 4**): add together 800 mL base medium, 200 mL glucose and SO_4 solution, and 1 mL trace element solution and 100 μL molybdate solution.

11. Ampicillin stock solution: 100 mg/mL. Final concentration in all media (liquid and solid) is 100 μg/mL.
12. Chloramphenicol stock solution: 25 mg/mL, dissolved in ethanol. Final concentration in all media (liquid and solid) is 25 μg/mL.
13. Kanamycin stock solution: 60 mg/mL. Final concentration in all media (liquid and solid) is 60 μg/mL.
14. Arabinose solution: 1 M L-arabinose.
15. CelLytic™ B Cell Lysis Reagent (Sigma-Aldrich).

2.2 Molecular Biology

1. TE buffer: 10 mM Tris, 1 mM EDTA, pH 8.0.
2. Assembly buffer (10×): 500 μL 1 M Tris-HCl (pH 7.5), 50 μL 2 M $MgCl_2$, 100 μL 100 mM ATP, 10 μL 1 M DTT, and 340 μL nuclease-free H_2O. Stored at −20 °C.
3. Oligonucleotides. Stock solutions at 500 μM and working solutions at 10 μM are prepared in nuclease-free H_2O. Both solutions are stored at −20 °C. An overview of used oligonucleotide sequences is provided in Table 1.
4. Synthetic transcription factor genes, codon optimized for *E. coli* (*see* **Note 5**). Design is described in Subheading 3.2.1. Stock solutions at 50 ng/μL are prepared by using TE buffer. Stored at −20 °C.
5. Promoter fragments. Design is described in Subheading 3.2.1. Stock solutions at 50 ng/μL are prepared by using TE buffer. Stored at −20 °C.
6. KAPA HiFi DNA Polymerase (*see* **Note 6**); Master mixes (2×), containing the appropriate amount of DNA polymerase, buffer, dNTPs, and $MgCl_2$, can be prepared according to manufacturer's guidelines.
7. GoTaq® G2 DNA Polymerase (*see* **Note 6**). Master mixes (2×), containing the appropriate amount of DNA polymerase, buffer, dNTPs, and $MgCl_2$, can be prepared according to manufacturer's guidelines.
8. Gel and PCR cleanup kit for purification of DNA fragments.
9. Plasmid preparations used as template for cloning purposes, listed in Table 2.
10. Plasmid DNA extraction kit.
11. Restriction enzymes. FastDigest restriction enzymes BamHI, EcoRV, KpnI, NcoI, SalI, and XbaI are used with the corresponding FastDigest buffer.
12. T4 DNA Ligase and its respective 10X buffer.

Table 1
Overview of the different oligonucleotides used in this article

Primer	Description	Sequence (5′–3′)
1	FW primer to create pPRC fragment 1	GACGATGGTAGTCAGCTGCCTACTAAGCTG**TCTAGA**TA CAATGC**GATATC**TCATGCTATGGAGGTTAAG**CATATG**GTT AGCG AGCTGATCAA AGAAAACATGC
2	RV primer to create pPRC fragment 1	CACGTTAAGGGATTTTGGTCATGAGATTATC**GGTACC**G TTCTAGTAA**GCGGCCGC**CCAGCGCTGGAGTTAATCACG TAAAAGAG
3	FW primer to create pPRC fragment 2	GATAATCTCATGACCAAAATCCCTTAACGTG
4	RV primer to create pPRC fragment 2	CACCCCAGCTTCAAAAGCGCTCTCTCACTCAAAGGCGG TAATACGGT
5	FW primer to create pPRC fragment 3	CAGCTTAGTAGGCAGCTGACTACCATCGT CAAAAAAAAACCCCGCTTCGGCGGGGTTTTTTTT**GGAT** **CC**CTTCAAGATCCCCCACGCTG
6	RV primer to create pPRC fragment 3	ACCGTATTACCGCCTTTGAGTGAGAGAGCGCTTTTGAAG CTGGGGT
7	FW primer to create pITC fragment 1	CTAGGTCTTACAGTATGCCATCATA**GCGGCCGC**TTAC TAGAAC**GGTACC**AGGAACTTCGAAGCAGCTCC
8	RV primer to create pITC fragment 1	AGCATCTGTACGCTAACGCAAGGTTGACAGTTG TTAGCCGATTACGA**CTCGAG**ACGAT**GTCGAC**TCACTCAG TAT**CCATGG**CTGTCTCCTTGTCTCCTAC
9	FW primer to create pITC fragment 2	CTGTCAACCTTGCGTTAGCGTACAGATGCTCTTGTTCAG AACGCTCGGTCTTGCACACCGGGCGTTTTTTCTTTGTGAG TCCACGACAGTAAGACGGGTAAGC
10	RV primer to create pITC fragment 2	GAACCATTGAGATCATACTTACTTTGCATGTC
11	FW primer to create pITC fragment 3	GACATGCAAAGTAAGTATGATCTCAATGGTTC
12	RV primer to create pITC fragment 3	GTAA**GCGGCCGC**TATGATGGCATACTGTAAGACC TAG**GATATC**GCATTGTA**GGATCC**TGCCGCAAGCACTCAG
13	FW primer to amplify synthetic DNA fragments	GAAGTGCCATTCCGCCTGAC
14	RV primer to amplify synthetic DNA fragments	CACTGAGCCTCCACCTAGC
15	FW primer for colony PCR pPRC plasmids	GTCCAGATAGCCCAGTAGCTGACATTC
16	RV primer for colony PCR pPRC plasmids	GTGGTTATTCACGGTGCCTTCC

(continued)

Table 1
(continued)

Primer	Description	Sequence (5′–3′)
17	FW primer for colony PCR pITC plasmids	TGAGCCTGGCCATGACAACG
18	RV primer for colony PCR pITC plasmids	TGACCACTTCGGATTATCCCGTGAC
19	FW primer for colony PCR 1-plasmid system	GTCCAGATAGCCCAGTAGCTGACATTC
20	RV primer for colony PCR 1-plasmid system	CCTTCGTAAATCTGGCGAGTGG
21	FW primer for cloning ProB/C/D in pPRC	GATGGTAGTCAGCTGCCTACTAAGCT**GTCTAGA**CACAGC TAACACCACGTCGTCC
22	RV primer for cloning ProB/C/D in pPRC	CTAACCATATGCTTAACCTCCATAGCATGA**G** **ATATC**ATACAGTTTCATGTGCATGTTTTCTTTGATC

Sequences in bold correspond with restriction sites, while underlined sequences belong to a terminator

Table 2
Overview of the different fragments generated by PCR, together with the used template, primers, the length, and the functionalities present on the respective fragment

Generated fragment	Template	Used primers	Length (bp)	Functionalities
pPRC fragment 1	pNatSens [23]	1 & 2	1039	*mkate2* reporter gene + terminator T2
pPRC fragment 2	pBR322 [25]	3 & 4	813	Origin of replication
pPRC fragment 3	pSC101-ProB-*mkate2* (previous work, own lab)	5 & 6	1442	Kanamycin resistance gene
pITC fragment 1	pNatSens [23]	7 & 8	1489	Naringenin inducible promoter
pITC fragment 2	pNatSens [23]	9 & 10	1714	Origin of replication
pITC fragment 3	pNatSens [23]	11 & 12	1813	Chloramphenicol resistance gene
Promoter ProB/C/D	pSC101-ProB/C/D-*mkate2* (previous work, own lab)	21 & 22	285	Constitutive promoter (control)

2.3 Plate Reader Measurements

1. Transparent 96-well plates.
2. Black 96-well plates (*see* **Note 7**).
3. Breathable membranes (*see* **Note 8**).
4. Naringenin stock solutions of 25 and 250 mg/L. Prepare by dissolving naringenin powder in pure ethanol.
5. Stock solutions of ligands (the concentration used will depend on the nature of ligand, e.g., solubility) in ddH_2O or in medium. Filter sterilized (0.22 μm) and stored at −20 °C.
6. Microplate reader for kinetic measurements (*see* **Note 9**).

3 Methods

In this work, an approach is described to conduct an in vivo screening of TFs and promoters of interest. A two-plasmid reporter system is designed and used for the cloning of TFs and their responsive promoter sequences to obtain a set of TF-specific reporters. Screening of strains bearing the different metabolite-inducible systems is conducted by measuring fluorescent reporter signals. Depending on the desired throughput, one can work with a microplate reader and manual pipetting, or one can use an automated handling and pipetting system to perform measurements of multiple plates at the same time. In addition, an overview of modifications to adapt the strains' performances is presented, using examples to illustrate this. These modifications are dependent on the chosen plasmid system, TF, and promoter sequences, and can therefore not be generalized.

3.1 Design and Construction of a Two-Plasmid Reporter System

The first step in constructing a reporter system is the selection of a plasmid backbone. Here, two plasmids are designed *de novo* to include all desirable properties while limiting plasmid size to a minimum (*see* Fig. 1). The two-plasmid system consisting of a promoter reporter construct (pPRC), and an inducible TF construct (pITC) has the advantage that numerous combinations of TFs and promoters can easily be tested by performing different co-transformations of plasmids into the cell without the need for additional cloning steps (*see* **Note 10**).

3.1.1 Promoter Reporter Construct (pPRC)

The pPRC plasmid is designed to have several essential characteristics (*see* Fig. 1). First, it needs an antibiotic resistance gene for selection and an origin of replication with low to medium copy number (*see* **Note 11**). Additionally, a reporter gene is integrated on the plasmid, more specifically a fluorescent reporter gene *mkate2* was chosen followed by a terminator (T2) to stop transcription (*see* **Notes 12** and **13**). A multiple cloning site (MCS1) is added directly upstream of the reporter gene to enable insertion of

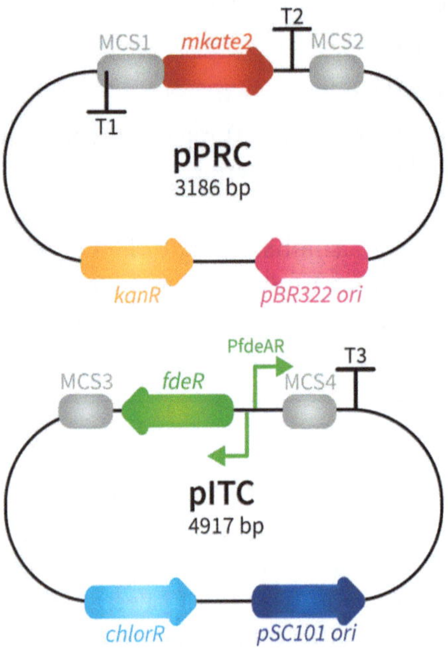

Plasmid	Part	Composition
pPRC	Origin of replication	pBR322 ori
	Antibiotic resistance gene	Kanamycin resistance
	MCS1	BamHI - XbaI - EcoRV - NdeI
	Terminator T1	Part: BBa_B1001
	Fluorescent reporter gene	*mkate2*
	Terminator T2	BioFab apFAB391
	MCS2	NotI - KpnI
pITC	Origin of replication	pSC101-RepA ori
	Antibiotic resistance gene	Chloramphenicol resistance
	MCS3	BamHI - EcoRV - NotI - KpnI
	Inducible system	*fdeR*-PfdeAR
	MCS4	NcoI - SalI - XhoI
	Terminator T3	T0 (part: BBa_K864600)

Fig. 1 Overview of the two-plasmid system consisting of a promoter reporter construct (pPRC) and an inducible TF construct (pITC) together with the different parts used to build the plasmids

promoter sequences. Upfront this MCS1, a second, bidirectional terminator (T1) is placed to avoid read-through transcription from promoters in both directions (*see* **Note 14**). Furthermore, additional restriction sites are inserted directly upstream of the T1 terminator and downstream of the T2 terminator (MCS2) (*see* Fig. 1) to enable the removal of the promoter reporter module. In this way, the promoter and reporter can be transferred to the pITC plasmid, obtaining a single-plasmid system.

3.1.2 Inducible TF Construct (pITC)

Analogous to the pPRC, the pITC plasmid requires an antibiotic resistance gene and an origin of replication resulting in a low to medium copy number (*see* **Note 11**), both compatible (*see* **Note 15**) with those of the pPRC to enable simultaneous existence of the two plasmids in the cell. Secondly, an inducible promoter is chosen to control expression of the TF genes. The naringenin-inducible system *fdeR*-PfdeAR is selected for this purpose (*see* **Note 16**). An MCS is inserted downstream of the PfdeAR promoter (MCS4) to allow insertion of TF genes (*see* Fig. 1). Finally, a second MCS is also added on the pITC (MCS3) to allow introduction of the promoter reporter module from the pPRC plasmid and thus the transition to a single-plasmid system.

3.1.3 Preparation of Cell Extract for Assembly Cloning

For the assembly of the pPRC and pITC plasmids, as well as for cloning of promoters and TF sequences in the respective plasmids, a cloning method is used that is a variation on the SLiCE (Seamless Ligation Cloning Extract) method described in [21] (*see* **Note 17**). In this assembly approach based on homologous recombination, cell extract is used that contains a high concentration of proteins from the λ Red recombinase system.

Preparation of cell extract:

1. Pick a colony of *E. coli* DH10B bearing pKD46 to inoculate 10 mL of 2YT medium, supplemented with ampicillin.
2. Grow the preculture overnight at 28 °C and 210 rpm (*see* **Note 18**).
3. Use this preculture to inoculate a 50 mL culture of 2YT medium, supplemented with ampicillin.
4. Grow the culture at 28 °C and 210 rpm until an OD_{600} of 2.5–3 is reached (*see* **Notes 18** and **19**).
5. Add 1 mL of arabinose solution to the culture to induce expression of the λ Red recombinase system and place the culture for 2 h at 28 °C and 210 rpm (*see* **Note 18**).
6. Take a 50 mL falcon tube and weigh it.
7. Transfer the culture to the falcon tube.
8. Centrifuge the culture for 20 min at 2000 × *g* and 4 °C. Resuspend the pellet in 50 mL ddH_2O.
9. Centrifuge the cells again for 20 min at 2000 × *g* and 4 °C.
10. Weigh the falcon containing the pellet. Resuspend the pellet in 300 μL CelLytic™ B Cell Lysis reagent per 0.25 g of pellet.
11. Vortex the mixture and transfer it to a 1.5 mL tube, incubate at room temperature for 10 min.
12. Centrifuge the lysate for 3 min at 13,000 rpm in a microcentrifuge. Mix the supernatant in a 1:1 ratio with 100% glycerol, aliquot, and store at −20 °C.

3.1.4 Construction of pPRC and pITC by Polymerase Chain Reaction and Assembly Cloning

The pPRC and pITC plasmids are each constructed by assembling 3 separate PCR fragments with overlapping ends. Polymerase chain reactions are used on different DNA templates to generate the fragments with the necessary characteristics and overhangs to form the pITC and pPRC plasmids. Primers and fragments generated by PCR are listed in Tables 1 and 2. Primers for PCR are designed so that they contain a priming part, extra nucleotides to build an MCS or a terminator, and an overlap with the other plasmid fragments to enable homologous recombination in the assembly step. The high-fidelity DNA polymerase KAPA (*see* **Note 6**) is used for amplification of the fragments. When using another type of high-fidelity polymerase, PCR conditions should be adjusted according to manufacturer's guidelines. After purification of the fragments, an assembly reaction is conducted to generate the full pITC and pPRC plasmids. Finally, a transformation by heat shock is performed in *E. coli* MG1655 competent cells.

1. Prepare a PCR reaction mix for every assembly fragment by adding 1.5 μL forward primer, 1.5 μL reverse primer, 1 μL template DNA, 21 μL nuclease-free H_2O, and 25 μL Kapa Master Mix (2×) to a PCR tube.
2. Spin down and place in a thermocycler.
3. Run the following program:

   95 °C – 3 min

   35 cycles of

   98 °C – 20 s

   55–60 °C – 15 s

   72 °C – 30 s/kb

   72 °C – 1 min/kb
4. Perform gel electrophoresis using a 0.8 or 1% agarose gel and analyze the result.
5. Cut bands at the correct height from the agarose gel and purify using the gel and PCR cleanup kit according to manufacturer's instructions (*see* **Note 20**).
6. For the construction of both the pPRC and pITC plasmid, pipet following ingredients on ice in separate 0.2 mL tubes:

   100 ng of each purified PCR fragment, in equimolar ratio (*see* **Note 21**)

   1 μL assembly buffer

   1 μL cell extract

   Add nuclease-free H_2O until a total volume of 10 μL
7. Incubate the reaction mixture for 1 h at 37 °C in a water bath or thermocycler.
8. After the incubation period, add 2–4 μL of the reaction mixture to *E. coli* MG1655 competent cells on ice.
9. Mix the cells by gently pipetting up and down and incubate the tubes on ice for 25 min.
10. Perform a heat shock by incubating the cells for 3 min at 37 °C.
11. Incubate the cells for 10 min at room temperature.
12. Add 1 mL of LB medium to the cells and incubate the tubes for 2 h at 37 °C while shaking.
13. Plate 100 μL on an LB agar plate, containing the appropriate antibiotics.

14. Centrifuge the remaining 900 μL, discard the supernatant and resuspend the pellet in 100 μL of fresh medium. Plate on an LB agar plate, containing the appropriate antibiotics.
15. Incubate the plates overnight at 37 °C.
16. Pick a few colonies grown on the plates to inoculate liquid cultures of 5 mL LB medium with the appropriate antibiotics. Incubate the cultures overnight at 37 °C and 170 rpm.
17. Extract plasmids from the liquid cultures using a plasmid extraction kit.
18. Perform an initial verification of the transformants by restriction analysis: digest the plasmids with enzymes that are predicted to cut the plasmid, by following the manufacturer's guidelines.
19. Load the restriction mixtures on an agarose gel and perform gel electrophoresis (*see* **Note 22**).
20. Analyze and select positive candidates, validate plasmids by Sanger sequencing.

3.2 Development and Screening of the Transcription Factor Library

Once the two plasmids are constructed and validated (*see* **Note 23**), TF genes and their corresponding promoter sequences can be inserted in pITC and pPRC, respectively. Subsequently, the resulting plasmids are co-transformed into *E. coli* cells, followed by an elaborate evaluation of the strains' performances using fluorescence measurements. The general workflow is depicted in Fig. 2.

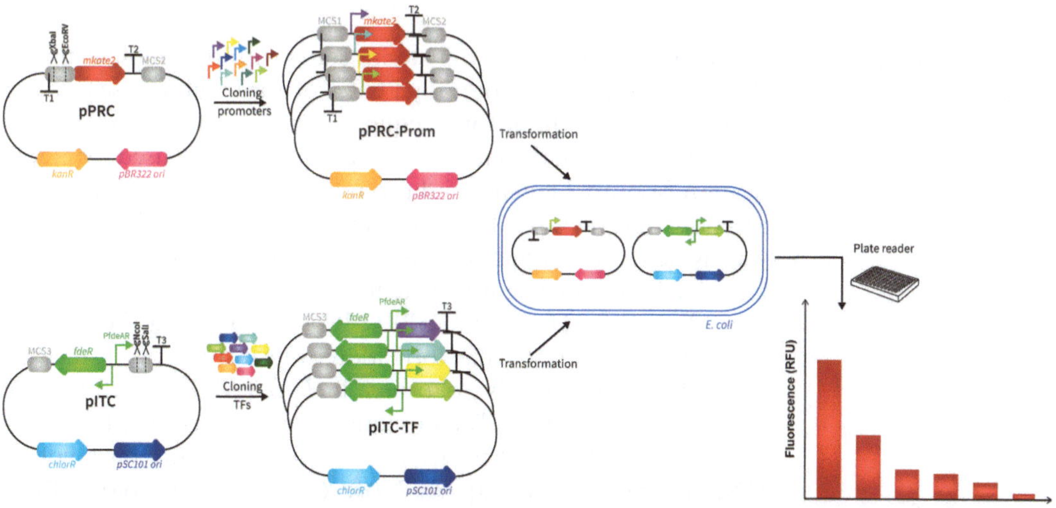

Fig. 2 Workflow describing the construction of a two-plasmid screen and the execution of in vivo fluorescence measurements. The pPRC and pITC plasmids are both linearized, respectively using XbaI and EcoRV, or NcoI and SalI as restriction enzymes. The restricted plasmids and promoters/TFs are used in a cloning reaction to create different sets of plasmids. Corresponding pPRC-Prom and pITC-TF plasmids are co-transformed by performing a heat shock transformation in *E. coli* competent cells. Finally, cultures are grown in 96-well microtiter plates, from which OD_{600} and mKate2 fluorescence are monitored over time in a plate reader and relative fluorescence units (RFU) are calculated

3.2.1 Selection of TFs and Corresponding Promoter Sequences

The selection of TFs and their promoters greatly depends on the aim of the experiment and on the knowledge available about the TFs or the TF family they belong to. Additionally, the size of the selected set of TFs and promoters relies upon the choice of screening method and its throughput. Different types of experiments can be conducted, ranging from a large screening of hitherto uncharacterized TF-promoter pairs, whether belonging to the same microorganism or the same TF family, to the characterization of one specific transcription factor or the optimization of an existing biosensor for a certain application. TFs and promoters can be selected by performing an extensive literature study, searches in databases, or even by performing an analysis of the genomic environment of genes of interest. For each TF, at least one promoter sequence should be selected for examination. In case promoter sequences are not known, either the promoter of the TF gene itself can be selected (in case the TF might be an autoregulator), the promoter of a gene located in a divergent operon, or the promoter of a (nearby located) gene with an interesting, predicted function. For promoter sequences that have not yet been characterized, a sufficiently large region (for example 300 base pairs) before the start of the gene of interest can be taken. Both TF genes as well as promoter sequences can be ordered as dsDNA fragments (*see* **Note 24**) and can already contain the necessary adjacent homology regions for assembly cloning on both ends. This allows the use of one set of primers (specific for either pPRC or pITC) to amplify all fragments for insertion into the plasmids.

3.2.2 Construction of Reporter Strains

The construction of reporter strains starts with the amplification of the TF genes and promoters by PCR. Forward and reverse primers, designed to bind the adaptor sequences (present at the ends of each synthetic fragment depending on the company), are used for amplification of both TF genes and promoters (*see* **Note 25**). The PCR reactions and their purification from gel are conducted in the same way as described in Subheading 3.1.4. Further, both plasmids are linearized, and an assembly reaction is conducted to generate the final plasmids pITC-TF and pPRC-Prom, containing the TFs and promoters of interest. Transformation by heat shock of the relevant plasmid combinations is performed in *E. coli* MG1655 competent cells. Finally, colony PCR and sequencing are performed to check whether the cloning has been successful. The non-high-fidelity DNA polymerase GoTaq (*see* **Note 6**) is used for colony PCR. When using another DNA polymerase, PCR conditions should be adjusted according to manufacturer's guidelines.

1. Pick colonies containing the pPRC or pITC plasmids, obtained in Subheading 3.1.4., to inoculate 5 mL of liquid LB medium, and supplemented with the appropriate antibiotic.
2. Grow the cultures overnight at 37 °C and 170 rpm.

3. Extract pPRC and pITC plasmids from liquid culture using a plasmid extraction kit.
4. Linearize the plasmids pPRC and pITC using restriction enzymes EcoRV and XbaI, and NcoI and SalI, respectively (*see* **Note 26**). FastDigest restriction enzymes are used. When using another type of restriction enzyme, reaction conditions should be adjusted according to manufacturer's guidelines. Prepare the reaction mixtures as follows:

   1 μg of plasmid DNA

   2 μL 10× FastDigest buffer

   1 μL restriction enzyme 1

   1 μL restriction enzyme 2

   Add nuclease-free H_2O until a total volume of 20 μL
5. Incubate the reaction for 1 h at 37 °C.
6. Inactivate the restriction enzymes by incubating the reaction for 15 min at 70 °C.
7. Load the reaction mixture (everything) on a 0.8% agarose gel and perform gel electrophoresis (*see* **Note 22**).
8. Cut out the band corresponding to the restricted plasmid from gel and purify using the gel and PCR cleanup kit according to manufacturer's instructions (*see* **Note 20**).
9. Use the amplified promoter sequence and restricted pPRC plasmid or the TF gene and pITC plasmid, in an assembly reaction, by adding the following components together to 0.2 mL tubes, work on ice:

   100 ng of restricted plasmid

   Promoter or TF gene in 1:5 (vector:insert) molar ratio

   1 μL assembly buffer

   1 μL cell extract

   Add nuclease-free H_2O until a total volume of 10 μL
10. Incubate the assembly mixture for 1 h at 37 °C.
11. Perform a heat shock transformation in *E. coli* MG1655 competent cells as described in Subheading 3.1.4.
12. Perform a colony PCR to screen for positive transformants. Start by picking colonies with a sterile toothpick and dissolving them in 50 μL of nuclease-free H_2O.
13. Place the tubes at 100 °C for 5 min and perform a short spin. Use the supernatant as template DNA in the PCR reaction (*see* **Note 27**).
14. Prepare the PCR reaction by adding 1.5 μL forward primer, 1.5 μL reverse primer, 1 μL template DNA, 8.5 μL nuclease-

free H_2O, and 12.5 μL GoTaq Master Mix (2×) to a 0.2 mL PCR tube.

15. Spin down and run the following PCR program:

    95 °C – 2 min

    35 cycles of

        95 °C – 30 s

        55–58 °C – 30 s

        72 °C – 1 min/kb

    72 °C – 5 min

16. Analyze the PCR fragments on a 0.8 or 1% agarose gel by performing gel electrophoresis.
17. Purify PCR fragments with a band at the expected height from the PCR mixture using the gel and PCR cleanup kit according to manufacturer's instructions (*see* **Note 20**).
18. Send the purified fragments to an external company for Sanger sequencing.
19. Analyze the sequencing results by comparing sequences *in silico* to the reference sequence.
20. Use colonies of the verified positive transformants to inoculate 5 mL LB medium, supplemented with the appropriate antibiotics. Incubate overnight at 37 °C and 170 rpm.
21. Extract the plasmids from liquid cultures using a plasmid DNA extraction kit.

3.2.3 Fluorescence Measurements

For each strain bearing a two-plasmid combination, two different experiments are performed. The first experiment tests the strains in the presence of a range of distinct naringenin concentrations to observe if there is a regulatory effect (activation/repression) of the TF on the selected promoter when the TF is expressed. It is important to also check the fluorescence values in absence of naringenin as a reference. Furthermore, an additional strain containing the two-plasmid system, but without TF present on the pITC plasmid, is included in these experiments. The latter ensures to have a condition where absolutely no TF is present, since the occurrence of leaky expression of the naringenin inducible promoter cannot be excluded (*see* **Note 16**). In a second experiment, the strains are investigated in the presence of variable ligand concentrations at fixed naringenin concentrations, to determine which compounds interact with the TF.

For both experiments, at least four biological replicates are recommended per condition (*see* **Note 28**). Furthermore, each microplate contains wells with only medium (background) and wells with a positive control (strain bearing the combination of pITC and pPRC-ProD), both in absence and in presence of the highest naringenin and/or ligand concentration to detect changes in fluorescence or optical density caused by either contamination or specific components of the medium. The positive control, having a validated *E. coli* promoter which results in a specific fluorescent outcome signal (*see* **Note 22**), enables comparison of results between separate microplates. A well-thought-out plate design, containing all necessary controls and replicates, should be designed for every 96-well plate in advance of each plate reader experiment.

Before starting a fluorescence measurement, it is important to perform fresh transformations of the plasmids in *E. coli* MG1655 competent cells (*see* **Notes 1** and **29**). The transformants can then be used for the inoculation of a preculture plate that is grown overnight. The day after, the final plate can be pipetted, and fluorescence measurements can be started.

1. Co-transform the pITC-TF plasmid and its corresponding pPRC-Prom plasmid by adding 10 ng of each plasmid to the same tube of 100 μL thawed *E. coli* MG1655 competent cells on ice.
2. Mix the cells by gently pipetting up and down and incubate the tubes on ice for 25 min.
3. Perform a heat shock by incubating the cells for 3 min at 37 °C.
4. Incubate the cells for 10 min at room temperature.
5. Add 1 mL of LB medium to the cells and incubate the tubes for 2 h at 37 °C while shaking.
6. Plate 100 μL of the cells on LB agar plates, containing the appropriate antibiotics.
7. Incubate the plates overnight at 37 °C.
8. For every plate to be measured, prepare first a preculture plate: pipet 200 μL of the defined medium, supplemented with the relevant antibiotics, to the wells of a transparent 96-well plate.
9. Inoculate corresponding to the plate design each well of the preculture plate with a separate colony, derived from the fresh transformation agar plates.
10. Incubate the preculture plate overnight in a thermal mixer, at a temperature of 37 °C and a rotation speed of 300 rpm.
11. The next day, pipet the correct volumes, according to the plate design, of naringenin solution in the wells of a black 96-well plate and incubate at 37 °C (*see* **Note 30**) to evaporate the ethanol, leaving the naringenin in the wells.
12. If the experiments are conducted to test a ligand, add 20 μL of ligand solution (*see* **Note 31**) to the appropriate wells after ALL the ethanol has evaporated.

13. Fill the black plate with 190 μL (or 170 μL, depending on whether ligand has been added or not) defined medium (*see* **Note 4**), supplemented with the relevant antibiotics.
14. Take a second transparent plate (referred to as dilution plate) and fill it with 140 μL of defined medium supplemented with the appropriate antibiotics.
15. Use the preculture plate for inoculation of this dilution plate, by pipetting 10 μL of the preculture plate to the corresponding wells of the dilution plate. Make sure to mix well by pipetting up and down. Immediately after mixing, pipet 10 μL of the wells of the dilution plate and transfer to the corresponding wells of the final, black 96-well plate.
16. Repeat **step 15** until the black plate has been completely inoculated, according to the plate design.
17. Seal the black plate using a breathable membrane (*see* **Note 8**). Make sure not to crease the seal, as this will change the degree of oxygen transfer through the membrane and therefore growth and measured values for optical density will be influenced.
18. Place the black 96-well plate in a microplate reader and conduct a kinetic measurement for a period of 48 h (or until all cultures have reached stationary phase) at a constant temperature of 30 °C with continuous shaking. Measure optical density at 600 nm (OD_{600}) and mKate2 fluorescence (excitation wavelength: 588 nm, emission wavelength: 633 nm, gain 110) (*see* **Note 32**) every 20 min.
19. Analyze the results with the software of your choice (e.g., R, Excel), preferences might change depending on the screening throughput. First, have a look at the general trends of the fluorescence and OD values. Determine the FL/OD values, corresponding to the relative fluorescence (RFU) for each condition at a certain time point in the exponential and/or stationary phase. When considering exponential phase, make sure that all cultures are in the same stage of the exponential phase at the chosen time point. If this cannot be achieved, a certain optical density interval, instead of a fixed time point, can be set at which values can be determined. Calculate mean average and standard deviation of the different replicates. If an outlier is observed among the different replicates of the same condition, that datapoint can be left out of the analysis. Visualize the data using scatter and/or bar plots.

3.3 Modifications to Tune Reporter Output and Allow Further Characterization

When screening heterologous TF-promoter pairs in the chosen host organism, it can be observed that not every strain is instantly characterized by the 'expected' response or even by a readable fluorescence output. This might be caused by various elements, such as the functionality of the heterologous promoter in the host organism, the occurrence of crosstalk/interference with the metabolism of the host, or the relative amounts of TF and promoter present in the cell. These situations are all illustrated below with an example in our host *E. coli* using the two-plasmid system expressing an LTTR-type TF (functioning as a repressor in absence of ligand) from *Corynebacterium glutamicum* on the pITC plasmid (pITC-TF), and two responsive promoters inserted in the pPRC plasmid (pPRC-Prom1 and pPRC-Prom2). A schematic overview of the promoter architecture of both Prom1 and Prom2, as well as of the additionally created promoter Prom1_ProD, is given in Fig. 3a.

3.3.1 Adaptation of a Heterologous Promoter to Improve Transcription in E. coli

The functionality of Prom1 is tested in a plate reader experiment by measuring the fluorescence of the strain containing the plasmid pPRC-Prom1. As depicted in Fig. 3b, no clear fluorescent signal could be observed. It is hypothesized that the heterologous promoter Prom1 is too weak to obtain a measurable fluorescent signal in *E. coli* or even non-functional. To test this hypothesis, a new promoter is built, replacing the promoter of *C. glutamicum* with the synthetic promoter ProD (*see* **Note 33**). Since the TF binding site (TFBS) is located downstream of the essential promoter elements (transcription start site, −35 and −10 box) in the native situation for Prom1, this could be easily done without hampering the TFBS. Furthermore, promoter ProD is a suitable candidate since it has already proven to function in the pPRC plasmid in *E. coli* (*see* **Note 23**). A new plate reader experiment is set up with the strain containing the adapted promoter Prom1_ProD, which results in a clear fluorescent signal and thus indicates a better functioning of the newly synthesized promoter in *E. coli* (*see* Fig. 3b). With a functional promoter in place, the repressing effect of the heterologous TF can still be validated in *E. coli*, however, caution must be taken toward interpretation of the native regulatory mechanism in the original host based on these results.

3.3.2 Changing the TF/Promoter Ratio by Transitioning to a Single-Plasmid System

Another adaptation that can be made to vary the output signal of a biosensor or TF-promoter pair is to change the ratio of TF versus corresponding promoter in the cell. As mentioned before, a two-plasmid system consisting of the pPRC and pITC plasmid is used in this work. Both plasmids are characterized by a different origin of replication and have a distinct copy number, which in this case means that the ratio of copies of the promoter sequence in the cell compared to the number of copies of the TF gene is high. To analyze the effect of copy number on fluorescence level, the ratio promoter sequence/TF gene can be decreased by employing the same origin of replication for both functionalities. This can be obtained by using one of the plasmid system's built-in features (*see* **Note 10**). A simple transition to a single-plasmid system can

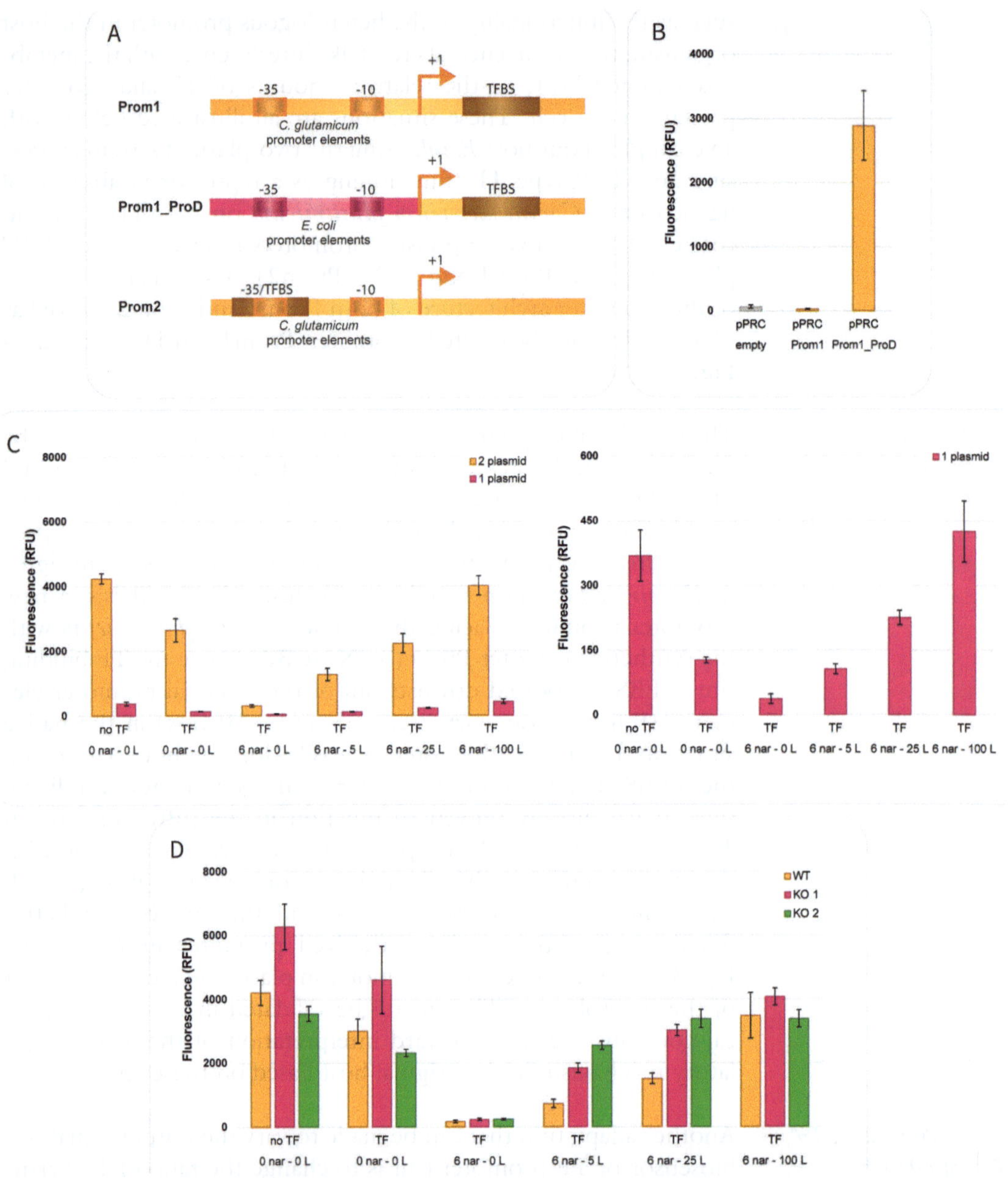

Fig. 3 Plate reader measurements conducted using a heterologous LTTR-based TF and responsive promoters from *C. glutamicum* in *E. coli* to illustrate which adaptations can be made to adapt a reporter system's functioning. For each experiment, an independent plate measurement is performed to determine fluorescence levels. Nar and L represent naringenin (in mg/L) and ligand concentration (in mM), respectively. (**a**) Overview of the promoter architecture of native *C. glutamicum* promoters Prom1 and Prom2 and of the artificially created promoter Prom1_ProD. The essential parts (−35 and −10 element, transcription start site) of the promoters are indicated, as well as the location of the TF binding site (TFBS). In Prom1_ProD, the Prom1 promoter is adapted by combining the TFBS with the core of the synthetic promoter ProD. (**b**) Fluorescence levels of the pPRC plasmid bearing heterologous promoter Prom1 or the adapted promoter Prom1_ProD. A strain containing the empty pPRC plasmid is used as a control. (**c**) Fluorescence levels of a two-plasmid biosensor, composed of pITC-TF and pPRC-Prom2, are compared to those of a single-plasmid biosensor (pITC-TF-Prom2) in different conditions (naringenin/ligand concentration). The no TF condition is

be accomplished by introducing the promoter and *mkate2* gene on the pITC plasmid. The latter is done for pPRC-Prom2 and pITC-TF. First, both plasmids are restricted using BamHI and KpnI restriction enzymes. Then, a ligation reaction is performed.

1. Prepare precultures of the pITC-TF and pPRC-Prom2 by inoculating 5 mL of liquid LB medium, supplemented with the appropriate antibiotics.
2. Grow the cultures overnight at 37 °C and 170 rpm.
3. Extract pITC-TF and pPRC-Prom2 plasmids from liquid culture using a plasmid extraction kit.
4. Linearize both plasmids pITC-TF and pPRC-Prom2 using restriction enzymes BamHI and KpnI (*see* **Note 26**). FastDigest restriction enzymes are used. When using another type of restriction enzymes, reaction conditions should be adjusted according to manufacturer's guidelines. Prepare the reaction mixtures as follows:

   1 μg of plasmid DNA

   2 μL 10× FastDigest buffer

   1 μL restriction enzyme 1

   1 μL restriction enzyme 2

   Add nuclease-free H_2O until a total volume of 20 μL
5. Incubate the reaction for 1 h at 37 °C.
6. Inactivate the restriction enzymes by incubating the reaction for 15 min at 70 °C.
7. Load the reaction mixture on a 0.8% agarose gel and perform gel electrophoresis (*see* **Note 22**).
8. Cut out the band corresponding to the restricted plasmid from gel and purify using the gel and PCR cleanup kit according to manufacturer's instructions (*see* **Note 20**).
9. Prepare the ligation reaction by adding the following ingredients to the same tube:

   100 ng of restricted pITC-TF

   × μL of restricted pPRC-Prom2 in a 1:3 (pITC:pPRC) molar ratio

   2 μL T4 DNA ligase buffer

Fig. 3 (continued) taken as a control and corresponds to the strain containing empty pITC and pPRC-Prom2, or the pITC-Prom2 plasmid in case of the 1 plasmid system. The right panel depicts the same results of the single-plasmid system, however on a smaller scale. (**d**) Fluorescence levels of two knock-out strains (KO 1 and KO 2) are compared to those of the WT strain, each containing the plasmids pITC-TF and pPRC-Prom2, in different conditions (naringenin/ligand concentration). The no TF condition is taken as a control and corresponds to the strain containing empty pITC and pPRC-Prom2

1 μL T4 DNA ligase

Add nuclease-free H_2O until a total volume of 20 μL

10. Mix the reaction and incubate overnight at room temperature.

Subsequently, a transformation by heat shock is performed, followed by a validation of the constructs by colony PCR and sequencing. The resulting strain bearing plasmid pITC-TF-Prom2 is used in a plate reader experiment, as well as the strain containing the two-plasmid system as a comparison (*see* Fig. 3c). Overall, a decrease in fluorescence levels can be observed when comparing the single- to the two-plasmid system (*see* **Note 34**). Nonetheless, the same general trends (decrease when TF is expressed, increase in the presence of ligand) in fluorescence are observed for both strains. Still, upon addition of the transcription factor in absence of inducer and cofactor, repression by the TF is more pronounced in the single-plasmid versus the two-plasmid system. The latter reflects the change of copy number for the promoters and thereby the altered ratio between TF and promoter sequence.

3.3.3 Elimination of Crosstalk with Native Metabolism by Using E. coli Knock-out Mutants

Finally, the effect of two transcriptional repressors from *E. coli*, homologous to the TF of *C. glutamicum*, is analyzed by using two *E. coli* MG1655 mutant strains where either one of the homologs has been deleted (referred to as KO 1 and KO 2) and by comparing the fluorescence levels to those of the wild type MG1655 strain. The two plasmids pITC-TF and pPRC-Prom2 are introduced in the WT and knock-out strains by performing a heat shock transformation, and plate reader measurements are conducted. A clear impact can be observed when comparing the fluorescence values of both knock-out strains to the WT (*see* Fig. 3d). In KO 1, all fluorescence levels are higher compared to the WT, which suggests that the knocked-out TF represses the expression of mKate2 both in the absence and presence of the *C. glutamicum* TF and its ligand. In KO 2 on the other hand, the effect is different: in absence of the heterologous TF, there is only a slight difference in fluorescence level between KO 2 and the WT, while when the heterologous TF is expressed and ligand is added, higher signals are observed for KO 2 compared to the WT.

4 Notes

1. For construction of the plasmids, other common cloning strains such as *E. coli* DH5α or DH10B can be used instead of *E. coli* MG1655. However, not all *E. coli* strains are suitable to conduct kinetic fluorescence measurements of reporter systems (especially in 96-well format). *E. coli* MG1655 has been

chosen because of its more robust growth in well-plate format displaying better reproducibility among replicates, compared to DH5α.

2. The pKD46 plasmid includes the genes γ, β, and *exo* from the λ Red recombinase system behind an arabinose-inducible promoter. When expressed, they inhibit the host's RecBCD system for recombination and promote a more efficient homologous recombination [20]. Furthermore, replication of the plasmid is thermosensitive, which enables the easy removal of the pKD46 plasmid after recombination has occurred.
3. The competent cells are prepared by using a standard protocol based on calcium chloride. Other methods can also be applied for the preparation of chemically competent cells, for example, by using rubidium chloride. Furthermore, electrocompetent cells can also be used. In that case, heat shock needs to be replaced by electroporation and it might be necessary to conduct an additional column purification of the assembly mixes before performing electroporation.
4. When performing in vivo measurements to characterize the functioning of a TF, it is recommended to use a minimal or defined medium (at least during the first set of tests). When using undefined rich medium (such as LB or 853), it is possible that certain components are present that can interact with the TF to induce an activating or repressing effect, which will lead to false conclusions.
5. Codon optimization is performed using the website tool of the supplier for synthetic DNA. When the start codon of the TF sequence is different from ATG (e.g., TTG, ATT, GTG), it is also adapted to ATG. Codon optimization can lead in exceptional cases to a negative effect on gene expression. When this occurs, it can be opted to order a non-codon optimized version (another option is to amplify the gene by PCR in case the genomic DNA is available) or to perform a milder mode of codon optimization by for example only adapting the rare codons.
6. During critical steps, such as the amplification of plasmid parts, TFs, and promoters, high-fidelity DNA polymerases are recommended to decrease the chances of inserting mutations in the sequences. However, when performing colony PCR, or for the amplification of very short fragments, a high-fidelity polymerase is not necessary and therefore other less expensive polymerases can be used.
7. The black 96-well plates are only used during fluorescence measurements, to minimize/eliminate the influence of neighboring wells on the fluorescence signal. During all other steps, transparent 96-well plates are utilized.

8. This type of membrane allows the transfer of oxygen through the pores, which is necessary for a good growth of the cultures.
9. The type of plate reader that is used has a great influence on certain aspects of fluorescence measurements. A monochromator-based reader has the advantage that a broad range of wavelengths can be measured. This is not the case when using a filter-based reader: here a special filter for each wavelength range is needed. When using a filter-based plate reader, the choice of reporter gene is thus dependent on the filters that you want to use and vice versa. However, filter-based plate readers also have some advantages worth considering, more specifically they are characterized with a higher sensitivity compared to the monochromator-based readers.
10. A switch can be made to a single-plasmid system by using the built-in multiple cloning sites in both plasmids. This transition will decrease the metabolic burden on the cells and results in a stable ratio between TF and promoter as they are both expressed from the same (low copy number) plasmid. The latter already resembles more to a genomic context.
11. When selecting plasmids for the purpose of building reporter systems, it is important to consider a plasmid's copy number. A high copy number is associated with a larger stress applied on the cells, which affects growth of the strain. In our case, two plasmids are used simultaneously, which already applies more burden to the cells than a single-plasmid system. Using plasmids with high copy numbers might lead to severe growth problems and instability of the plasmids. Furthermore, when the expression system is characterized with leakiness, a high copy number will lead to a higher level of background fluorescence, which can complicate the analysis of results. Also, when working with two plasmids concurrently as is the case here, it is important to be aware that copy numbers of both plasmids differ. This might lead to an excess of either available promoters or TFs, which is however not necessarily a bad property. In our case, the pPRC has a somewhat higher copy number than the pITC.
12. Besides a fluorescent reporter, also other types of reporter genes, such as antibiotic resistance genes, can be used for the construction of reporter systems. In our case, a fluorescent reporter gene is more appropriate since the final aim is to perform a screening of TF-promoter interactions. Antibiotic resistance can be used when the aim is to select strains with a specific property or to select the best performing strain by increasing selective pressure.
13. The red fluorescent reporter gene *mkate2* is used to circumvent background signals (from the cells or the medium) that can

occur during growth at wavelengths used for measuring yellow or green fluorescence. Furthermore, mKate2 has been reported to have a higher pH resistance and brightness compared to other red fluorescent proteins [22].

14. Depending on the type of experiment, different types of promoters can be selected for cloning in the pPRC plasmid. When using synthetic promoter sequences, promoters can be designed to prevent them from being functional in their reverse direction. However, when naturally occurring promoters are used, for example, a promoter that controls a gene located in a divergent operon, the promoter can be functional in both directions. To avoid that, the promoter (in its reverse direction) influences transcription of the upstream located kanamycin resistance gene, a terminator is placed upfront the promoter sequence.

15. Plasmids with an identical system of replication cannot be used in the same cell: both plasmids will compete for the same apparatus, which increases the probability that one of the plasmids will be lost during replication of the cells. Furthermore, plasmids need to have at least one different antibiotic resistance to enable the simultaneous selection for both plasmids.

16. The naringenin-inducible promoter from pNatSens [23] is chosen instead of the more commonly used LacI or AraC inducible promoters, mainly because of the following reasons: the *fdeR*-PfdeAR promoter is characterized by a low leaky expression, a broad dynamic and operational range, and since naringenin is a plant metabolite, cross-reactions with the native *E. coli* metabolism and/or the TFs of interest are very unlikely.

17. When the assembly method is not resulting in positive transformants, Gibson assembly can most of the times be used instead, since both methods depend on homologous ends of minimally 15 nucleotides. However, since our assembly method is based on the functioning of a recombinase compared to an exonuclease and polymerase in Gibson assembly, the method has the additional advantage that it can be used when heterologous sequences are flanking the overlapping ends. A disadvantage of this is however that sequences located off-target on the plasmid with unintentional similarity can be recombined with the insert, which will lead to unsuccessful cloning results. A third cloning method that could be used is restriction-ligation, but only when restriction sites are available at both ends of the fragments that are not internally present in the sequence of the plasmid, TF gene, or promoter.

18. As already mentioned in **note 2**, plasmid pKD46 is characterized by a temperature-sensitive replication system. Do not

grow at temperatures above 28 °C to prevent loss of the plasmid.

19. Grow the culture until the late exponential phase is reached. When the culture volume is increased/decreased, the target OD also needs to be adapted.
20. In general, it can be stated that whenever an obtained product is not 'pure', an agarose gel should be run and purification from gel should be performed. This is always the case after a DNA restriction, since the reaction mixture still contains non-digested DNA, which will impede the following cloning steps. After a PCR however, a simple purification on column suffices in most cases. PCR products amplified from a template containing the same antibiotic resistance gene as the final construct should nevertheless be purified from gel, again to minimize the occurrence of negative background. Furthermore, it is important to acknowledge that a large fraction of the fragment is lost when purifying from gel. Therefore, different reactions and purifications can be performed in parallel to make sure enough fragment is obtained for the generation of all constructs.
21. If a PCR fragment is included that is much larger or smaller compared to the other fragments, it is important to take this into account. For example, if a fragment is added that is 5–10 times smaller compared to the other two fragments, this smallest fragment should be added in a 5–10 times molar excess.
22. When loading restriction reactions on gel for purification, it is advised to also load a sample of nondigested plasmid on gel to enable comparison with the digested fragment. Make sure that the digested and nondigested plasmids are migrating at different heights at the time of cutting out the bands of the linearized fragments.
23. An additional validation of the constructs is recommended and can be done as follows. Validation of the pPRC construct can be performed by inserting known (synthetic) promoters, as, for example, ProB, ProC, and ProD [24] in the plasmid. The pPRC-ProB/C/D plasmids are transformed in *E. coli* cells, followed by an evaluation of the promoter's performances using fluorescence measurements. All three promoters result in a fluorescence signal, with ProB being the weakest promoter and ProD the strongest (*see* Fig. 4), as was expected from previous experiments using other plasmids. A similar validation of the pITC plasmid can be done by cloning a fluorescent reporter gene downstream of the naringenin inducible promoter and by performing measurements in the presence of different naringenin concentrations. In this manner, proper

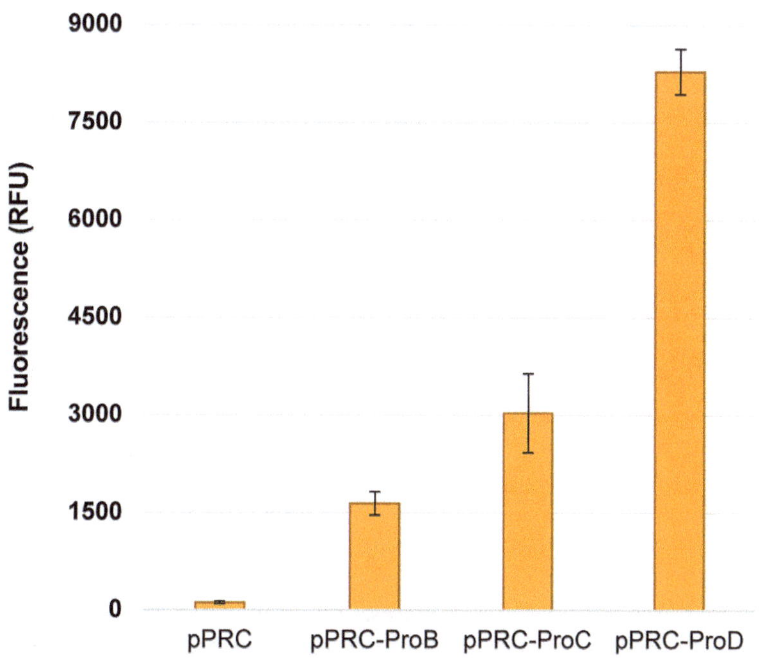

Fig. 4 Validation of the pPRC construct. Well-known, synthetic promoters ProB, ProC, and ProD are cloned in the pPRC plasmid. The resulting strains are used for the performance of fluorescence measurements

induction of the gene of interest by addition of naringenin can be verified. When a fluorescent reporter is chosen with different wavelength requirements compared to mKate2 one can also verify the compatibility and stability of the two plasmids together expressed in *E. coli.*

24. Depending on the supplier's method of DNA synthesis, it is possible that some sequences cannot be synthesized due to their complexity, for example, sequences with an extremely low/high GC content or sequences containing many repeats. Sometimes repeats are introduced through codon optimization and these could be circumvented or corrected by adaptation of some codons in the sequence.
25. When ordering synthetic genes, adaptor sequences might be present at the end of the synthesized DNA fragment. In this case, as an assembly technique based on homologous recombination is used, the adaptor sequences can remain present on the DNA fragments used for cloning as they will be removed during the assembly reaction.
26. Make sure to test both enzymes separately before using them together in a restriction reaction, to avoid having an excess of non-cut (or only restricted by one enzyme) plasmids that will negatively influence the consecutive cloning steps and the number of positive colonies after transformation. If one of

the enzymes is not working efficiently, two successive restriction reactions can be performed instead of one, using one restriction enzyme at a time (starting with the least efficient one), each time followed by purification from the gel.

27. Too many cells negatively influence a PCR reaction. Therefore, the picked colony is first diluted in water, after which a short heat-treatment is performed to lyse the cells. Using 1 μL of the supernatant as the template is enough starting material and generally leads to a better PCR efficiency.
28. Besides the four replicates on each plate, every plate is measured independently at least two times to make sure the results are reproducible.
29. Depending on the number of transformations that need to be performed, the transformation protocol can be adapted to allow a higher throughput. Smaller volumes of chemically competent *E. coli* MG1655 cells, for example, 10 μL, can be aliquoted in advance and stored at −80 °C in 96-well PCR plates. The following steps can be undertaken to perform multiple transformation reactions at the same time:
    1. Co-transform the pITC-TF plasmid and its corresponding pPRC-Prom plasmid by adding 0.5 μL of each plasmid to the same well containing 10 μL thawed *E. coli* MG1655 competent cells on ice.
    2. Mix the cells by gently pipetting up and down and incubate on ice for 25 min.
    3. Perform a heat shock by incubating the cells for 3 min at 37 °C in a thermocycler.
    4. Incubate the cells for 10 min at room temperature.
    5. Add 100 μL of LB medium to the cells, mix well and incubate the PCR plate for 1 h in a thermocycler at 37 °C.
    6. Plate 30 μL of the cells on LB agar plates, containing the appropriate antibiotics. 6-well plates can be used instead of normal Petri dishes, together with sterile glass beads to allow the homogenous distribution of cells on the surface of the well.
    7. Incubate the plates overnight at 37 °C.
30. It does not need to be exactly 37 °C, much higher temperatures (e.g., 75 °C) or lower temperatures (e.g., room temperature) can also work, as long as all the ethanol is evaporated before continuing to pipette ligand solutions or medium in the wells.

31. The same volume of 20 μL is always added, even when a 0 mM concentration is aimed for. In that case, 20 μL of nuclease-free water (or medium when the stock solutions have been prepared in medium) is used instead of ligand. Furthermore, it is important to state that the ligand concentrations that are tested are dependent on the ligand itself and the sensitivity of the reporter strain for this ligand. The highest possible ligand concentration that can be tested is determined by factors such as solubility of the ligand in water and acidity of the ligand (more specifically the effect of acidity on the growth of the used *E. coli* strain).
32. Depending on the type of microplate reader used, it might be necessary to set a certain gain value. Since we are conducting a kinetic type of experiment, the option to perform an 'autogain' is not achievable because of the low fluorescence levels at the beginning of the experiment compared to higher levels toward the end of the experiment. One possibility is to do a point measurement at the end of growth (or at a certain time point), then the autogain function can be used. For each new combination of plasmid system and fluorescent protein, it is advised to test different gain settings before conducting elaborate experiments.
33. Changing the promoter sequence has a direct effect on the efficiency of transcription. However, it is also possible that a low fluorescence signal is due to a low translation rate. In that case, the ribosome binding sites can be adapted to improve expression of the reporter gene.
34. By shifting from the two-plasmid system to a single-plasmid system, the origin of replication used to replicate the promoter is changed to have a lower copy number. Therefore, an overall decrease is observed in fluorescence values.

Acknowledgments

This work was supported by the Vrije Universiteit Brussel and funded by Research Foundation Flanders (FWO-Vlaanderen) (FWO SB PhD fellowship 1S90319N to Amber Bernauw, FWO Junior Post-doc fellowship 1204921N to Indra Bervoets).

References

1. Browning DF, Busby SJW (2004) The regulation of bacterial transcription initiation. Nat Rev Microbiol 2:57–65. https://doi.org/10.1038/nrmicro787
2. Bervoets I, Charlier D (2019) Diversity, versatility and complexity of bacterial gene regulation mechanisms: opportunities and drawbacks for applications in synthetic biology. FEMS

Microbiol Rev 43:304–339. https://doi.org/10.1093/femsre/fuz001

3. Perez-Rueda E, Hernandez-Guerrero R, Martinez-Nuñez MA et al (2018) Abundance, diversity and domain architecture variability in prokaryotic DNA-binding transcription factors. PLoS One 13:1–16. https://doi.org/10.1371/journal.pone.0195332
4. Zhao EM, Zhang Y, Mehl J et al (2018) Optogenetic regulation of engineered cellular metabolism for microbial chemical production. Nature 555:683–687. https://doi.org/10.1038/nature26141
5. Immethun CM, Ng KM, Delorenzo DM et al (2016) Oxygen-responsive genetic circuits constructed in Synechocystis sp. PCC 6803. Biotechnol Bioeng 113:433–442. https://doi.org/10.1002/bit.25722
6. Valdez-Cruz NA, Caspeta L, Pérez NO et al (2010) Production of recombinant proteins in E. coli by the heat inducible expression system based on the phage lambda pL and/or pR promoters. Microb Cell Factories 9. https://doi.org/10.1186/1475-2859-9-18
7. Soma Y, Hanai T (2015) Self-induced metabolic state switching by a tunable cell density sensor for microbial isopropanol production. Metab Eng 30:7–15. https://doi.org/10.1016/j.ymben.2015.04.005
8. Gupta A, Reizman IMB, Reisch CR, Prather KLJ (2017) Dynamic regulation of metabolic flux in engineered bacteria using a pathway-independent quorum-sensing circuit. Nat Biotechnol 35:273–279. https://doi.org/10.1038/nbt.3796
9. Hanko EKR, Paiva AC, Jonczyk M et al (2020) A genome-wide approach for identification and characterisation of metabolite-inducible systems. Nat Commun 11:1213. https://doi.org/10.1038/s41467-020-14941-6
10. Shi S, Ang EL, Zhao H (2018) In vivo biosensors: mechanisms, development, and applications. J Ind Microbiol Biotechnol 45:491–516. https://doi.org/10.1007/s10295-018-2004-x
11. Mustafi N, Grünberger A, Kohlheyer D et al (2012) The development and application of a single-cell biosensor for the detection of L-methionine and branched-chain amino acids. Metab Eng 14:449–457. https://doi.org/10.1016/j.ymben.2012.02.002
12. Goers L, Ainsworth C, Goey CH et al (2017) Whole-cell Escherichia coli lactate biosensor for monitoring mammalian cell cultures during biopharmaceutical production. Biotechnol Bioeng 114:1290–1300. https://doi.org/10.1002/bit.26254
13. Watstein DM, Styczynski MP (2018) Development of a pigment-based whole-cell zinc biosensor for human serum. ACS Synth Biol 7:267–275. https://doi.org/10.1021/acssynbio.7b00292
14. Liu D, Mannan AA, Han Y et al (2018) Dynamic metabolic control: towards precision engineering of metabolism. J Ind Microbiol Biotechnol 45:535–543. https://doi.org/10.1007/s10295-018-2013-9
15. Zhang F, Carothers JM, Keasling JD (2012) Design of a dynamic sensor-regulator system for production of chemicals and fuels derived from fatty acids. Nat Biotechnol 30:354–359. https://doi.org/10.1038/nbt.2149
16. Moser F, Espah Borujeni A, Ghodasara AN et al (2018) Dynamic control of endogenous metabolism with combinatorial logic circuits. Mol Syst Biol 14:1–18. https://doi.org/10.15252/msb.20188605
17. Doong SJ, Gupta A, Prather KLJ (2018) Layered dynamic regulation for improving metabolic pathway productivity in Escherichia coli. Proc Natl Acad Sci U S A 115:2964–2969. https://doi.org/10.1073/pnas.1716920115
18. Wu Y, Chen T, Liu Y et al (2020) Design of a programmable biosensor-CRISPRi genetic circuits for dynamic and autonomous dual-control of metabolic flux in Bacillus subtilis. Nucleic Acids Res 48:996–1009. https://doi.org/10.1093/nar/gkz1123
19. Yang Y, Lin Y, Wang J et al (2018) Sensor-regulator and RNAi based bifunctional dynamic control network for engineered microbial synthesis. Nat Commun 9. https://doi.org/10.1038/s41467-018-05466-0
20. Datsenko KA, Wanner BL (2000) One-step inactivation of chromosomal genes in Escherichia coli K-12 using PCR products. Proc Natl Acad Sci U S A 97:6640–6645. https://doi.org/10.1073/pnas.120163297
21. Zhang Y, Werling U, Edelmann W (2012) SLiCE: a novel bacterial cell extract-based DNA cloning method. Nucleic Acids Res 40:e55. https://doi.org/10.1093/nar/gkr1288
22. Shcherbo D, Murphy CS, Ermakova GV et al (2009) Far-red fluorescent tags for protein imaging in living tissues. Biochem J 418:567–574. https://doi.org/10.1042/BJ20081949.Far-red
23. de Paepe B, Maertens J, Vanholme B, de Mey M (2018) Modularization and response curve engineering of a naringenin-responsive

transcriptional biosensor. ACS Synth Biol 7: 1303–1314. https://doi.org/10.1021/acssynbio.7b00419

24. Davis JH, Rubin AJ, Sauer RT (2011) Design, construction and characterization of a set of insulated bacterial promoters. Nucleic Acids Res 39:1131–1141. https://doi.org/10.1093/nar/gkq810

25. Bolivar F, Rodriguez RL, Greene PJ et al (1977) Construction and characterization of new cloning vehicles. II. A multipurpose cloning system. Gene 2:95–113

Chapter 9

Application of Special Electron Microscopy Techniques to the Study of DNA – Protein Complexes in *E. coli* Cells

Andrey Moiseenko, Nataliya Loiko, Olga S. Sokolova, and Yurii F. Krupyanskii

Abstract

Various electron microscopy techniques were applied recently to the study of DNA condensation in dormant bacterial cells. Here, we describe, in detail, the preparation of dormant *Escherichia coli* cells for electron microscopy studies and electron tomography and energy dispersive spectroscopy (EDS) approaches, which were used to reveal the structures of DNA-protein complexes in dormant *Escherichia coli* cells.

Key words *Escherichia coli*, Dormant cells, Electron tomography, Energy Dispersive Spectroscopy, DNA condensation, DNA-Dps structures

1 Introduction

Bacterial genomic DNA interacts with nucleoid-associated proteins (NAPs) and is located in a highly condensed and functional organized form in the nucleoid of the cell. The structure of the bacterial nucleoid is still awaiting its determination in high resolution. However, five decades of intensive research showed [1, 2] that condensed DNA in the bacterial nucleoid has a hierarchical structure. Several distinct tiers of structural organization characterize the compact bacterial DNA [2, 3] in the bacterial cell. The lowest tier (small scale $\sim$1 kbp) is provided by the interaction of DNA with histone-like NAPs. The next, intermediate tier (scale $\sim$ 10 kbp) is characterized by DNA plectonemic supercoil loops. On the third, highest tier (megascale $\sim 10^6$ bp) the bacterial chromosome becomes organized in the form of six spatial macrodomains [2, 3]. Such architecture only exists as a result of dynamic structural rearrangements, which characterize actively growing bacteria [4].

Changes in environmental conditions are perceived by bacteria as stress. In response to stress, bacteria turn on universally

Eveline Peeters and Indra Bervoets (eds.), *Prokaryotic Gene Regulation: Methods and Protocols*,
Methods in Molecular Biology, vol. 2516, https://doi.org/10.1007/978-1-0716-2413-5_9,

hereditary adaptive strategies, based on structural, biochemical, and genetic rearrangements that allow to preserve part of the population to survive in adverse conditions. Importantly, these strategies are aimed to protect the DNA vital structures [5]. One of the adaptive strategies for the constantly changing conditions of the environment utilized in bacterial cells involves the condensation of DNA in complex with Dps. In the stationary phase caused by nutrient depletion, energy production processes become inefficient. Bacteria in the prolonged stationary phase use another energy-independent mechanism for maintaining an order to protect the DNA: the creation of stable structures, like those in inanimate nature (crystals, liquid crystals, etc). Most of the cells (up to 99.98%) in starving populations are exposed to autolysis. The remaining cells develop into dormant forms that differ significantly in structural organization from growing cells [6, 7]. For dormant cells, one can expect completely new structures of condensed DNA, compared to actively growing cells.

The study of DNA condensation in dormant cells is an important task, since it may shed light on the mechanisms of bacterial resistance to antibiotics. Bacterial resistance to antibiotics represents one of the most important medical problems in the world to date and often coincides with the mechanism of bacterial response to starvation. To solve these important problems of bioengineering and medicine, one needs to study the peculiarities of dormant bacterial cell morphology.

Various EM analytical techniques, including EDS, as well as tomographical studies, were applied to the study of bacterial cells. In our recent work [8], we used electron tomography and EDS techniques to visualize the structures of condensed DNA in long-term (up to 7 months) dormant *E. coli* cells. Two of them: *nanocrystalline* and *liquid crystalline* (*see* Fig. 1), are also typical for inanimate nature [4]. The novel *folded nucleosome-like* structure (*see* Fig. 2) is likely the result of the interaction and the multiple folding of the nucleoid DNA around Dps dodecamers and their associates. This structure, to our knowledge, never has been observed in starving bacterial cells before. We hypothesized that the heterogeneity of bacterial cells allows for a flexible response to environmental changes and to surviving stress situations. Multiple types of DNA condensation within the same dormant *E. coli* cell increase the chances for rapid resumption of growth when conditions turn back to favorable.

Here, we aim to describe the experimental procedures that could be used to visualize nanostructures on the lower hierarchical level of DNA compactization found in dormant *E. coli* cells. Experimental details of transmission electron microscopy, electron tomography, and analytical methods (EDS) that clearly reflect protein-DNA packing in dormant cells on a small scale ($\leq$1 kbp) are provided.

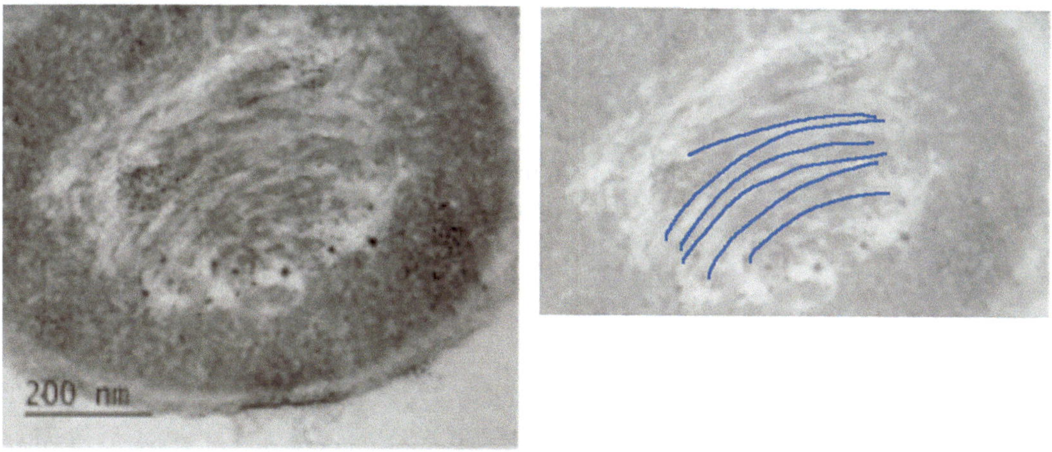

Fig. 1 Morphology of the DNA-Dps *liquid crystalline* assemblies in *E. coli* BL21-Gold (DE3)/pET-DPS growing on M9 media with induced DPS production in the linear growth phase

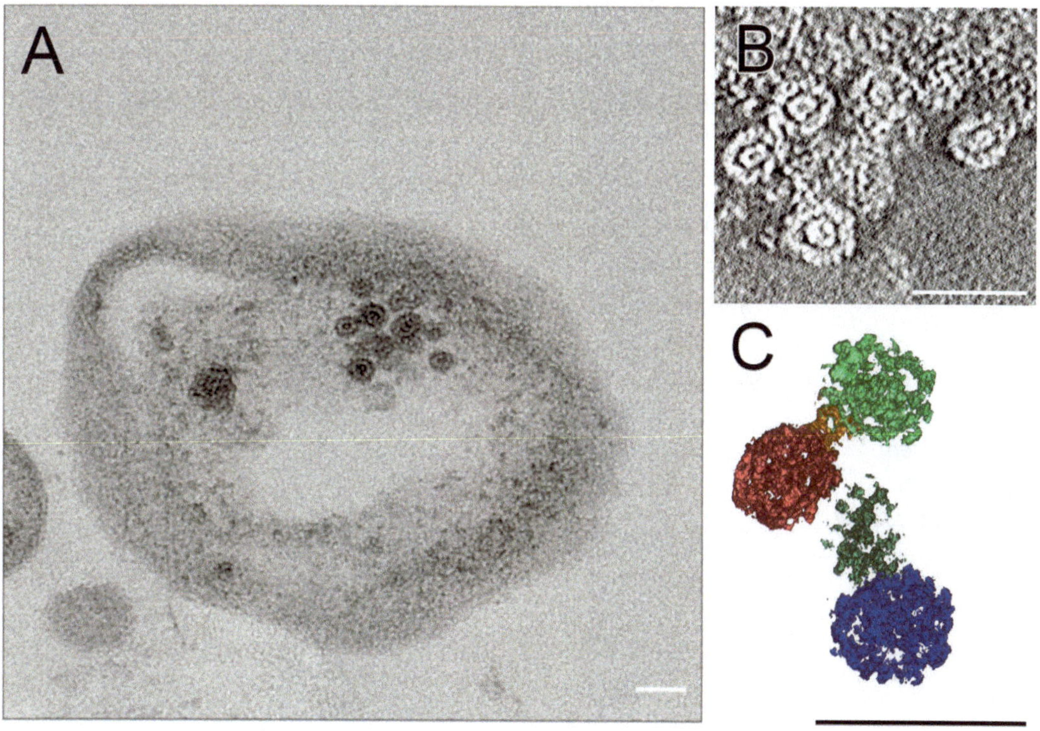

Fig. 2 Morphology of the DNA-Dps *folded nucleosome-like* structure in dormant *E. coli* cells. (**a**) TEM image of the cell, (**b**) Tomogram slice revealing the DPS spherical associates, (**c**) 3d model of the spherical associates. Bar size 55 nm

2 Materials

2.1 Preparation of Dormant E. coli Cells BL21-Gold (DE3)/pET-DPS

1. Bacterial strain *E. coli* BL21-Gold (DE3)/pET-DPS. Dps overproducer mutant [8].
2. Modified medium M9: Na_2HPO_4 – 6 g, KH_2PO_4 – 3 g, NaCl – 0.5 g, NH_4Cl – 0.2 g, $MnSO_4$ – 0.0004 g, $MgSO_4$ – 0.0025 g, $CaCl_2$ – 0.0002 g, glucose – 10 g. Place it in a 1 l glass beaker. Add 1 l of distilled water and stir until dissolved. Bring the pH to 7.0 by adding the desired amount of 1 M NaOH (1.0–1.5 ml). Pour 50 ml of modified M9 medium into a 250 ml glass flask. Close with cotton plugs and autoclave at 107°C for 40 min.
3. Ampicillin solution: dry ampicillin powder – 1 g, sterile distilled water – 10 ml, stir until smooth. Prepare a new solution each time, immediately before use.
4. Lactose solution (1 M): lactose – 34.2 g, distilled water – 100 ml, stir in 250 ml flask, close with cotton plugs, and autoclave at 105°C for 30 min.

2.2 Equipment Used for Preparation of Bacterial Culture

1. Incubating light-duty orbital shakers model ISLD04HDG (OHAUS, USA)
2. Electronic Balance, PX84 AM (OHAUS, USA)
3. Autoclave 2540 ML
4. pH-meter
5. Safe 2020 Class II Biological Safety Cabinets (Thermo Fisher, USA)
6. Centrifuge Eppendorf MiniSpin plus (Eppendorf, Germany)

2.3 Specimen Preparation for Electron Microscopy

1. 1.5% glutaraldehyde
2. 0.1 M cacodylate buffer: pH 7.4
3. 0.5% paraformaldehyde
4. 1% osmium tetroxide solution (OsO_4) in 0.1 M cacodylate buffer: pH 7.4
5. Ethanol solutions: 50%, 60%, 70%, 80%, and 96%.
6. 2% uranyl acetate
7. Epoxy resin kit: Epon 812, DDSA (dodecyl amber anhydride), MNA (methylandyric anhydride), acetone, and DMP-30 (tridimethylaminophenol).

2.4 Equipment for Specimen Preparation for Electron Microscopy

1. Tweezers with fine tips (e.g., Dumont #5)
2. Diamond knife (Diatome Knife Ultra 45°)
3. TEM grids 300 mesh copper with formvar support film
4. Grid storage box
5. Liquid nitrogen
6. Carbon rods for carbon coating

2.5 Equipment Used for Electron Microscopy and Electron Tomography

1. Transmission Electron Microscope JEM-2100 with LaB6 electron source (Jeol, Japan)
2. High-Tilt specimen holder EM-21311HTR (Jeol, Japan)
3. Cooling Holder EM-21090 (Jeol, Japan)
4. Gatan Orius SC200D CCD camera (Gatan, USA)
5. Gatan Ultrascan 1000XP CCD camera (Gatan, USA)
6. Gatan GIF Quantum ER imaging filter (Gatan, USA)
7. Gatan HAADF STEM detector model 806 (Gatan, USA)
8. Oxford Instruments X-Max 80 EDS detector
9. Hitachi HUS-5GB vacuum evaporator
10. Ultramicrotome Leica UC7

2.6 Software Used for Electron Microscopy and Electron Tomography

1. Gatan Digital Micrograph ver. 2.32 (Gatan, USA)
2. SerialEM ver. 3.8 [9]
3. UCSF Chimera ver. 1.14 [10]
4. IMOD ver. 4.11 X [11]

3 Methods

3.1 Bacterial Culture Preparation

3.1.1 Preparation of a Bacterial Inoculum

1. All work with bacterial culture should be carried out under sterile conditions, using sterile equipment.
2. Add 75 μl ampicillin solution to the prepared sterilized flask with 50 ml of M9 medium.
3. Take an aliquot of the bacterial biomass with a sterile microbiological loop and transfer it to flask.
4. Place the flasks in a shaker and incubate at 28°C for 24 h.

3.1.2 Bacteria Cultivation

1. Add 75 μl ampicillin solution to the prepared sterilized flask with 50 ml sterile M9 medium.
2. Add 0.5 ml of bacterial inoculum to the prepared flask.
3. Place the flasks in a shaker and incubate at 28 °C for 3.5 h.
4. After 3.5 h of incubation, add 0.5 ml of lactose solution to each flask (*see* **Note 1**).
5. Continue to incubate for 44 h.

3.1.3 Storage of Bacterial Cultures

1. Remove the flasks with the two-day bacterial culture from the shaker.
2. Close with paraphilic cotton plugs tightly to avoid drying out.
3. Store at room temperature (23 °C) in a dark place for up to 7 months.

3.2 Specimen Preparation for Electron Microscopic Studies

1. Take cell aliquots (1 ml) from flasks into 2 ml plastic tubes (Eppendorf).
2. Place plastic tubes in a microcentrifuge and centrifuge for 8 min at 10,000 g.
3. Drain the supernatant and repeat the process three times to ensure that a sufficient biomass remains at the bottom of the tube.
4. Discard the supernatant and add 1 ml of 1.5% glutaraldehyde in 0.1 M cacodylate buffer, pH 7.4, and resuspend the pellet using Vortex.
5. Fix cells in 0.5 ml of 1.5% glutaraldehyde for 5 h at room temperature, following with post-fixation in 0.5 ml of 0.5% paraformaldehyde for 2 h.
6. Wash cells in 0.5 ml of 0.1 M cacodylate buffer pH 7.4.

3.2.1 Washing Procedure

1. Place plastic tubes in a microcentrifuge and centrifuge for 2 min at 10,000 g.
2. Discard the supernatant and add 1 ml of washing solution.
3. Repeat three times.

3.2.2 Preparation for Transmission Electron Microscopy and Tomography Experiments

1. Add 2 ml of 1% osmium tetroxide solution (OsO_4) in 0.1 M cacodylate buffer, pH 7.4 to the content of the tube, and keep at RT for 1 h.
2. Wash the samples with cold (4 °C) 50% ethyl alcohol solution 4 times for 5 min each until the solution stops darkening; discard all supernatants that contain OsO_4 and dispose them according to your Lab regulations for chemical hazardous waste.
3. Treat samples with 1 ml of 60% ethanol solution 2 times for 20 min each (4 °C).
4. Add 1 ml of 2% uranyl acetate in 70% ethanol solution for 12 h (may be left overnight in refrigerator).
5. Treat samples with cold (4 °C) 80% and 96% ethyl alcohol solutions for 15 min each at RT.
6. Dehydrate the samples in acetone 3 times for 45 min at room temperature.

7. Impregnate the samples with epoxy resin using a mixture of Epon 812, DDSA (dodecyl amber anhydride), and MNA (methylandyric anhydride) in a ratio of 13:8:7.
8. For impregnation, place the samples at room temperature for 1 h in a 1:3 mixture of resin and acetone, followed by 1:1 mixture of resin and acetone for 1 h, and finally, 3:1 mixture of resin and acetone for 1 h.
9. Add DMP-30 (tridimethylaminophenol) at required concentration to the resin before polymerization.
10. Place the material in the resin overnight.
11. Polymerization of the material should be performed for 24 h at 37 °C and then for 48 h at 60 °C.
12. Slices up to 100 nm thick should be obtained on an ultramicrotome (Reichert-Jung, Austria or similar) using a diamond knife (Diatom).
13. Slices should be transferred to formvar-coated copper mesh (Ted Pella, USA or similar) and, if desired, additionally contrasted by a lead citrate solution, according to the published protocol.

3.2.3 Preparation for EDS

Skip osmium tetroxide solution and uranyl acetate steps described above (**steps 1–4**), instead treat samples with 1 ml of 60% and 70% ethanol solution 2 times for 20 min each (4 °C). Proceed from **step 5** from above. Do not use lead citrate for contrasting.

3.2.4 Thin-Section Carbon Coating

Coat the specimen with a thin carbon layer to reduce the beam-induced motion and mass loss rate during data acquisition (*see* **Note 2**).

1. Put the sample grids on a clean glass slide and mount it inside the vacuum evaporator chamber.
2. Prepare a clean carbon rod by cutting it in a rod grinder. Set up the carbon rod in the rod holder and gently adjust the rod tension with the holder spring. Place the carbon rod holder as far as possible from the sample surface to reduce the possible heating damage. Close the shutter between the carbon rod and the sample.
3. Put the small piece of porcelain or Teflon, partially shadowed with a metal plate, in the evaporator chamber at the level of the sample. This piece will be used as a rough thickness controller.
4. Evacuate the vacuum chamber up to 10^{-5} Torr.
5. Slowly increase the current through the rod until it glows yellow. Leave it for 1–2 min to bake out the residual contamination from the rod.
6. Increase the current until the rod shines brightly white. Open the shutter for 5–10 s, then turn off the rod current.

7. Examine the thickness controller: the evaporated carbon layer should be clearly visible as a grey surface in contrast to the white shadowed area of the porcelain/Teflon. If the thickness controller remains white, repeat the coating.

3.3 Transmission Electron Microscopy and Electron Tomography

Electron tomography is routinely used to obtain the local 3D structure of unique objects. In our recent study [8], we use electron tomography to demonstrate that the newly identified round structures are not toroids, as described previously [6], but represent nearly spherical formations. This was unclear from the study of the 2D sections. For that, we used the single-axis tomographical approach described below.

3.3.1 Single-Axis Tomography Data Acquisition

1. Fill the anti-contaminator device (cold trap) with liquid nitrogen at least half an hour before imaging.
2. Turn on the accelerating voltage. It is preferred to use at least 200 kV for tomography acquisition (*see* **Note 3**).
3. Load the sample into the microscope high-tilt sample holder. Before transferring the holder to the microscope airlock, ensure that the holder O-rings are dust-free and clean.
4. Turn on the LaB_6 electron source and check that beam current is in the range of 4–6 μA.
5. Align the microscope optics in TEM mode using Alpha 2, Spot size 2, 70 μm condenser aperture, and 60 μm objective aperture. During the alignment, check the beam centering, beam tilt, condenser astigmatism, beam-tilt condenser pivot point, condenser and objective aperture centering, and objective astigmatism. Aligning should be performed at 40 k magnification and a standard objective lens focus value. Set up parallel beam illumination conditions by focusing the beam at the objective aperture plane with a CL3 lens.
6. Locate the thin sections at low magnifications and move them to the stage field of view.
7. Switch the microscope to EFTEM mode (Imaging Filter mode) and check the TEM alignments.
8. Move the stage to the region of interest and choose the appropriate magnification for tomography (*see* **Note 4**).
9. Pre-illuminate the region of interest with a high electron dose of >5000 e/A^2.
10. Tune the energy filter. Start from the manual pre-slit GIF alignment in EELS mode (FX, FY, SX, SY) to obtain the sharpest possible ZLP. Align ZLP to zero energy shift. Then, perform auto-tuning in EFTEM mode, which includes slit centering, achromaticity tuning, magnification and distortion tuning, and isochromaticity tuning. Insert a 20 eV slit and check that the image intensity remains uniform.

11. Adjust the exposure in Record, Focus, and Trial mode in SerialEM to check that the CCD is not saturating and has a moderate number of counts. Set Focus area to the center quarter of the image or to the wide elongated area along the holder tilt axis to minimize focusing errors at large field of view.
12. Perform the Check Autofocus task in SerialEM to verify that the autofocusing error is small enough. The measured defocus should be at least 0.90 of the set one.
13. Adjust the eucentric height with an automatic task. This procedure is carried out at low magnifications. If the energy selecting slit is misaligned at low magnifications, perform the mag-dependent energy shift calibration in SerialEM.
14. Tilt the holder to 60° with the Walk Up task in SerialEM and then reset the image shift.
15. Check that the region of interest did not shift and center it in the stage movement field of view, if necessary.
16. Perform initial autofocusing with a −0.8 μm defocus and check that the region of interest is well-focused.
17. Set up automated data acquisition with a 2° tilt step, one-directional tilt range from 60 to −60°. Set exposure to be variable with 1/cos of the current tilt angle.

3.3.2 Tomography Data Processing

1. Import the tilt series into the IMOD software.
2. Perform hot pixel replacement with a pre-processing task.
3. Coarsely align the stack with cross-correlation. Note, that sometimes it is useful to perform cross-correlation only on the central quarter of the image.
4. For the tilt series obtained without the gold fiducials, the fine alignment is performed with patch tracking. Generate patches with the size of 150 with 0.33 fractional overlap and use the boundary model to perform patch tracking only on areas with contrast features. When calculating rotations and changes in magnification, one needs to group the images of the tilted series by angles and not take distortion into account. Check the patch tracking model after solving for rotation and exclude any track paths that look fuzzy. Go from lowest to highest tilt and delete any patch contours that are misplaced during tilting. Iterate over patch contours and delete all patches with residue more than 1.0. Repeat the fine alignment and contour deleting until the mean residual will be less than 1.0.
5. Perform tomogram positioning; manually draw contours over thin-section surfaces on three sample tomogram slices. Inspect the tomogram slices. If the densities look elongated or "moon-shaped", it may be necessary to redo the fine alignment. Compute the final Z shift and angles.

6. Generate tomogram with the Back Projection algorithm, with GPU processing enabled. Check the final tomogram positioning: if the plastic section is not parallel with the tomogram XY plane, repeat the tomogram positioning using the whole tomogram as input.
7. Trim the tomogram based on the region of interest in XY and on the plastic section boundaries in Z. The final volume can also be squeezed with reduction factor of 1.25 without quality loss, in case of data from the CCD detector.
8. Visualize the tomogram in 3dmod. Draw the contours around the objects and either visualize the contours as 3d models or use the contours as volume-limiting masks to show only densities inside them. Adjust threshold and filtering to optimize features display.

A representative tomographic reconstruction is shown in Fig. 2c.

3.3.3 Fourier-Filtering the Crystal Lattice TEM Images

In cells with Dps overproduction, most of the ordered structures exist in nanocrystalline form. To enhance the contrast in the TEM images of crystalline arrays in cells, we oriented the plastic section under the electron beam to obtain clear FFT peaks from the crystal and performed Fourier-filtering with the following protocol:

1. Proceed with the protocol listed in Subheading 3.1.1 to align the microscope. Use double-tilt holder for precise positioning of the sample under the electron beam.
2. Locate the region of interest and turn on the live FFT from the crystal area.
3. Tilt the double-tilt holder in the range of $\pm 10°$ by both axes, optimizing the FFT peaks.
4. Take an image and create a lattice mask on its FFT transform by manually aligning the mask spots with the FFT peaks. Create the mask with a 5 pix. Soft edge and set the intensity outside of it to zero. Perform inverse FFT on the masked image.

The representative crystal lattice of DNA-Dps co-crystal from bacterial cytoplasm is presented in Fig. 3b.

3.4 Energy Dispersive Spectroscopy (EDS)

Energy Dispersive Spectroscopy (EDS) was recently used to detect and map desired elements on thin slices. For example, the Ka peak (2.307 keV) of S reflects the existence of a DNA-binding protein, Dps (each dodecamer of Dps contains 48 Methionines), while the Ka peak (2.013 keV) of P corresponds to the phosphorus in the nucleoid DNA. The simultaneous presence of both peaks on the spectra of the same structure reveals the formation of the DNA-Dps complex.

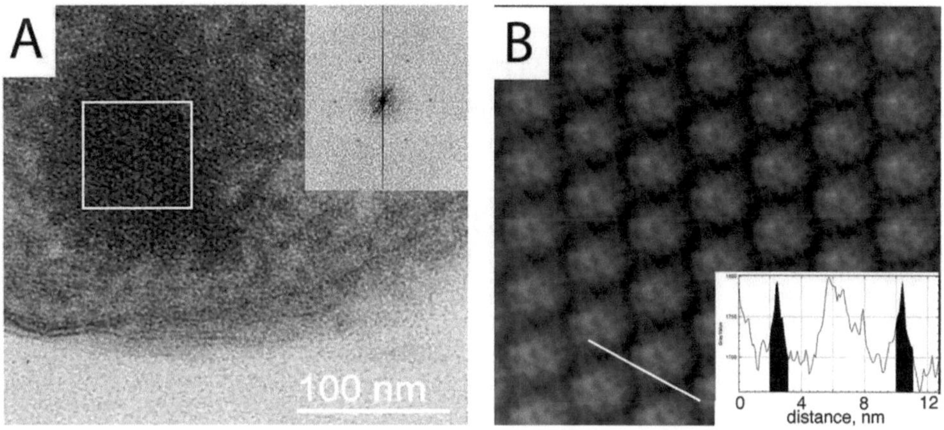

Fig. 3 DNA-Dps *nanocrystal* in dormant *E. coli* cell. (**a**) Area of the crystal exhibits peaks on the FFT, (**b**) Fourier-filtered image of the crystal area. Maximums on the cross-section (white line) reveal the positions of DNA

1. Fill the microscope anti-contaminator device (cold trap) with liquid nitrogen at least half an hour before imaging. Turn on the accelerating voltage.
2. Load the sample into the microscope beryllium sample holder. Before transferring the holder to the microscope airlock, ensure that the holder O-rings are dust-free and clean. In case the sample exhibits strong contamination in STEM mode, use the cooling holder instead of the beryllium one (*see* **Note 5**).
3. Switch on the cooling of the EDS detector.
4. Turn on the LaB_6 electron source and check that the beam current is in the range of 4–6 μA.
5. Roughly align microscope in TEM mode to locate the region of interest. The EDS spectrum data is obtained in STEM mode, which requires independent alignment.
6. Take an image of the region of interest in TEM before any spectrum acquisition because specimen damage, due to mass loss and contamination, may deteriorate the contrast.
7. Switch the microscope to STEM mode and align it with Spot Size 1 and 100 μm condenser aperture. Do not use the objective aperture or high-contrast aperture. Obtain the image with a BF detector and locate the region of interest.
8. Insert the EDS detector and then fine-focus the image on the BF detector.
9. Enable drift correction with the shortest time interval possible (e.g., every line of pixels or every 20 s). The region for the drift correction should contain some contrast features.

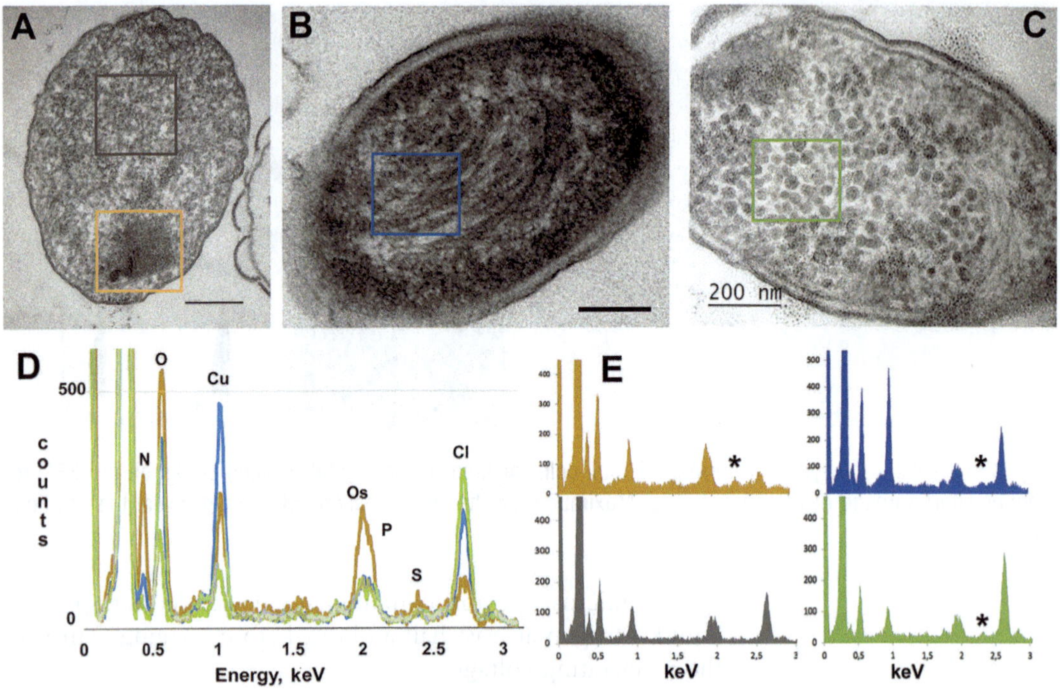

Fig. 4 Three types of condensed DNA-Dps structures found in dormant *E. coli* cells, starving for 7 months. (**a**) *nanocrystalline*; (**b**) *liquid crystalline*; (**c**) *folded nucleosome-like* type. Bar – 200 nm. Colored frames mark the specific areas where the EDX spectra were obtained; (**d**) Superimposed EDX spectra from the selected areas, marked in (**a–d**), normalized to the C peak. Line colors reflect the corresponded frame colors in (**a–d**). (**e**) EDX spectra from the selected areas marked in (**a–d**) colored by the specific area. Asterisks mark positions of the S peak

10. Mark the area for spectra acquisition and start the acquisition. Monitor the spectrum data (*see* **Note 6**).
11. To compare the relative atomic compositions of different samples, use Excel plot to overlap the spectra from regions of interest and the control region, aligning them by the length of the carbon peak.

The representative spectra are demonstrated in Fig. 4d, e.

4 Notes

1. Lactose solution should be added at the beginning of the linear growth stage of the bacterial culture. Since the growth of a culture depends on many factors, the time for adding lactose can vary from 2.5 to 6 h from the beginning of the experiment. Thus, a preliminary experiment should be performed: record the growth curve of *E. coli*, record the change in optical density (OD) of the culture suspension at 450 nm. The beginning of the linear stage will be in accordance with OD 2.5–3.0.

2. The best results are achieved if the plastic section is "sandwiched" between two carbon layers. Thus, it is necessary to carbon coat the sample from both sides when using pure formvar support film. When using carbon-coated formvar grids, the ultrathin section should be put on the carbon side of the support film and then the sample is carbon-coated from the section side.
3. The maximum sample thickness that can be imaged with electron tomography depends on the accelerating voltage of the microscope. At 200 kV, the electron mean free path in typical biological specimen is about 260 nm, which corresponds to the electron path through the 130 nm section tilted to 60°. Using EFTEM to filter out the inelastically scattered electrons allows to use at least 1.5 thicker sections.
4. The magnification for the tomography data acquisition should be chosen so that the region of interest occupies around 3/4 of the field of the view area; some image shifting is allowed without loss of information. The CCD has limited detective efficiency at high spatial frequency, so the pixel size of the magnified image should be at least three to four times smaller than the desired resolution.
5. In STEM mode, the beam-sensitive samples (i.e., plastic sections) may exhibit beam-induced contamination, which may significantly affect the spectrum data [12]. Using the cooling sample holder at −180 °C allows to reduce the contamination effects and to increase the sample stability. The cooling holder liquid nitrogen Dewar has to be evacuated overnight with the oil-free dry pumping station down to 5e–6 Torr pressure.
6. In some cases, the EDS spectrum contains a very low signal from the sample, but a high signal from the grid (e.g., copper). This may happen if the spectrum is collected too close to the grid square border, or the support film is bent too much. In such cases, it is often better to move to another region of interest.

Acknowledgments

Bacterial strain with Dps overexpression was kindly provided by Prof. Vassili N. Lasarev, Head of the Laboratory of Gene Engineering, Federal Research and Clinical Center of Physical-Chemical Medicine, Federal Medical Biological Agency, Moscow, Russia. We thank Yana Danilova for her help with EM analysis and Lisa Trifonova for proofreading the manuscript.

The electron microscopy of DNA-protein complexes was performed on the Unique equipment setup "3D-EMC" of Moscow State University (under support from Russian Science Foundation, grant #19-74-30003). O.S.S. acknowledges the partial support from the Interdisciplinary Scientific and Educational School of Moscow Lomonosov University «Molecular Technologies of the Living Systems and Synthetic Biology».

Y.F.K. and A.M. are grateful for the support from The Ministry of Science and Higher Education of the Russian Federation (state assignments 122040400089-6), N.L. for the support from The Ministry of Science and Higher Education of the Russian Federation (state assignments 122040800164-6).

References

1. Stonington OG, Pettijohn DE (1971) The folded genome of Escherichia coli isolated in a protein-DNA-RNA complex. Proc Natl Acad Sci 68:6–9
2. Verma SC, Qian Z, Adhya SL (2019) Architecture of the Escherichia coli nucleoid. PLoS Genet 15:e1008456
3. Valens M, Penaud S, Rossignol M et al (2004) Macrodomain organization of the Escherichia coli chromosome. EMBO J 23:4330–4341
4. Minsky A, Shimoni E, Frenkiel-Krispin D (2002) Stress, order and survival. Nat Rev Mol Cell Biol 3:50–60
5. Tkachenko AG (2018) Stress responses of bacterial cells as mechanism of development of antibiotic tolerance (review). Appl Biochem Microbiol 54:108–127
6. Frenkiel-Krispin D, Ben-Avraham I, Englander J et al (2004) Nucleoid restructuring in stationary-state bacteria: nucleoid restructuring in stationary-state bacteria. Mol Microbiol 51:395–405
7. Krupyanskii YF, Loiko NG, Sinitsyn DO et al (2018) Biocrystallization in bacterial and fungal cells and spores. Crystallogr Rep 63: 594–599
8. Loiko N, Danilova Y, Moiseenko A et al (2020) Morphological peculiarities of the DNA-protein complexes in starved Escherichia coli cells. PLoS One 15:e0231562
9. Mastronarde DN (2005) Automated electron microscope tomography using robust prediction of specimen movements. J Struct Biol 152: 36–51
10. Pettersen EF, Goddard TD, Huang CC et al (2004) UCSF chimera – a visualization system for exploratory research and analysis. J Comput Chem 25:1605–1612
11. Kremer JR, Mastronarde DN, McIntosh JR (1996) Computer visualization of three-dimensional image data using IMOD. J Struct Biol 116:71–76
12. Egerton RF (2013) Control of radiation damage in the TEM. Ultramicroscopy 127: 100–108

Chapter 10

High-Speed Atomic Force Microscopy Visualization of Protein-DNA Interactions Using DNA Origami Frames

Ronnie G. Willaert and Sandor Kasas

Abstract

Direct, live imaging of protein-DNA interactions under physiological conditions is invaluable for understanding the mechanism and kinetics of binding and understanding the topological changes of the DNA strand. The DNA origami technology allows for precise placement of target molecules in a designed nanostructure. Here, we describe a protocol for the self-assembly of DNA origami frames with 2 stretched DNA sequences containing the binding site of a transcription factor, i.e., the Protein FadR, which is a TetR-family tanscription factor regulator for fatty acid metabolism in the archaeal organism *Sulfolobus acidocaldarius.* These frames can be used to study the dynamics of transcription factor binding using high-speed AFM and obtain mechanistic insights into the mechanism of action of transcription factors.

Key words DNA nanotechnology, DNA origami frames, Self-assembly, Protein-DNA interaction, Atomic force microscopy (AFM), High-speed AFM

1 Introduction

Since its invention in the late 1980s by Binning Quate and Gerber [1], atomic force microscopy (AFM) has led to unique insights through the visualization and analysis of nanoscale features, events, and processes [2]. AFM techniques have turned out to be a suitable and versatile tool to directly observe biomolecules at single-molecule spatial resolution under physiological conditions. Over the last 30 years, AFM has made many significant contributions to the study of protein-DNA interactions since it can reveal protein binding specificity and affinity, protein-induced DNA binding, protein-binding stoichiometry, and protein-DNA interaction forces [3, 4]. Due to the development of high-speed (HS) AFM [5, 6] and the commercial availability of these AFMs, a unique tool can now be used to investigate dynamic molecular processes. HS AFM can directly visualize single protein molecules as well as their complexes during their functional activity at nanometer sub-second

Eveline Peeters and Indra Bervoets (eds.), *Prokaryotic Gene Regulation: Methods and Protocols*,
Methods in Molecular Biology, vol. 2516, https://doi.org/10.1007/978-1-0716-2413-5_10,

(video rate) resolution [6–13]. HS AFM molecular dynamics movies can provide mechanistic insights that are hard to be attained with other approaches.

Scaffolded DNA origami allows the arrangement of thousands of nucleotides with sub-nanometer precision at specified locations in space to yield a diversity of objects with nanometer-scale dimensions [14–17]. The concept of DNA origami relies on folding a long ssDNA called the scaffold (such as for example the M13mp18 viral genome of the M13 bacteriophage), with hundreds of designed short ssDNAs called staples. Because AFM visualizes all molecules in a scanning area, an observation scaffold is very helpful to facilitate the target-specific imaging of molecules in the dynamic state [18–20]. DNA origami frames have been specifically designed and constructed to study protein-DNA interactions [21–26]. Here, we describe a protocol for the design and construction of a DNA origami frame to study the interaction of the TetR-family transcription factor FadR from the thermoacidophilic crenarchaeon *Sulfolobus acidocaldarius* [27]. FadR is a transcriptional regulator of genes involved in fatty acid metabolism. Furthermore, we describe a protocol to perform HS AFM visualization at high spational and temporal resolution in physiological (liquid) conditions.

2 Materials

Prepare all solutions using ultrapure water (prepared by purifying deionized water, to attain a sensitivity of 18 MΩ-cm at 25 °C) and molecular biology grade reagents. All solutions were filtered through a 0.20 μm non-pyrogenic filter. Store DNA origami solution at 4 °C and all other stock and working buffers at room temperature. Discard the waste solutions and materials appropriately.

2.1 DNA Origami Design and Synthesis

1. The DNA origami frame was designed using caDNAno [28] (*see* **Note 1**); cadnano v0.2.3 (honeycomb lattice) and cadnanoSQ v0.2.4 (square lattice) were installed (cadnano.org) according to the instructions (http://douglaslab.org/tutorials/). As the scaffold, we selected the M13mp18 single-stranded viral genome of the M13 bacteriophage (100 nM in 10 mM TRIS-BASE and 1 mM EDTA, type p7249, Tilibit Nanosystems, Germany) since its length of 7249 nucleotides fits the frame dimensions of the frame that was constructed.
2. The sequences of the staples were generated by caDNAno and ordered.
3. Ds 74-mer DNA sequences that contain the FadR binding site: top and bottom strands (1 μM in water).

4. 10x origami folding buffer: 50 mM tris(hydroxymethyl)-aminomethane (TRIS), 50 mM NaCl, 10 mM ethylenediaminetetraacetic acid (EDTA), and pH 8.
5. Stock magnesium buffer: 200 mM $MgCl_2$.

2.2 Protein Solution

1. We selected FadR, which is a TetR-family transcriptional regulator that is involved in the regulation of fatty acid metabolism in the crenarchaeon *Sulfolobus acidocaldarius* [27], as a model protein to analyze its binding with its DNA-binding site.
2. Protein buffer: 150 mM NaCl and 50 mM TRIS (pH 7.2).

2.3 DNA Origami AFM Sample Preparation and AFM Imaging

1. Fluid cell cleaning solutions: 2% (m/v) sodium dodecyl sulfate, ethanol (analytical grade).
2. AFM USC-F0.3-K0.3 (NanoWorld) or Fastscan-D (Bruker) probes are used for HS imaging in liquid.
3. HS AFM is performed using a NanoWizard 4 Ultraspeed 2 AFM (JPK Bruker).
4. The AFM fluid cell consisted of a mica disk (diameter of 10 mm, NanoAndMore, Germany) which was glued with epoxy glue to a microscopy glass. A plastic ring (radius of 25 mm, height of 6.5 mm) was glued to the microscope glass using silicon glue (Type 1, TFC TrollFactory, Germany) with the mica disc in the center.
5. DNA origami adsorption and imaging buffer (TE Mg 12.5 mM buffer): 20 mM TRIS (pH 7.6), 1 mM EDTA, and 12.5 mM $MgCl_2.6H_2O$ (*see* **Note 2**).
6. AFM image analysis processing software: JPKSPM Data Processing software (Bruker).

3 Methods

3.1 DNA Origami Design

1. Select the helices (from 0 to 31) in the "slice" panel of the caDNAno interface, which gives each selected helix with a three-base long scaffold path in the "path" panel (*see* Fig. 1) [28]. Use the path-panel editing tool to extend the scaffold in the z direction and connect neighboring helices with Holliday junction crossovers until the target shape is filled with the scaffold strand (*see* **Note 3**).
2. Use the "auto-staple" tool to create staple paths. Staple paths that do not fall in the preferred length range (18–49 bases) are manually broken into shorter segments using the "break" tool.
3. When all staples are designed in the preferred layout, the "add sequence" tool is used to populate the scaffold path with a DNA sequence such as the M13mp18 scaffold (*see* **Note 4**). Generate the list with staple sequences, which can be next ordered.

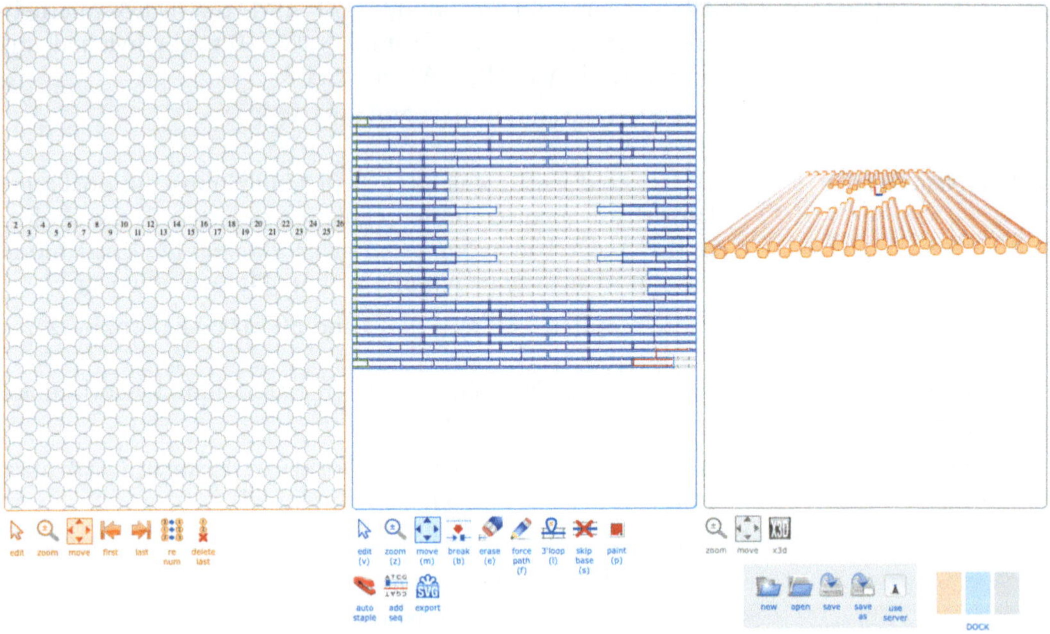

Fig. 1 Screenshot of designing the DNA origami frame in caDNAno. The left panel is the "slice" panel showing a cross-sectional view of the honeycomb lattice of the design. The middle "path" panel provides an interface to edit the unrolled 2D schematic of the scaffold and staple strands. The right "render" panel shows the 3D model of the design

3.2 Construction of the 74-mer Duplex DNA Fragments

1. The 74-mer ds DNA sequences that contain the binding site of the FadR protein [27] were designed (e.g., DNASTAR Laser-Gene SeqBuilder Pro) with unique sticky ends such that they can only hybridize to one of the four inner DNA origami frame extensions A, B, C or D (*see* Fig. 2).
2. The sequences are produced by a polymerase chain reaction (PCR) amplification (KAPA Taq PCR kit) from a DNA fragment from *S. acidalcarius* containing this binding site (see [27]) (*see* **Note 5**).

3.3 DNA Origami Synthesis

1. Mix 10 μl of M13mp18 scaffold solution (100 nM), 20 μl of staple solution (475 nM), 5 μl 10x origami folding buffer, 5 μl Mg buffer (200 mM), 5 μl top strand (1 μM), 5 μl bottom strand (1 μM), and 1.5 μl ultrapure water.
2. Incubate the mixture in a thermocycler using the following temperature sequence: 80 °C during 5 s, 65 °C during 15 min, cool down from 60 °C to 40 °C, at 1 °C/h. The total time for this sequence takes 20 h (*see* **Note 6**).

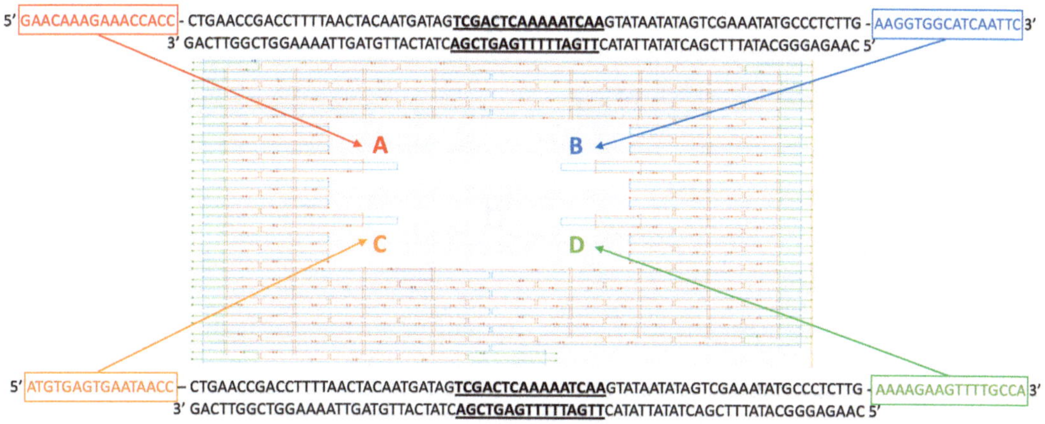

Fig. 2 The 74-mer FadR binding DNA sequences that fit into the DNA origami frame. The FadR binding site is indicated in bold and underlined

3.4 Sample Preparation for AFM Imaging

1. Wash a microscope glass sequentially with a detergent solution (e.g., 2% (m/v) sodium dodecyl sulfate), 100% ethanol, and ultrapure water. Dry the surface with compressed air (*see* Fig. 3).
2. Glue the mica disc in the middle of a microscope glass (*see* Fig. 3).
3. Cleave the mica surface using single-sided adhesive tape. Check that a complete mica layer is cleaved from the surface before continuing.
4. For a total volume of 20 μl of working solution, mix 0.5 μl DNA origami solution with 19.5 μl TE Mg 12.5 mM buffer (*see* **Note 7**).
5. Drop 20 μl of working solution on the mica surface and incubate for 5–15 min (*see* **Note 6**).
6. Gently wash the mica surface by dropping 70 μl of TE Mg 12.5 mM buffer on the surface and withdraw the drop immediately by touching the drop at the edge with a paper towel. Repeat at least ten times.
7. Glue the plastic fluid cell ring to the microscope slide with the mica disc in the center and leave for 10–15 min to dry and cure the silicone glue.
8. Fill the liquid cell with 1.5 ml TE Mg 12.5 mM buffer.

3.5 Pre-imaging Setup for High-Speed AFM of Protein-DNA Interactions in fluid

1. Position the microscopy slide with fluid cell on the AFM-optical microscope stage and, next, the AFM head.
2. Prior to imaging, soak the chosen cantilever in ethanol (or isopropanol) for a minute and next in ultrapure water.
3. Place the cantilever in a UV ozone cleaner (ProCleaner, Bio-Force Nanosciences) for 30 min.

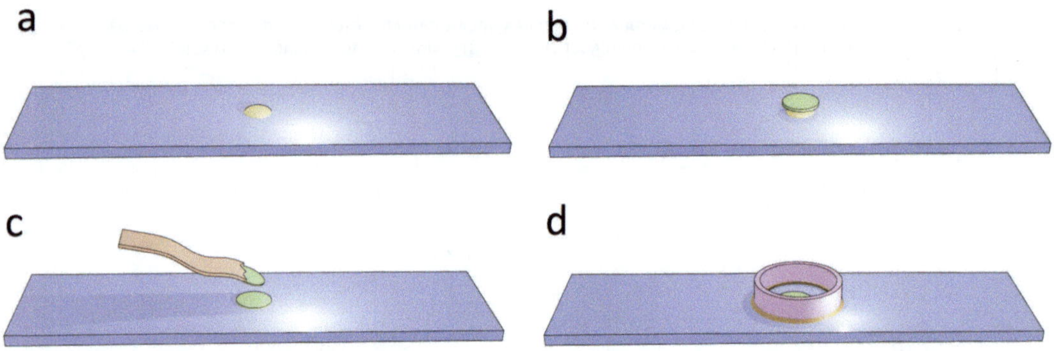

Fig. 3 DNA origami sample preparation and constructing the fluid cell using the plastic ring. (**a**) Place a drop of 2-component epoxy glue on the microscope slide. (**b**) Place a mica disc on the drop of glue. (**c**) Cleave a layer of mica using adhesive tape. (**d**) After DNA origami adsorption and washing, glue the plastic ring using 2-component rubber glue on the microscope glass to build the AFM fluid cell

4. Mount the cantilever on the NanoWizard Ultraspeed 2 cantilever holder and connect this to the AFM head.
5. Put the AFM head on the microscope stage and approach the cantilever to within a few hundred micrometers of the mica surface using the motors.
6. Once the cantilever is immersed in the liquid, align the laser on the cantilever of the AFM probe and position the detector in the center.
7. Select as imaging mode "AC Fast Imaging" and calibrate the cantilever.
8. Select the high-speed Fast KD scanner.
9. Select a drive frequency around the resonance frequency of the selected cantilever and a target amplitude of 4–5 nm.

3.6 High-Speed AFM Imaging of Protein-DNA Interactions in fluid

3.6.1 Imaging of the DNA Origami Frames

1. Set baseline adjustment to "dynamic". Select an initial scan size, e.g., 2–4 μm, and pixels at 256 × 256 (or 512 × 512), and a gain of 150.
2. Approach the surface. Once approached, press run.
3. Optimize the imaging by increasing the gain to 1000 and more (*see* Fig. 4).
4. Adjust the setpoint if needed: increasing the setpoint will reduce the cantilever force, decreasing will increase the force (*see* **Note 6**).
5. In the setting manager, put the baseline adjustment to "none".
6. Zoom in on a DNA origami frame.
7. To increase the imaging speed:
   (a) increase the line rate in steps until the desired frame rate is obtained

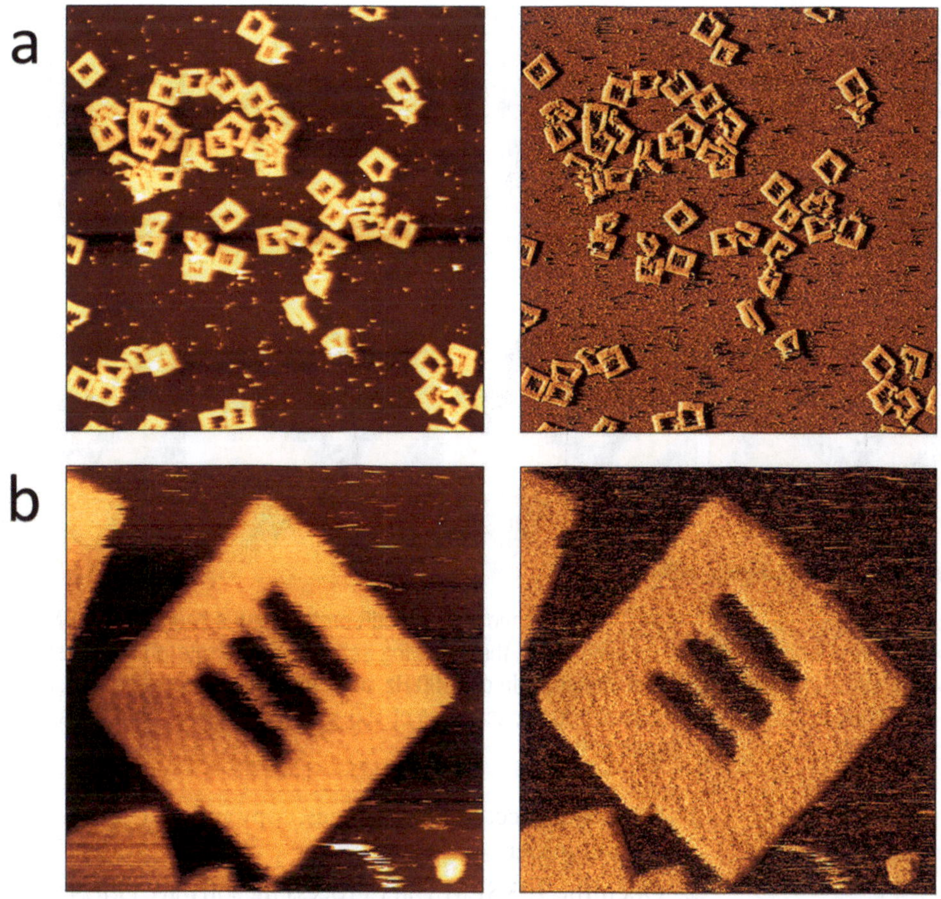

Fig. 4 AFM imaging of DNA origami frames. (**a**) A representative image of DNA origami frames adsorbed on mica. Height (left) and phase image (right) obtained using the following settings: a line rate of 40 Hz, scan size of 1.5 × 1.5 μm, 512 × 512 pixels, an integral gain of 1500 Hz, and proportional gain of 0.0239. (**b**) High-resolution height (left) and phase (right) image of a single DNA origami frame that was obtained using the following settings: a line rate of 10 Hz, scan size of 150 × 150 nm, 512 × 512 pixels, an integral gain of 2000 Hz and proportional gain of 0.0318

(b) decrease the number of pixels to 128 × 128

(c) Activate bidirectional scan in the settings manager.

3.6.2 Imaging of Protein – DNA Interactions

1. After imaging the DNA origamis and selecting an area on the mica surface for subsequent study of protein interaction, running is stopped and disengaged. The AFM head is removed.
2. Add the protein solution to the buffer in the liquid cell using a pipette, e.g., add 50 μl of the FadR solution (concentration of 60 μg/ml) (*see* **Note 8**).
3. Mount the AFM head again and approach.
4. Image the sample. Adjust (increase) the setpoint to minimize the force exerted by the tip on the sample (*see* Fig. 5).

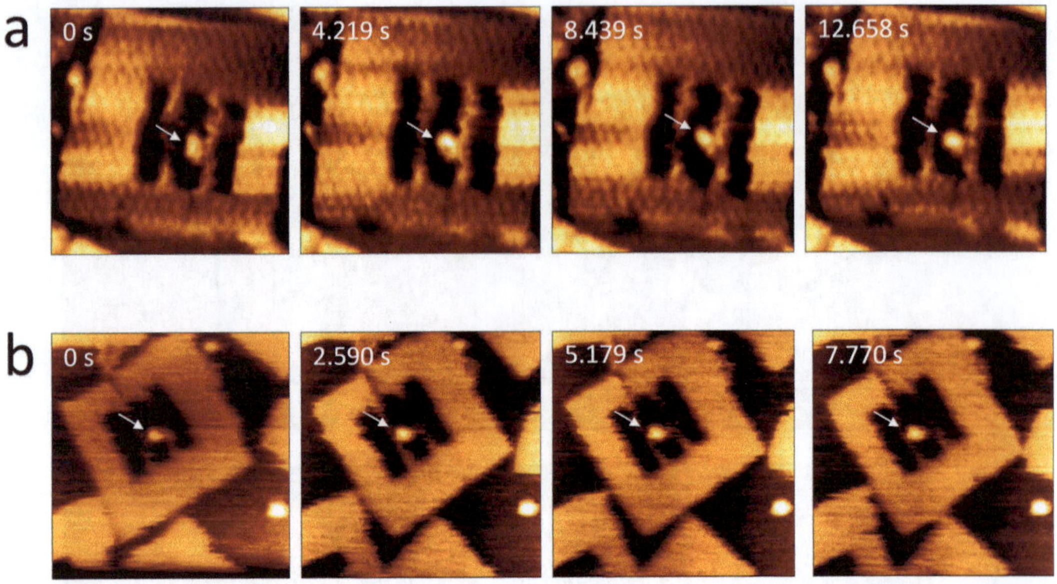

Fig. 5 Time-lapse imaging of protein-DNA interactions. The FadR protein that is bound to the DNA strand is indicated with a white arrow. (**a**) Imaging using the following settings: a line rate of 30 Hz, scan size of 100 × 100 nm, 128 × 128 pixels, an integral gain of 1800 Hz, and proportional gain of 0.0286. (**b**) Imaging using the following settings: a line rate of 100 Hz, scan size of 150 × 150 nm, 256 × 256 pixels, an integral gain of 1860 Hz, and proportional gain of 0.0296

3.6.3 Image Analysis

1. Select the selected set of images, copy and paste them in a newly created folder.
2. Open the JPKSPM Data Processing software (*see* **Note 9**).
3. Select "Open a batch of images", select the created folder, and click "Load".
4. Select the height channel and click "Next".
5. Select "Save images" and "Export as picture", and click on "Finish".
6. Plane leveling: click on the "Plane fit" icon; select some areas on the mica surface that are not covered by the DNA origamis.
7. Click on "Low-pass Filter" and select "Gaussian Smoothing"; adjust the values of Sigma X and Sigma Y to obtain the desired smoothing.
8. Optional: apply "Line leveling" with a degree of 2.
9. Click on "Keep and apply to all".
10. Create a video by clicking on the "Video" button.

4 Notes

1. The DNA origami design should translate the constructed shape using a DNA origami design software into the folding route of the scaffold and generate corresponding staple sequences that fulfill the folding. Besides caDNAno, different DNA origami design software can be used such as Tiamat [29], SARSE-DNA [30], oxView [31], Hex-Tiles [32], vHelix [33], DAEDALUS [34], TALOS [35], ATHENA [36], and Adenita [37]. A comparison of these software can be found in [17].
2. Since $MgCl_2$ is hygroscopic, it is recommended to use $MgCl_2.6H_2O$ to ensure the correct concentration.
3. The DNA origami frame was constructed using M13mp18 single-stranded DNA scaffold and sequence-designed DNA staple strands based on previous designs to analyze protein-DNA interaction [22, 23].
4. Optionally, the DNA origamis can be purified and enriched from the folding reaction mixture. Protocols for this can be found among others in these references [15, 38].
5. Alternatively, the designed DNA fragments can be ordered and custom produced.
6. The degree of coverage of the mica surface depends on the dilution of the origami solution and the incubation time on the mica surface.
7. For high-resolution imaging, the setpoint needs to be decreased until the high-resolution image appears. For high-speed imaging, a higher setpoint value needs to be selected to avoid a too strong interaction of the tip with the sample.
8. A suitable dilution of the protein solution is best determined experimentally by evaluating different dilutions.
9. Image analysis is performed using the JPKSPM Data Processing software. Other software packages can also be used such as the free and open-source Gwyddion software [39].

Acknowledgment

We thank Dr. Andreas Kraus and Dr. Heiko Haschke (BRUKER JPK, Berlin, Germany) for their help with the sample preparation and imaging; prof. Eveline Peeters for providing the protein FadR and collaborative work on FadR-DNA interaction; prof. Giovanni Dietler for collaborative work; Timothy Januarius, Yeseren Kayacan and Charlotte Yvanoff for their contribution to this DNA origami research topic. This work was supported by the Belgian Federal Science Policy Office (Belspo) and the European Space Agency

(ESA) PRODEX, FWO Flanders funded research equipment (AUGE/13/19; I002620N), FWO Flanders, and the Swiss National Fund (SNF) are funding the lead agency project (G068121N-310030L_197946). The Swiss National Grants 200021-144321, 407240-167137, and CRSII5_173863; the Gebert Rüf Stiftung GRS-024/14; and NASA NNH16ZDA001N-CLDTCH.

References

1. Binnig G, Quate CF, Gerber C (1986) Atomic force microscope. Phys Rev Lett 56:930–933. https://doi.org/10.1103/PhysRevLett.56.930
2. Bian K, Gerber C, Heinrich AJ et al (2021) Scanning probe microscopy. Nat Rev Methods Prim 11(1):1–29. https://doi.org/10.1038/s43586-021-00033-2
3. Beckwitt EC, Kong M, Van Houten B (2018) Studying protein-DNA interactions using atomic force microscopy. Semin Cell Dev Biol 73:220–230. https://doi.org/10.1016/J.SEMCDB.2017.06.028
4. Kasas S, Dietler G (2018) DNA-protein interactions explored by atomic force microscopy. Semin Cell Dev Biol 73:231–239. https://doi.org/10.1016/J.SEMCDB.2017.07.015
5. Ando T, Kodera N, Takai E et al (2001) A high-speed atomic force microscope for studying biological macromolecules. Proc Natl Acad Sci U S A 98:12468–12472. https://doi.org/10.1073/pnas.211400898
6. Ando T, Uchihashi T, Kodera N et al (2007) High-speed AFM and nano-visualization of biomolecular processes. Pflügers Arch - Eur J Physiol 456:211–225
7. Ando T (2019) High-speed atomic force microscopy. Curr. Opin. Chem. Biol. 51: 105–112
8. Ando T, Uchihashi T, Kodera N (2013) High-speed AFM and applications to biomolecular systems. Annu Rev Biophys 42:393–414. https://doi.org/10.1146/annurev-biophys-083012-130324
9. Uchihashi T, Scheuring S (2018) Applications of high-speed atomic force microscopy to real-time visualization of dynamic biomolecular processes. Biochim Biophys Acta – Gen Subj 1862:229–240. https://doi.org/10.1016/J.BBAGEN.2017.07.010
10. Ando T, Uchihashi T, Scheuring S (2014) Filming biomolecular processes by high-speed atomic force microscopy. Chem Rev 114: 3120–3188. https://doi.org/10.1021/CR4003837
11. YI L (2018) Direct AFM visualization of the nanoscale dynamics of biomolecular complexes. J Phys D Appl Phys 51. https://doi.org/10.1088/1361-6463/AAD898
12. Casuso I, Redondo-Morata L, Rico F (2020) Biological physics by high-speed atomic force microscopy. Philos Trans R Soc A 378: 20190604. https://doi.org/10.1098/RSTA.2019.0604
13. Uchihashi T, Ganser C (2020) Recent advances in bioimaging with high-speed atomic force microscopy. Biophys Rev 122(12):363–369. https://doi.org/10.1007/S12551-020-00670-Z
14. Rothemund PWK (2006) Folding DNA to create nanoscale shapes and patterns. Nat Cell Biol 440:297–302
15. Castro CE, Kilchherr F, Kim DN et al (2011) A primer to scaffolded DNA origami. Nat Methods 8:221–229. https://doi.org/10.1038/nmeth.1570
16. Chandrasekaran AR (2016) Programmable DNA scaffolds for spatially-ordered protein assembly. Nanoscale 8:4436–4446. https://doi.org/10.1039/C5NR08685J
17. Dey S, Fan C, Gothelf KV et al (2021) DNA origami. Nat Rev Methods Prim 11(1):1–24. https://doi.org/10.1038/s43586-020-00009-8
18. Tørring T, Voigt NV, Nangreave J et al (2011) DNA origami: a quantum leap for self-assembly of complex structures. Chem Soc Rev 40: 5636–5646. https://doi.org/10.1039/C1CS15057J
19. Rajendran A, Endo M, Sugiyama H (2012) Single-molecule analysis using DNA origami. Angew Chemie Int Ed 51:874–890. https://doi.org/10.1002/ANIE.201102113
20. Endo M, Yang Y, Sugiyama H (2013) DNA origami technology for biomaterials applications. Biomater Sci 1:347–360. https://doi.org/10.1039/C2BM00154C
21. Nakata E, Liew FF, Uwatoko C et al (2012) Zinc-finger proteins for site-specific protein positioning on DNA-origami structures.

Angew Chemie Int Ed 51:2421–2424. https://doi.org/10.1002/ANIE.201108199

22. Endo M, Sugiyama H (2014) Single-molecule imaging of dynamic motions of biomolecules in DNA origami nanostructures using high-speed atomic force microscopy. Acc Chem Res 47:1645–1653. https://doi.org/10.1021/AR400299M
23. Yamamoto S, De D, Hidaka K et al (2014) Single molecule visualization and characterization of Sox2-Pax6 complex formation on a regulatory DNA element using a DNA origami frame. Nano Lett 14:2286–2292. https://doi.org/10.1021/nl4044949
24. Endo M (2019) AFM-based single-molecule observation of the conformational changes of DNA structures. Methods 169:3–10. https://doi.org/10.1016/J.YMETH.2019.04.007
25. Lee AJ, Endo M, Hobbs JK, Wälti C (2017) Direct single-molecule observation of mode and geometry of RecA-mediated homology search. ACS Nano 12:272–278. https://doi.org/10.1021/ACSNANO.7B06208
26. Kurokawa T, Kiyonaka S, Nakata E et al (2018) DNA origami scaffolds as templates for functional tetrameric Kir3 K+ channels. Angew Chemie Int Ed 57:2586–2591. https://doi.org/10.1002/ANIE.201709982
27. Wang K, Sybers D, Maklad HR et al (2019) A TetR-family transcription factor regulates fatty acid metabolism in the archaeal model organism Sulfolobus acidocaldarius. Nat Commun 101(10):1–16. https://doi.org/10.1038/s41467-019-09479-1
28. Douglas SM, Marblestone AH, Teerapittayanon S et al (2009) Rapid prototyping of 3D DNA-origami shapes with caDNAno. Nucleic Acids Res 37:5001–5006. https://doi.org/10.1093/nar/gkp436
29. Williams S, Lund K, Lin C et al (2008) Tiamat: a three-dimensional editing tool for complex DNA structures. Lect Notes Comput Sci (including Subser Lect Notes Artif Intell Lect Notes Bioinf) 5347 LNCS:90–101. https://doi.org/10.1007/978-3-642-03076-5_8
30. Andersen ES, Dong M, Nielsen MM et al (2008) DNA origami design of dolphin-shaped structures with flexible tails. ACS Nano 2:1213–1218. https://doi.org/10.1021/nn800215j
31. Poppleton E, Bohlin J, Matthies M et al (2020) Design, optimization and analysis of large DNA and RNA nanostructures through interactive visualization, editing and molecular simulation. Nucleic Acids Res 48:E72–E72. https://doi.org/10.1093/nar/gkaa417
32. Matthies M, Agarwal NP, Poppleton E et al (2019) Triangulated wireframe structures assembled using single-stranded DNA tiles. ACS Nano 13:1839–1848. https://doi.org/10.1021/acsnano.8b08009
33. Benson E, Mohammed A, Gardell J et al (2015) DNA rendering of polyhedral meshes at the nanoscale. Nature 523:441–444. https://doi.org/10.1038/nature14586
34. Veneziano R, Ratanalert S, Zhang K et al (2016) Designer nanoscale DNA assemblies programmed from the top down. Science 80(352):1534–1534. https://doi.org/10.1126/science.aaf4388
35. Jun H, Shepherd TR, Zhang K et al (2019) Automated sequence design of 3D polyhedral wireframe DNA origami with honeycomb edges. ACS Nano 13:2083–2093. https://doi.org/10.1021/ACSNANO.8B08671
36. Jun H, Wang X, Bricker W et al (2020) Rapid prototyping of wireframe scaffolded DNA origami using ATHENA. bioRxiv 2020.02.09.940320. https://doi.org/10.1101/2020.02.09.940320
37. de Llano E, Miao H, Ahmadi Y et al (2020) Adenita: interactive 3D modelling and visualization of DNA nanostructures. Nucleic Acids Res 48:8269–8275. https://doi.org/10.1093/NAR/GKAA593
38. Wagenbauer KF, Engelhardt FAS, Stahl E et al (2017) How we make DNA origami. ChemBioChem 18:1873–1885. https://doi.org/10.1002/CBIC.201700377
39. Nečas D, Klapetek P (2012) Gwyddion: an open-source software for SPM data analysis. Open Phys 10:181–188. https://doi.org/10.2478/S11534-011-0096-2

Chapter 11

Separation and Characterization of Protein–DNA Complexes by EMSA and In-Gel Footprinting

Daniel Charlier and Indra Bervoets

Abstract

In-gel footprinting enables the precise identification of protein binding sites on the DNA after separation of free and protein-bound DNA molecules by gel electrophoresis in native conditions and subsequent digestion by the nuclease activity of the 1,10-phenanthroline-copper ion $[(OP)_2\text{-}Cu^+]$ within the gel matrix. Hence, the technique combines the resolving power of protein–DNA complexes in the electrophoretic mobility shift assay (EMSA) with the precision of target site identification by chemical footprinting. This approach is particularly well suited to characterize distinct molecular assemblies in a mixture of protein–DNA complexes and to identify individual binding sites within composite operators, when the concentration-dependent occupation of binding sites, with a different affinity, results in the generation of complexes with a distinct stoichiometry and migration velocity in gel electrophoresis.

Key words Protein–DNA binding, Native gel electrophoresis, Electrophoretic mobility shift assay, Denaturing gel electrophoresis, Chemical nucleases, In-gel footprinting, 1,10-phenanthroline, Sequencing ladders

1 Introduction

Numerous major cellular processes including replication, transcription, recombination, and modification of DNA are governed by sequence-specific protein–DNA interactions, and various genetic, biochemical, and biophysical methods or combinations thereof can be applied or have been specifically developed to study different aspects of these interactions in a qualitative or quantitative manner in vivo and/or in vitro [1]. Just to cite a few: chromatin immunoprecipitation (ChIP) either in combination with DNA hybridization (ChIP-chip) or sequencing (ChIP-seq), systematic evolution of ligands by exponential enrichment (SELEX), methylation protection, electrophoretic mobility shift assay (EMSA), affinity chromatography, nitrocellulose filter binding, analytical ultracentrifugation, circular dichroism (CD), isothermal titration calorimetry (ITC), surface plasmon resonance (SPR), bio-layer

Eveline Peeters and Indra Bervoets (eds.), *Prokaryotic Gene Regulation: Methods and Protocols*, Methods in Molecular Biology, vol. 2516, https://doi.org/10.1007/978-1-0716-2413-5_11,

interferometry (BLI), thermophoresis, enzymatic and chemical footprinting, base- and phosphate-specific chemical premodification binding interference, groove-specific ligand-binding interference, atomic force microscopy (AFM), X-ray crystallography, and small angle X-ray scattering (SAXS). Each one of these techniques has its advantages and limitations/shortcomings; the choice will largely depend on which aspect(s) of the complex formation—affinity, stoichiometry, cooperativity, equilibrium and/or rate constants, precise binding site identification, groove-, base- or backbone specific contacts—one wants to address/determine, the purity and amount of available biochemical material, the characteristics of the individual interacting components and the complex (solubility, affinity, stability), and the availability/access to equipment.

EMSA is a very easy, yet robust, reliable, and powerful method [2–6]. It has the advantage that it does not require sophisticated (expensive) equipment and consumes very small amounts of biological material only. DNA-binding proteins used in EMSA are preferentially purified to electrophoretic homogeneity, but in some instances, partially purified protein samples or even total or fractionated cell-free extracts may be used. As a consequence, the technique is widely used since decennia, frequently for screening of potential target sites or as the very first step in the characterization of a particular protein–DNA interaction. But, importantly, EMSA can also be used in combination with other techniques such as chemical in-gel footprinting (see below) and premodification binding interference (see Chapter 12) for the simultaneous identification of multiple binding sites in composite operators and the identification of local base and backbone-specific contacts [7–10].

EMSA is based on the early observation that a protein–DNA complex migrates with a different velocity (generally more slowly) through a gel matrix (agarose or polyacrylamide) than the same unbound DNA [11, 12]. This automatically implies that EMSA is not a real equilibrium method since bound and unbound DNA fractions are separated. However, this limitation is equally valid for the vast majority of the other techniques enumerated above, including filter binding, SPR, BLI, chromatography, and ultracentrifugation, which all rely on the physical separation of distinct molecular species. Despite this limitation, EMSAs may provide essential qualitative and quantitative information on major aspects of the protein–DNA interaction such as apparent equilibrium constants, number of target sites on the DNA probe, stoichiometry, and specificity of the binding, cooperativity, requirement/interference of ligands, protein-induced DNA deformations (circular permutation assay). For examples see [10, 13–20].

EMSA can be performed with unlabeled DNA probes (restriction fragments, PCR amplicons, oligonucleotide duplexes) but is much more sensitive with radioisotope or fluorescently labeled

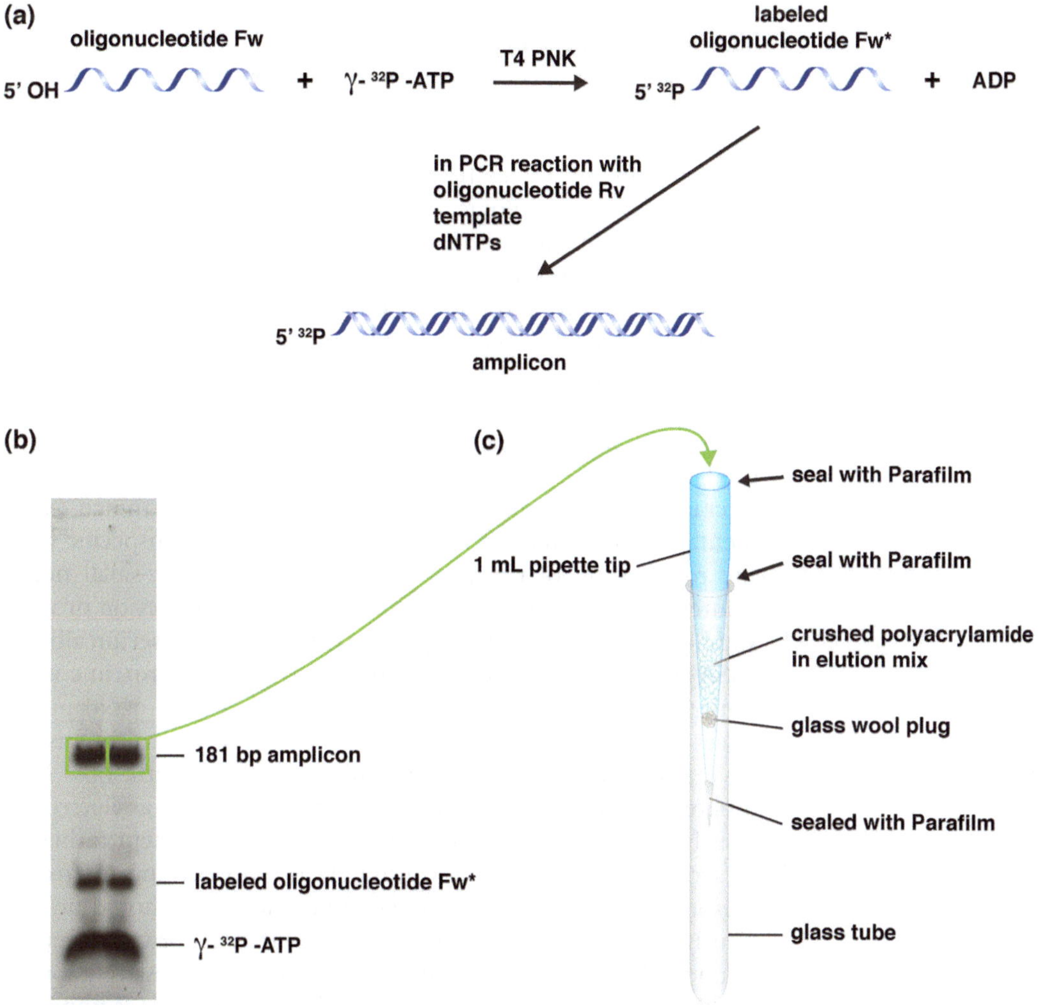

Fig. 1 Outline of the synthesis, purification by gel electrophoresis, and recovery from gel of a [5′-^{32}P] single-end labeled DNA amplicon. (**a**) 5′-end labeling of a synthetic oligonucleotide by T4 PNK and [γ-^{32}P]-ATP and its use in amplification of a single-end labeled DNA fragment by PCR. (**b**) Autoradiograph of a typical PCR amplification reaction with one labeled and one nonlabeled primer separated by gel electrophoresis on a nondenaturing 6% polyacrylamide gel. (**c**) Elution set-up for the recovery of labeled DNA from an acrylamide gel

molecules and when the technique is used in combination with in-gel footprinting, single-end labeled molecules are a prerequisite (*see* Fig. 1). The composition (agarose or polyacrylamide) and concentration of the gel must be adapted in function of the size of the DNA probe, the amplitude of the protein-induced difference in migration velocity, and the potential need of separating different protein–DNA complexes. Generally, 5–10% polyacrylamide gels are used with probes ranging from a few tens to a few hundred bps, whereas agarose is used for larger fragments, up to a few kb. Protein–DNA interactions generally comprise base-specific and backbone contacts and rely on the formation of hydrogen bonds (most sequence-specific contributions), ionic, hydrophobic, and Van der Waals interactions. Hence, complex formation is sensitive to pH, temperature, and salt concentration. Consequently, buffer composition and incubation temperature have to be adapted and ideally mimic physiologic conditions. As the formation of ionic interactions, major contributors to the free energy of complex formation, releases heat (exotherm), binding is generally more efficient and complexes more stable at low temperatures. Some salt is required to decrease the contribution of nonsequence-specific interactions, but too much salt may significantly inhibit or even completely abolish complex formation. Furthermore, an excess nonsequence-specific nonlabeled competitor DNA may be used to further reduce the contribution of nonsequence-specific interactions. Again, its concentration has to be determined experimentally in the function of the ratio of specific/nonspecific target site recognition of the protein. Glycerol, or other small neutral solutes such as sucrose, often exert a stabilizing effect on proteins, and hence may improve complex formation. Bovine serum albumin (BSA) is frequently added to standardize the total protein concentration among different samples in an assay and to avoid protein losses during sample manipulation, especially with proteins that have a tendency to stick to the tube wall and pipette tips. Many different buffers can be used for gel preparation and as electrophoresis buffer, but the ionic concentration should be kept rather low to avoid too much heating during electrophoresis and to ensure a reasonable migration velocity. The migration time should be sufficiently long to allow separation of the different forms of free and bound DNA molecules in the sample, but not excessively long to avoid too much dissociation during the migration. It is important to notice that the amplitude of the reduction in migration velocity of a protein–DNA complex compared to free DNA will not only depend on the mass of the protein (more precisely on the ratio of protein/DNA masses) but also on its charge (pH dependent), and potential protein-induced DNA deformations, whereby bend or kinked DNA migrates more slowly through a polyacrylamide (but not agarose, where the pores are too large) gel than straight DNA. The latter may be explained by the worm-like movement (also called reptation) of DNA through the gel matrix and the reduced end-to-end distance of curved as compared to straight DNA (*see* **Note 1**). It is worth noticing that some degree of DNA curvature is frequently induced upon protein binding (induced fit) and that both the amplitude of the bend and its position within the fragment can be determined by EMSA in the so-called circular permutation assay [13].

As developed above, EMSA may reveal many aspects of a protein–DNA interaction but it does not allow the precise identification of one or more binding sites within the limits of the DNA probe. This information may, however, be gathered from various enzymatic and chemical footprinting methods. The principle of these "protection type" of techniques (as opposite to the premodification binding interference type) relies on the observation that protein binding will protect zones of tight interaction on the DNA against digestion by an enzymatic (generally DNase I) [21] or chemical nuclease (generally a highly reactive radical) [21–26]. DNase I interacts in a nonsequence specific manner (but yet is sensitive to minor groove width and generates a ladder of heterogeneous intensity) with the minor groove of B-form DNA, and in the presence of Mg^{2+} ions introduces a single-stranded nick. Partial digestion (statistically one cut per molecule) of single-end-labeled DNA molecules will result in the generation of a series of labeled DNA fragments of different sizes (a ladder) that can be separated by gel electrophoresis under denaturing conditions. Local protection against digestion will result in a blank (or region of decreased band intensity) in the ladder with respect to the control lane with unbound DNA treated in the same conditions. Migration of the reaction products alongside sequencing reference lanes generated from the same DNA fragment allows the precise delimitation of the zone of protection on the sequence. Potential limitations of the technique result from the low cutting activity of DNase I in A+T rich regions (narrow minor groove) and at TpA steps, which both frequently occur in promoter regions, and the rather large size of the nuclease which results in some oversizing of the region of interaction due to steric hindrance between DNase I and DNA-bound protein. This limitation is overcome by the use of chemical nucleases, such as the highly reactive hydroxyl radical (OH•), which are much smaller and allow a more precise delimitation of the zones of tight contact [22]. Hydroxyl radicals are generated directly in the test tube containing the protein–DNA mix by the Fenton reaction (H_2O_2 reacting with the ferrous iron–EDTA complex $[Fe(EDTA)^{2-}]$) but notice that the binding buffer should not contain glycerol, as it is a potent scavenger of OH• radicals. Hydroxyl radicals cleave the DNA backbone from the minor groove, with nearly no sequence specificity. The main primary target is the H1' proton of the deoxyribose ring; secondary reactions emanating from the reactive intermediates eventually result in breakage of the phosphodiester bond. Due to the accessibility of the "backside" of the DNA–protein complex (which is generally not the case for DNase I) and its limited permeability for the small nuclease, the binding site is generally split up in several small zones of protection (3–5 nucleotides long) corresponding to tight contacts, separated by short unprotected regions. With proteins that bind to one face of the DNA helix only, such as numerous prokaryotic transcription regulators with a helix-turn-helix DNA-binding motif do, this results in a typical pattern with the centers of the patches of protection separated by approximately one helical turn and slightly shifted toward the 3′-end with respect to the patches on the complementary strand [22]. It is evident that in order to obtain a clear footprint of complexes in solution by treatment with enzymatic or chemical nucleases, a major fraction of the DNA probe should be bound, which is not always possible to reach due to low-binding affinity and/or low solubility of the protein. Furthermore, with mixed populations of complexes the resulting footprint will be dominated by the most abundant and stable species but provide no information on the distinct components.

These latter limitations are overcome in the in-gel footprinting technique, which combines the advantages of the resolving power of gel electrophoresis to separate free from bound DNA by EMSA with the accuracy of chemical footprinting with the 1, 10-phenanthroline-copper ion [$(OP)_2$-Cu^+] [23, 24]. In this approach, different molecular species (free DNA, different complexes) with a distinct migration velocity are first separated by EMSA and subsequently treated with the chemical nuclease, directly within the gel matrix (*see* Fig. 2a, b). Finally, the DNA is extracted from gel and the reaction products separated by gel electrophoresis under denaturing conditions (*see* Fig. 2c). This approach allows the simultaneous characterization of all different protein–DNA complexes present in the binding assay, including minor forms, as long as a sufficient amount of DNA can be extracted from the native EMSA gel to proceed to the final gel electrophoresis under denaturing conditions. For an example see [16]. Chemical footprinting with the [$(OP)_2$-Cu^+] ion does not produce free floating radicals as in the OH• radical footprinting described above. Instead, the radicals are formed at the very site of binding of the [$(OP)_2$-Cu^+] ions in the minor groove of DNA, which occurs with nearly no sequence specificity, though DNA conformation may have an effect and some hyperreactivity has been observed at adenine residues from TAT triplets [27]. Other advantages of the method are that glycerol can be used in the binding buffer (but instead it should not contain mM concentrations of DTT or EDTA) [28] and the caging effect of the gel matrix may exert a stabilizing effect on the protein–DNA complexes all along the cutting reaction. Furthermore, as the cleavage reaction produces predominantly 3’- and 5’-phosphorylated ends, footprints can be accurately calibrated with chemical sequencing ladders generated from the same single-end labeled DNA probe [29]. The procedure starts by adding appropriate concentrations of $CuSO_4$ and 1,10-phenanthroline to the submerged EMSA gel. This results in the formation of a coordinated cupric complex with 2:1 stoichiometry ($[OP]_2$-Cu^{2+}) that is subsequently reduced to its cuprous form ($[OP]_2$-Cu^+) with 3-mercaptopropionic acid. The coordinated tetrahedral cuprous complex binds reversibly to the minor groove of the DNA and undergoes an *in situ* one-electron oxidation with hydrogen peroxide (H_2O_2), an

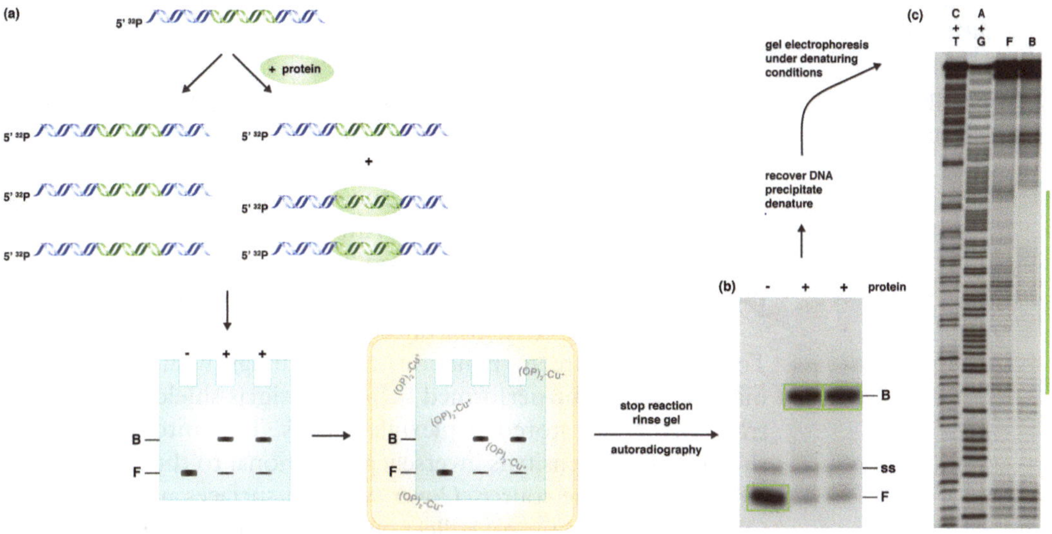

Fig. 2 Schematic outline and example of the in-gel footprinting technique. (**a**) Schematic outline. A constant amount of single [5′-^{32}P] end-labeled DNA is first incubated with and without protein in appropriate conditions (binding buffer composition, temperature, time). When the equilibrium of the interaction has been reached, the samples are loaded on a polyacrylamide gel and free (F) and bound (B) DNA populations separated by electrophoresis in native conditions. Then, the whole gel is treated with the chemical nuclease, the reaction stopped, bands visualized by autoradiography, the various DNA populations recovered from the polyacrylamide matrix, and the reaction products separated by gel electrophoresis in denaturing conditions alongside chemical sequencing reactions. (**b**) Autoradiograph of an EMSA with protein binding to a single [5′-^{32}P] end-labeled DNA fragment. (**c**) Autoradiograph of the separation of the reaction products of the in-gel footprinting by gel electrophoresis in denaturing conditions on 8% polyacrylamide. A+G and C+T are the chemical sequencing reactions of the labeled probe

essential co-reactant for the nuclease activity of the complex. The resulting highly reactive intermediate will then attack the H1′ protons of the exposed deoxyribose rings and the reaction culminates into local cleavage of the phosphodiester backbone. Notice that H_2O_2 can be added to the reaction mix, but this is not required as in the presence of molecular oxygen (in air), H_2O_2 will automatically be produced *in situ* by oxidation of part of the cuprous complex. As in the other "protection type" of footprinting methods, binding of the protein to specific regions on the DNA will interfere with binding of the cuprous complex to these sites, and hence protect them from the nuclease activity and be visible as a region of reduced intensity in the ladder as compared to the control reaction performed with free DNA (*see* Fig. 2). See [10] for a comparison of footprints obtained with DNase I, OH•, and in-gel footprinting of the same protein–DNA interaction. How individual binding sites, cognate binding constants, and cooperativity parameters of protein oligomers binding to a multipartite operator can be determined by the combination of in-gel footprinting and EMSA is illustrated in [16, 17].

2 Materials

2.1 Safety Instructions

Many chemicals, including acrylamide/*bis*-acrylamide, ammoniumpersulfate, TEMED, EDTA, 3-mercaptopropionic acid, 1–10 phenanthroline, hydrazine, and piperidine, are toxic, neurotoxic, potentially carcinogenic, and/or flammable. Avoid inhalation and all contact with the skin. Always wear gloves and protective clothing, and work in a fume hood when necessary. Always consult the accompanying safety sheets prior to handling a chemical.

[^{32}P] is a high energy β-ray emitter with a half-life of 14 days. To limit exposure to radiation, all handlings with [^{32}P]-labeled material should be performed behind plexiglas shields and samples transported and stored in plexiglas boxes. Take into account that the radiation intensity is inversely proportional to the square of the distance from the source. Cover working surfaces with benchcoat (absorbing face on top); collect and store contaminated liquid and solid waste in separate authorized containers; and regularly verify gloves, working areas, and equipment for potential spill and contamination with a Geiger counter and decontaminate as soon as possible. Working exclusively in a dedicated and specifically equipped room with dedicated racks, pipette sets, etc., reserved for manipulation of isotopes, is highly recommended. At any rate, authorization to manipulate nonencapsulated radionuclides should be obtained from the institutional health and safety office and its instructions/guidelines strictly applied. This generally includes wearing of a personal dosimeter to monitor cumulative exposure to radiation.

2.2 Preparation of [5′-^{32}P] Single-End Labeled DNA Fragments

2.2.1 Solutions, Enzymes, Chemicals

1. Freshly delivered [γ-^{32}P]-ATP (3000 Ci/mmol). Take into account that the half-life of [^{32}P] is 14 days. Store at −20 °C but make sure that the solution is completely thawed and homogenous before removing an aliquot.
2. T4 polynucleotide kinase (10 U/μL) (store at −20 °C, use on ice).
3. 10× commercial kinase buffer (store at −20 °C, thaw on ice).
4. Nuclease-free H_20.
5. Probe-specific oligonucleotides (dissolved in nuclease-free H_20) as primers for PCR amplification. Make stock solutions at 500 pmol/μL, conserved at −20 °C for long-term storage, and working solutions at 10 pmol/μL.
6. Template DNA (genomic or plasmid) (store at −20 °C in small aliquots to avoid multiple cycles of freeze and thaw).
7. 2x commercial PCR amplification mix or separate heat-resistant DNA polymerase and dNTPs.

2.2.2 Other Materials and Equipment

1. Sterile disposables: 1.5 mL Eppendorf tubes, 0.25 mL PCR tubes, and pipette tips
2. Heating thermostatic water baths or heating blocks at 37 °C and 65 °C
3. Microcentrifuge
4. Thermocycler

2.3 Native Gel Electrophoresis

2.3.1 Solutions and Chemicals

1. 5x Electrophoresis buffer. TBE: 445 mM Tris-base, 445 mM boric acid, 12.5 mM EDTA. TBE is commonly used but other low ionic strength buffers are equally possible (*see* **Note 2**).
2. 100 mg/mL ammoniumpersulfate stock solution dissolved in ddH_20. This can be stored up to a few weeks at 4 °C.
3. *N*,*N*,*N*′,*N*′-tetramethylenediamine (TEMED). Store in dark, at 4 °C.
4. 30% acrylamide/*N*, *N*′-methylene *bis*-acrylamide solution (ratio 29:1). Store in dark at 4 °C.
5. ddH_2O.
6. Repel solution (e.g., Blue Slick spray, Serva, to siliconize the glass plates).

2.3.2 Other Materials and Equipment

1. Pair of glass plates (19.5 × 16.0 cm and 19.5 × 19.0 cm). Other dimensions are possible, but take into account that they should be compatible with the dimensions of the electrophoresis tank and sufficiently long to allow good separation of all bound DNA forms without loss of the free DNA by migration out of the gel.
2. Pair of spacers (0.1 cm thick) and cognate 14 well comb.
3. Commercial rubber sealing system for glass plate assembly (adapted to the size of glass plates) or tape and clamps.
4. Syringe (50 mL).
5. Büchner funnel and rubber tubing to connect to the vacuum pump.
6. Filter paper for Büchner funnel.
7. Vacuum pump.
8. Power supply and electric cabling.
9. Electrophoresis tank for vertical gel electrophoresis

2.4 Autoradiography

2.4.1 Solutions

1. Developer for photography (5x diluted in H_20). Store at room temperature but protected from light in a brown bottle. Can be used repeatedly but refresh when becoming brownish colored.
2. Fixer for photography (5x diluted in H_20). Store at room temperature but protected from light. Can be used repeatedly but refresh when becoming slightly pink colored.

2.4.2 Other Materials and Equipment

1. Radioisotope and chemiluminescent imaging film
2. Intensifying screen
3. Plastic foil (e.g., Saran Wrap or cling film)
4. Black plastic film or black plastic bags (sufficiently large to pack the gel + film and protect from light)
5. Dark room equipped with red light bulbs and a sink with running tap water and demineralized water
6. Two trays slightly larger than the dimensions of the autoradiograph for developer and fixer solutions

2.5 Recovery of DNA from Polyacrylamide Gel

2.5.1 Solutions and Chemicals

1. Elution buffer: 0.5 M ammonium acetate, 0.01 M magnesium acetate, 0.1 mM EDTA, 0.1% SDS. Sterilize by autoclaving, store at room temperature.
2. Ice-cold absolute ethanol. Store at −20 °C.
3. 80% ice-cold ethanol. 80% absolute ethanol, 20% nuclease-free H_2O. Store at −20 °C.

2.5.2 Other Materials and Equipment

1. Sterile glass tubes (L = 10 cm, d = 0.8 cm).
2. Sterile 1 mL pipette tips.
3. 150 mm Pasteur pipettes.
4. Drawn-out 150 mm Pasteur pipettes with thin tip to allow removal of a maximum of supernatant without touching the DNA pellet upon DNA precipitation in Eppendorf tubes.
5. Sterile glass wool. Sterilize by autoclaving.
6. Self-sealing tape.
7. Nuclease-free scalpel.
8. Water bath at 45 °C.
9. Microcentrifuge.
10. Vacuumcentrifuge.
11. Vortex.

2.6 EMSA

2.6.1 Solutions and Products

1. [5′-^{32}P] single-end labeled DNA (*see* Subheading 2.2); 10 000–15 000 cpm per lane
2. Nonlabeled, nonspecific competitor DNA (e.g., 0.5 mg/mL sonicated salmon or herring sperm DNA) (*see* **Note 3**)
3. DNA-binding protein, preferably purified to electrophoretic homogeneity (*see* **Note 4**)
4. Protein–DNA binding buffer: 20 mM Tris-HCl pH 8.0, 0.4 mM EDTA, 0.1 mM DTT, 50 mM NaCl, 1 mM $MgCl_2$
5. 10× retardation loading dye mix (25% ficoll, 0.1% xylene cyanol, 0.1% bromophenol blue)

6. All solutions and products for native gel electrophoresis (*see* Subheading 2.3)
7. All solutions for autoradiography (*see* Subheading 2.4)

2.6.2 Other Materials and Equipment

1. All materials and equipment for native gel electrophoresis (*see* Subheading 2.3)
2. All materials and equipment for autoradiography (*see* Subheading 2.4)
3. Heating thermostatic water bath

2.7 In-Gel Footprinting

2.7.1 Solutions and Chemicals

1. All solutions and chemicals for EMSA (*see* Subheading 2.6), but take into account that at least 100,000 cpm of labeled DNA per lane should be used.
2. All solutions and chemicals used for autoradiography (*see* Subheading 2.4).
3. All solutions used for recovery of DNA from polyacrylamide gel (*see* Subheading 2.5).
4. 200 mL 10 mM Tris-HCl, pH 8.0.
5. 1, 10-phenantroline solution. Freshly prepared by dissolving 40 mg in 5 mL absolute ethanol.
6. $CuSO_4$ solution. Freshly prepared by dissolving 14.4 mg in 10 mL nuclease-free H_2O.
7. 100x-diluted 3-mercaptopropionic acid. Freshly prepared by mixing 100 μL 3-mercaptopropionic acid with 9.9 mL nuclease-free H_2O. Work in chemical fume hood.
8. Neocuproine solution. Freshly prepared by dissolving 125 mg in 20 mL absolute ethanol.

2.7.2 Other Materials and Equipment

1. Pyrex tray of dimensions slightly larger than the glass plate of the EMSA gel
2. All materials and equipment used for EMSA (*see* Subheading 2.6)
3. All materials and equipment used for autoradiography (*see* Subheading 2.4)
4. All materials and equipment used for recovery of DNA from polyacrylamide gel (*see* Subheading 2.5)

2.8 Preparation of A+G and C+T Sequencing Ladders

2.8.1 Solutions and Chemicals

1. [5′-^{32}P] single-end labeled DNA (about 100,000 cpm per reaction, *see* Subheading 2.2) (*see* **Note 5**).
2. Carrier DNA: 0.5 mg/mL sonicated salmon or herring sperm DNA dissolved in nuclease-free H_20.
3. Nuclease-free H_20.

4. Citrate solution: 50 mM Na-citrate, 0.1 mM EDTA, pH 4.0, adjusted with NaOH.
5. Hydrazine (NH_2-NH_2).
6. 1 M piperidine. Freshly prepared on ice by 10× dilution of a 10 M piperidine stock solution in nuclease-free H_20. Work in a chemical fume hood. First add the H_20 to the Eppendorf tube, then add the piperidine and pipette a couple of times up and down. Close the tube and mix thoroughly by vortexing.
7. 0.3 M sodium acetate. Sterilize by autoclaving, store in the refrigerator.
8. 5 mg/mL yeast-tRNA dissolved in nuclease-free H_20. Store at −20 °C or small working aliquots at 4 °C.
9. Ice-cold absolute ethanol. Store at −20 °C.
10. 80% ice-cold 80% ethanol: 80% absolute ethanol, 20% nuclease-free H_2O. Store at −20 °C.
11. Formamide dye mix: 100 mL deionized formamide, 0.03 g xylene cyanol, 0.03 g bromophenol blue, 20 mM EDTA.

2.8.2 Other Materials and Equipment

1. 150 mm Pasteur pipettes
2. Drawn-out 150 mm Pasteur pipettes with a thin tip to allow removal of a maximum of supernatant without touching the DNA pellet upon DNA precipitation in Eppendorf tubes
3. Microcentrifuge
4. Vacuumcentrifuge
5. Vortex

2.9 Denaturing Gel Electrophoresis

2.9.1 Solutions and Chemicals

1. 5× Electrophoresis buffer TBE: 445 mM Tris-base, 445 mM boric acid, 12.5 mM EDTA.
2. 100 mg/mL ammoniumpersulfate in ddH_20. Can be stored for a few weeks at 4 °C.
3. *N*,*N*,*N*′,*N*′-tetramethylenediamine (TEMED). Store in dark at 4 °C.
4. 40% acrylamide/*N*, *N*′-methylene *bis*-acrylamide solution (ratio 19:1). Store in dark at 4 °C.
5. Urea.
6. ddH_2O.
7. Repel solution (e.g., Blue Slick spray, Serva).
8. Sticky solution: 48.5 mL ethanol, 1.5 mL 10% acetic acid, 150 μL γ-methacryloxypropyltrimethoxysilane.
9. 2 L of 10% acetic acid (can be used for five gels, then refresh).

2.9.2 Other Materials and Equipment

1. A pair of heat-resistant sequencing type glass plates (33.5 × 39.5 cm and 33.5 × 41.5 cm). Other dimensions are possible, but take into account that they should be compatible with the dimensions of the electrophoresis tank.
2. Pair of spacers (0.25 mm thick) and cognate 20 well comb.
3. Commercial rubber sealing system for plate assembly (adapted to the size of glass plates) or tape and clamps.
4. 50 mL syringe.
5. Büchner funnel and rubber tubing to connect to vacuum pump.
6. Filter paper for Büchner funnel.
7. Vacuum pump.
8. Power supply allowing electrophoresis at a constant power of at least 50 W (±1 500 V).
9. Electrophoresis tank for sequencing-type gel electrophoresis (ensuring homogenous heat distribution during migration).
10. Plastic tray slightly larger than dimensions of the glass plates.
11. Small electric heating system to blow hot air.
12. Heating magnetic stirrer

3 Methods

3.1 Preparation of [5′-^{32}P] Single-end Labeled DNA Probes by PCR Amplification

3.1.1 DNA Probe Characteristics

In-gel footprinting requires highly pure single-end labeled DNA molecules and is ideally performed with rather short probes (100–200 bp), whereby the binding site should not be situated too far from the labeled end to ensure high resolution of the reaction products in the zone of interest upon gel electrophoresis in denaturing conditions (depending on the polyacrylamide concentration, but ideally not exceeding 50–60 bp on 6% polyacrylamide). When the DNA sequence is known, such fragments can be most easily generated by PCR amplification with one labeled and one nonlabeled oligonucleotide as primers and genomic or plasmid DNA as a template, according to standard PCR amplification reaction conditions and purified by gel electrophoresis and recovery from gel (*see* Fig. 1).

3.1.2 Oligonucleotide Labeling and PCR Amplification

1. Mix in a 1.5 mL sterile Eppendorf tube 60 pmol of oligonucleotide with 3.0 μL of commercial 10x kinase buffer, 1.0 μL T4 polynucleotide kinase, 2.5 μL [γ-^{32}P]-ATP, and adjust the volume to 30 μL with nuclease-free H_20.
2. Mix gently by pipetting a couple of times up and down, spin briefly to collect all material at the bottom of the tube.
3. Incubate at 37 °C for 40 min.

4. Transfer the tube to 65 °C for 10 min to inactivate the enzyme and stop the reaction.
5. Let it cool down and use aliquots of the labeled oligonucleotide as primer in normal PCR amplification reactions with parameters of template denaturation, annealing and elongation adapted to the nature of the template, the melting temperature of the primers and the nature of the DNA polymerase.
6. Proceed to the purification of the amplicon of interest by gel electrophoresis and recovery from gel (see below) or store the reaction mixes overnight at 4 °C or at −20 °C for longer periods.

3.2 Purification of [^{32}P]-Labeled DNA by Nondenaturing Polyacrylamide Gel Electrophoresis

3.2.1 Aim

Polyacrylamide gel electrophoresis is used to separate the labeled amplicon of interest from potential aberrant side products generated during the PCR amplification, an excess nonelongated labeled primer, and nonincorporated labeled [γ-^{32}P]-ATP (*see* Fig. 1b).

3.2.2 Treatment of Glass Plates

1. Start by thoroughly cleaning the front and back glass plates of the gel casting system with detergent, rinse abundantly with tap water, then with demineralized water, and wipe dry with absorbing paper.
2. Siliconize the inside face of the large front plate by spraying some repel solution on the plate, spread equally all over the plate with absorbing paper, and let evaporate for a few minutes.
3. Assemble the back and front plates with the spacers with the help of a commercial rubber assembly system or seal the edges and bottom with tape and clamps to avoid leaks. Deposit the assembly slightly tilted on a protected working surface (easily cleanable surface such as a Teflon plate or disposable benchcoat).

3.2.3 Preparation of a 6% Nondenaturing Polyacrylamide Gel (see **Note 6**)

1. Mix in a 50 mL cylinder 10 mL of 5x TBE buffer with 10 mL of 30% acrylamide/*bis*-acrylamide (ratio 29:1) stock solution, 0.5 mL ammoniumpersulfate solution and adjust to 50 mL with ddH$_2$0.
2. Filter the solution on a Büchner funnel connected to a vacuum pump and degas for 10 min.
3. Remove a 10 mL aliquot and add 10 μL TEMED, mix and use a syringe to let 5 mL of the solution flow down along the spacers, on either side of the assembly.
4. Allow to polymerize (takes a few minutes only and can be verified by tilting the assembly). This reduces the risk of further leakage while pouring the bulk of the gel.

5. Add 20 µL TEMED to the remaining 40 mL of polyacrylamide solution and use a syringe to carefully fill up the plates till the top of the small back plate. Let the liquid flow down between the glass plates slowly but regularly to avoid trapping air bubbles in the gel solution.
6. Insert the comb and avoid trapping very small air bubbles beneath the teeth of the comb by starting with a tilted comb, insert one side, then progressively push down the remaining part of the comb.
7. Let it polymerize for at least 30–45 min.
8. Remove sealing, comb, and potentially spilled polyacrylamide on the outer surface of the glass plates and mount the gel in the electrophoresis tank.
9. Add 1x TBE running buffer to the upper and lower chamber of the electrophoresis tank.
10. Use a syringe to rinse the wells and ensure that no polyacrylamide remains in the wells.
11. Switch on the current and prerun the gel for 30 min at 180 V (11 V/cm) to remove excess of nonpolymerized reagents.

3.2.4 Sample Preparation and Gel Electrophoresis

1. Mix 50 µL aliquots of the labeled PCR reaction mix (*see* Subheading 3.1) with 5 µL of 10x retardation loading dye mix. Centrifuge briefly.
2. Switch off the current and use long, thin pipette tips to load the samples in the wells. Avoid blowing air bubbles in the samples.
3. Switch on the current and let migrate at 180 V for 15 min to allow fast penetration of the material in the gel matrix.
4. Then reduce the current to 150 V and continue the migration to obtain a good separation, depending on the length of the labeled amplicon, using the dyes as a guide. On a 6% polyacrylamide gel, bromophenol blue (the fastest migrating dye) co-migrates with an approximately 58 bp fragment (45 bp on 8%, 33 bp on 10%), and xylene cyanol with a 230 bp (160 bp on 8%, 117 bp on 10%).

3.2.5 Imaging by Autoradiography

1. Stop the electrophoresis.
2. Remove the assembly from the electrophoresis tank and wipe the glass plates dry. Make sure to remove all running buffer from the wells before disassembling the glass plates (aspire with a Pasteur pipette or shake heavily above the sink with the wells turned downward).
3. Put the assembly on a flat surface (behind a plexiglas shield) with the large siliconized front plate on top.

4. Use a spacer or spatula as a lever to carefully pry the plates apart and remove the top plate. Avoid distorting the gel and remove the spacers.
5. Cover the gel still staying on the small back plate with a plastic film (fold back on the edges to avoid disturbance upon later manipulations) to avoid any direct contact of the radioactive material in the gel with the X-ray-sensitive film.
6. Put a sheet of paper or cardboard on top of the Saran Wrap (to avoid that the latter would stick later to the front plate) and cover with the front plate to protect from radiation.
7. Move the assembly to the darkroom.
8. In the dark (red light) remove the front plate and the paper or cardboard.
9. Cut a piece of X-ray-sensitive film sufficiently large to cover the area of interest and deposit such that the film perfectly aligns with the bottom of the back plate and slightly sticks out on at least one side and the top of the back plate.
10. Place the front plate on top of the X-ray sensitive film and fix the whole together with clamps.
11. Use the tip of scissors to scratch the contour of the small back plate on the film and indicate the orientation by making a small incision at one of the top corners of the film. This will allow a perfect realignment of the film and gel after development.
12. Expose for 10 min, then remove the X-ray-sensitive film from the assembly.
13. Pour a sufficient volume of 5x diluted developer to cover the entire surface of the film in a tray and develop the film for 2–3 min (gently shake the tray regularly).
14. Abundantly rinse the film with tap water.
15. Then move it to a second tray with 5x diluted fixer. Allow to stand with gentle shaking for a few minutes.
16. Rinse abundantly with tap water, then with demineralized water.
17. Suspend the film with a clamp and let air dry.
18. In the meantime, remove the running buffer from the electrophoresis tank. Be aware that the bottom chamber may be heavily contaminated with [γ-^{32}P]-ATP that may have migrated out of the gel, depending on the electrophoresis conditions. Verify with the Geiger counter. Rinse the electrophoresis apparatus abundantly with tap water and verify for potential remaining contamination with the Geiger counter.

3.2.6 Recovery of Labeled DNA from the Polyacrylamide Gel Matrix

1. Prepare one assembled elution set-up for every DNA band that has to be recovered (*see* Fig. 1c). Start by inserting a small amount of sterile glass wool in a sterile 1 mL pipette tip and very gently push (don't press, *see* **Note 7**) the glass wool toward the bottom of the tip with a Pasteur pipette to form a plug that will retain the polyacrylamide particles but allows buffer to flow through. Then seal the tip of the 1 mL pipette tip with self-sealing tape and insert the lower part in a sterile glass tube.
2. Work behind a plexiglas shield and deposit the autoradiography on a piece of benchcoat. Take the orientation into account and put the gel with the front plate on top of the autoradiography. Align the small back plate with the marked contour on the autoradiography and remove the front plate.
3. Cut and remove the plastic with a scalpel.
4. Use the scalpel to cut out a slice of acrylamide corresponding to the band of labeled DNA on the autoradiography. Keep the size of the gel slice as small as possible.
5. Use the scalpel to transfer the gel slice to the 1 mL pipette tip and use a Pasteur pipette to push it further down without touching the glass wool plug.
6. Crush the polyacrylamide between the wall of the 1 mL pipette tip and the Pasteur pipette by making a rotating movement (be careful not to touch or displace the plug; do it gently to avoid breaking the tip of the Pasteur pipette) till the slice has been transformed into small bead-like particles (this will improve the recovery by facilitating diffusion of the DNA) that stick to the Pasteur pipette.
7. Let 200 μL of elution mix flow down along the tip of the Pasteur pipette to collect the beads at the bottom of the 1 mL pipette tip, on top of the plug.
8. Repeat this operation to collect all acrylamide beads. Use a total of 400–600 μL elution mix per gel slice for elution, depending on the size of the gel slice. Make sure that all the acrylamide is submerged in the elution buffer. Use the Pasteur pipette to push all polyacrylamide in the liquid and to remove potential air bubbles in the solution.
9. Seal the top of the 1 mL pipette tip and the connection between the pipette tip and the glass tube with self-sealing tape.
10. Put the assembly in a floater and incubate the whole in a water bath at 45 °C, overnight.
11. Remove the tube from the water bath after overnight incubation.
12. Wipe dry, remove the self-sealing tape from the connection between the pipette tip and the glass tube. Verify that no water

from the bath has penetrated the glass tube (otherwise replace with a clean tube).

13. Carefully remove the self-sealing tape at the tip of the 1 mL pipette tip and immediately insert it in the glass tube (*see* **Note 8**).
14. Subsequently, remove the self-sealing tape at the top of the 1 mL pipette tip and collect the liquid (containing labeled DNA) that will flow through the glass wool plug, whereas the polyacrylamide particles are retained.
15. When no more liquid comes out of the tip, add another 100 μL of elution mix to the 1 mL pipette tip to rinse the polyacrylamide and glass wool. Collect the flow through in the same glass tube.
16. Transfer the collected liquid to a clean sterile 1.5 mL Eppendorf tube, add 3 μL yeast-tRNA (5 mg/mL) and 1 mL ice-cold absolute ethanol (*see* **Note 9**).
17. Mix well by inverting the tube a couple of times, and incubate at −20 °C for at least 1 h.
18. Centrifuge for 15 min at maximum speed in a microcentrifuge to pellet the precipitated DNA. Immediately remove the supernatant with a drawn-out Pasteur pipette (use a different one for every DNA fragment).
19. Add 1 mL of 80% ice-cold ethanol to the pellet, spin 3 min at maximum speed, remove the supernatant.
20. Dry the pellet for a few minutes in the vacuum centrifuge.

3.3 Optimization of Protein–DNA Complex Formation by EMSA

Before proceeding to the in-gel footprinting experiment, protein-binding conditions should be optimized (protein concentrations, buffer composition, amount of nonspecific competitor DNA, ligand concentration, etc.). This can be done by EMSA as explained below for protein concentrations to determine the apparent equilibrium dissociation constant (*see* **Notes 3** and **10**).

Notice that for a correct K_D determination, the EMSA should be performed at low DNA concentration (below the K_D of the interaction) and that a sufficiently wide range of protein concentrations should be used. Inversely, to determine the stoichiometry of the interaction, the binding assay should be performed at high DNA concentration (significantly above the K_D of the interaction to immediately titrate all protein added) and with various protein concentrations, ranging from below to far above the stoichiometric equivalent, in order to reach saturation and a stable plateau in the graphical representation (a rectangular hyperbole). For examples of both K_D and stoichiometry determinations by EMSA, see [9, 20].

1. Start by preparing and installing a nondenaturing polyacrylamide gel in the electrophoresis tank as described above (*see* Subheading 3.2.3).
2. Mix in sterile 1.5 mL Eppendorf tubes a constant amount of single-end-labeled DNA (10,000–15,000 cpm) with 1.0 μL of nonlabeled, nonspecific competitor DNA (*see* **Note 3**) and various amounts of protein (always include a reference without protein) in binding buffer (*see* **Note 10**) and adjust the volumes to 20 μL (*see* **Note 11**).
3. Mix gently to homogenize the samples (don't vortex as it may denature the protein), centrifuge very briefly (few seconds) to collect all material at the bottom of the tube.
4. Incubate the samples for 25–30 min at an appropriate temperature (*see* **Note 12**). At this point, also start the prerun of the gel (30 min at 180 V, 11 V/cm).
5. Remove the samples from the water bath.
6. Add 2.0 μL of retardation loading dye mix to each tube, homogenize gently, and centrifuge briefly.
7. Stop the prerun and load the samples on the gel.
8. Switch on the current at 180 V for 15 min allowing fast penetration of the material in the gel (*see* **Note 11**), then reduce the current to 130 V (8 V/cm) to avoid heating of the gel.
9. Let it migrate for 2–3 h (*see* **Note 13**).
10. Stop the electrophoresis and proceed to imaging by autoradiography as described above (*see* Subheading 3.2.5) but extend the exposure time to approximately 10–14 h (generally done overnight).

3.4 In-Gel Footprinting (see Fig. 2 for a Schematic Outline and Example)

The first step in the procedure consists of an EMSA, but here a much larger amount of labeled DNA per lane is required since the different DNA populations (free and bound DNA) have to be recovered from gel (*see* Fig. 2b), and the material corresponding to a single band on the EMSA gel will be distributed over as many bands as there are nucleotides in the DNA fragment on the denaturing gel (*see* Fig. 2c). Hence, it may even be necessary to perform identical reactions in duplicate or triplicate to recover enough material per molecular species (especially of minor protein–DNA forms) when the binding affinity is low and/or if the DNA is spread over multiple complexes with a different migration velocity. Also take into account that for DNA probes harboring multiple binding sites spread over a relatively large distance, the final gel electrophoresis under denaturing conditions may have to be performed in more than one condition (loading of the same footprinting reaction products on the same % of gel but with a different migration time or migration on gels with a different polyacrylamide concentration) (*see* also **Note 5**).

1. Start by preparing a nondenaturing polyacrylamide gel (*see* Subheading 3.2.3). Use a suited electrophoresis buffer composition and gel concentration, as determined in Subheading 3.3.
2. Perform an EMSA with 100,000–150,000 cpm per lane using a suited incubation temperature, binding buffer composition, and protein concentration, as determined in Subheading 3.3. Use at least one aliquot without protein (free DNA) and various protein concentrations aiming for at least 50% bound DNA. Adapt the migration time to ensure a good separation of all bands of interest, as determined in Subheading 3.3.
3. Stop the electrophoresis and remove the gel assembly from the electrophoresis apparatus.
4. Remove the large siliconized front plate as described in Subheading 3.2.5 points 3–4 but do not cover the gel and immediately proceed with the in-gel cleavage reactions prior to imaging by autoradiography (*see* **Note 14**). See Fig. 2a for a schematic outline of the experiment.
5. Move the gel still staying on the small back plate to a clean Pyrex tray with 200 mL of 10 mM Tris-HCl, pH 8.0. Make sure that the whole gel is submerged in the solution and work in a chemical fume hood.
6. Mix 1 mL freshly prepared 1, 10 phenanthroline solution with 1 mL freshly prepared $CuSO_4$ solution and 18 mL of ddH_2O.
7. Add this 20 mL mixture to the tray. Homogenize by gentle shaking of the tray. Allow the chemicals to penetrate the gel for 5 min.
8. Add 10 mL of freshly diluted 3 mercaptopropionic acid to the tray. Mix gently and allow the reaction to proceed for 10–20 min (*see* **Note 15**).
9. Stop the reaction by adding 20 mL of neocuproine solution. Mix by gentle shaking and allow to stand for 5 min.
10. Remove the liquid from the tray and abundantly rinse the gel with demineralized water (at least 5x with 500 mL). Be careful not to damage the gel, especially if it has detached from the back plate during the treatment.
11. Remove the gel laying on the back plate from the tray.
12. Wipe dry without touching the gel and cover the gel with a plastic film.
13. Proceed to imaging by autoradiography as described in Subheading 3.2.5 (points 5–17) but expose the X-ray-sensitive film for 2–3 h (*see* Fig. 2b for an example).

14. Recover all bands of interest from the gel matrix and precipitate the DNA as described in Subheading 3.2.6.
15. Dissolve the dry DNA pellets in formamide dye mix (use a vortex) or store the dry pellets at −20 °C.
16. Separate the reaction products by gel electrophoresis under denaturing conditions alongside chemical sequencing ladders (*see* Subheading 3.5) as described below (*see* Subheading 3.6).

3.5 Preparation of Chemical Sequencing Ladders (Maxam and Gilbert Reactions) [29]

When the sequence of the DNA probe is known, two chemical sequencing ladders with A+G and C+T reactions, respectively, are sufficient to unambiguously identify each band on the sequencing gel. Furthermore, as with combination of reactions, a band is generated for every nucleotide present in the DNA probe this also allows for an optimal alignment of the reaction products of the in-gel footprinting with the reference sequencing ladders (*see* Fig. 2c for an example).

3.5.1 A+G Sequencing Ladder

1. Use about 100,000 cpm (*see* **Note 5**) of single-end labeled DNA per reaction.
2. Add 2.5 μL of carrier DNA and adjust the volume to 10 μL with nuclease-free H_2O.
3. Add 10 μL citrate solution to the mix of labeled and carrier DNA, mix gently, spin briefly to collect all the material at the bottom of the tube.
4. Immediately incubate for 4 min at 80 °C (*see* **Note 16**).
5. Immediately put on ice.
6. Centrifuge briefly to collect the condensate from the cover.
7. Add 200 μL 0.3 M sodium acetate, 3 μL yeast-tRNA, and 1 mL ice-cold absolute ethanol.
8. Mix well by inverting the tube a couple of times and allow to precipitate at −20 °C for at least 1 h.
9. Spin down for 15 min at maximum speed in a microcentrifuge to pellet the precipitated DNA.
10. Immediately remove the supernatant with a drawn-out Pasteur pipette (use a different one for every reaction).
11. Add 1 mL of 80% ice-cold ethanol to the pellet and centrifuge 3 min at maximum speed.
12. Remove the supernatant and dry the pellet for a few minutes in the vacuum centrifuge.
13. Subsequently proceed to the cutting reaction of the DNA backbone with piperidine (*see* **Note 16**). Dissolve the dry pellet in 100 μL of 1 M piperidine (work in a chemical fume hood and use a vortex to ensure complete dissolution of the pellet, short DNA fragments are very resistant to mechanical shearing) and

centrifuge briefly to collect all the material at the bottom of the tube.

14. Incubate in a heating block at 90 °C for 30 min.
15. Centrifuge briefly to collect the condensate from the cover.
16. Open the tube and evaporate all the piperidine in a vacuum centrifuge.
17. Resuspend (vortex) the dry pellet in 40 μL nuclease-free H_2O.
18. Evaporate the water in the vacuum centrifuge.
19. Repeat **steps 17** and **18** another two times (*see* **Note 17**).
20. Dissolve the dry pellet in formamide dye mix (vortex).
21. Proceed to the gel electrophoresis under denaturing conditions (*see* Subheading 3.6) or store the dry pellet at −20 °C.

3.5.2 C+T Sequencing Ladder

1. Use about 100,000 cpm (*see* **Note 5**) of single-end labeled DNA per reaction.
2. Add 2.5 μL of carrier DNA and adjust the volume to 10 μL with nuclease-free H_2O.
3. Add 30 μL hydrazine (work in a chemical fume hood) to the mix of labeled and carrier DNA.
4. Mix and centrifuge briefly to collect all the material at the bottom of the tube.
5. Incubate for 10 min at 25 °C (*see* **Note 18**).
6. Proceed further as described above for the A+G ladder to stop the reaction, precipitate the modified DNA, and cleave the phosphodiester backbone at modified pyrimidines (*see* Subheading 3.5.1, **steps 5–21**).

3.6 Analysis of the Reaction Products by Gel Electrophoresis Under Denaturing Conditions

3.6.1 Prepare a Denaturing Acrylamide/Bis-Acrylamide Stock Solution of Suited Concentration.

1. For 500 mL of a 6% polyacrylamide gel: weigh 210 g urea in a 1 L glass beaker.
2. Add 100 mL 5x TBE electrophoresis buffer, 100 mL ddH_20 and 75 mL of 40% acylamide/*bis*-acrylamide solution (ratio 19/1) (*see* **Note 19**).
3. Dissolve the urea on a heating magnetic stirrer (endothermic reaction).
4. Transfer to a 500 mL cylinder when all the urea is dissolved and adjust the volume to 500 mL with ddH_20.
5. Filter on a Büchner funnel and degas for 10 min.
6. Transfer the solution to a glass bottle and store at 4 °C in the dark (can be kept for a few months).

3.6.2 Treatment and Assembly of the Glass Plates

1. Start by thoroughly cleaning the front and back glass plates with detergent, rinse abundantly with tap water, then with demineralized water, wipe dry with absorbing paper and degrease the inside face with technical grade ethanol. Ensure that the plates are extremely clean since the slightest contamination will result in the formation of air bubbles in the gel.
2. Treat the inside face of the large front plate with "sticky" solution to ensure tight adhesion of the gel during the fixing and drying procedures subsequent to gel electrophoresis. Pour a few mL of sticky solution on the plate, spread all over the plate with absorbing paper, and let it dry for at least a couple of hours (or overnight) (*see* **Note 20**).
3. Siliconize the inside face of the small back plate by spraying some repel solution, spread equally with absorbing paper, and let it evaporate a few minutes before assembling the plates with the spacers.
4. Assemble the small and large plates with 0.25 mm thick spacers (use a commercial rubber sealing system or seal the bottom and edges with tape and clamps) and deposit the assembly slightly tilted on the protected working surface (easily cleanable surface such as Teflon plate or disposable benchcoat).

3.6.3 Gel Preparation

1. Filter 60 mL of cold denaturing polyacrylamide stock solution (*see* Subheading 3.6.1) on a Büchner funnel and degas for 10 min.
2. Transfer the solution to a beaker.
3. Add 600 μL ammoniumpersulfate solution and 30 μL TEMED.
4. Mix by gentle swirling and use a 50 mL syringe to fill up the assembled glass plates. Let the solution flow down slowly but constantly (without interruption) from one side of the assembly and avoid trapping air bubbles.
5. Insert the comb (avoid trapping small air bubbles underneath the teeth of the comb by starting at one end and having some solution on top of the border of the small back plate).
6. Let it polymerize for at least 45 min.

3.6.4 Sample Preparation, Gel Electrophoresis Under Denaturing Conditions and Visualization by Autoradiography

1. Start by measuring the radioactivity of every dry pellet from the in-gel footprinting and sequencing reactions with a Geiger counter (or by Cerenkov counting in a beta counter).
2. Dissolve (use vortex) the pellets in an appropriate volume of formamide dye mix. Use as a guide so that 50,000 cpm will be required to obtain sufficiently strong signals after overnight exposure of the X-ray-sensitive film and that no more than 7 μL should be loaded in a well.

3. Transfer aliquots with an equal number of counts of all in-gel footprinting reactions and 50% thereof for the sequencing ladders to a sterile 1.5 mL Eppendorf tube and adjust all samples to the same number of counts in the same volume with formamide dye mix (*see* **Note 21**).
4. Remove the sealing and comb from the gel assembly. Remove all remaining acrylamide above the wells on the front plate and use a capillary (made from the tip of a Pasteur pipette) to clean the wells, if required. Clean the outer surface of the glass plates.
5. Install the assembly in an appropriate electrophoresis tank that allows for a homogenous distribution of the heat developed during the electrophoresis (metal back plate or buffer layer behind the back plate).
6. Fill the upper and lower chamber with 1x TBE buffer and use a syringe to rinse and remove air bubbles from the wells.
7. Connect to the power supply and perform a prerun of 30 min at constant power of 60 W.
8. Immediately prior to loading, heat the samples for 3 min at 90 °C in a heating block to completely denature the DNA strands and remove any secondary structures in single-stranded DNA to ensure that separation of the reaction products will only rely on the length.
9. Stop the prerun, use a syringe to rinse the wells, and load the samples.
10. Start the electrophoresis (constant power of 60 W). Take into account that bromophenol blue and xylene cyanol co-migrate with approximately 26 nt and 106 nt long fragments, respectively, on a 6% polyacrylamide gel (19 nt and 75 nt on 8%; 12 nt and 55 nt on 10%) to calculate the required migration distance for an optimal resolution of the bands in the zone of interest.
11. Stop the electrophoresis.
12. Remove the gel assembly from the electrophoresis tank and deposit it with the small back plate (siliconized) on top on a flat surface.
13. Use a small wedge to pry open the plates and remove the back plate and spacers.
14. Incubate the front plate with the gel sticking to it for 10 min in a plastic tray with 2 L of 10% acetic acid and shake regularly (work in a chemical fume hood).
15. Remove the plate from the tray, let a maximum of liquid drain off.
16. Wipe the back side of the plate dry with absorbing paper.
17. Put on a flat surface and remove liquid from the gel by rolling an absorbing paper roll over the gel (don't be afraid, the gel

should not detach or get damaged because it sticks very well to the plate if the "sticky" treatment has been performed correctly).

18. Use a small electric heating device blowing hot air to completely dry the gel that should become even thinner and very hard (blow from different sides to ensure homogenous drying). This will inhibit diffusion during exposure of the X-ray-sensitive film and generate sharp bands on the autoradiograph.
19. Let the plate cool down.
20. Cover the gel with plastic film (put it very flat, without wrinkles, fold back, and attach with tape on the back face of the plate) to avoid any direct contact of the radioactive material in the gel with the X-ray-sensitive film.
21. Cover with the back plate to protect from radiation and move to the dark room.
22. Proceed to imaging by autoradiography as described Subheading 3.2.5 but use a longer, appropriate exposure time. As a guide, take into account that about 50,000 cpm per lane are required to have sufficiently strong signals after overnight exposure. If the band intensity is not appropriate, a new exposure can be made for an adapted period of time. If the signals are very low, an intensifying screen placed on top of the X-ray-sensitive film and exposure at −80 °C can be used. This will increase the signal intensity by approximately a factor two.

4 Notes

1. The migration velocity (v) of a DNA fragment through a non-denaturing polyacrylamide gel is calculated as follows: $v = h^2.Q.E./L^2.f$ (with E: the electric field; Q: the effective charge; L: the contour length of the DNA; h: the end-to-end distance of the DNA; f: the friction coefficient of the gel).
2. Some alternative running buffers for EMSA:
3. 0.5x TBE: 45 mM Tris 45 mM boric acid, 1.25 mM EDTA
4. TE: 10 mM Tris, 1.0 mM EDTA
5. TAE: 40 mM Tris-acetate, 1 mM EDTA, pH 7.9
6. Tris-glycine: 50 mM Tris, 0.4 M glycine, 2.5 mM EDTA
7. Always use the same buffer in the preparation of the gel solution and as running buffer in the electrophoresis apparatus. Notice that the pH of the gel and running buffer may have a significant impact on the migration velocity and resolving

power of complexes, also dependent on the pI of the protein. As an example, see [30].

8. Different types of nonspecific competitors can be used in EMSA and other protein-DNA binding techniques, including commercially available sonicated salmon or herring sperm DNA, sonicated calf thymus DNA, synthetic polymers such as poly (dI:dC), or an unrelated DNA fragment. However, notice that the use of naturally occurring DNA better reflects the physiologic situation. Furthermore, synthetic polymers do not have exactly the same structure as naturally occurring DNA and hence are not truly equivalent. At any rate, do not use genomic DNA from a species that is too closely related and might bear binding sites resembling the specific target. The ideal concentration of a nonspecific competitor should be determined experimentally such that the binding specificity is enhanced without having a too strong negative effect on binding affinity. For examples, see [31, 32].
9. Depending on the research questions, EMSAs may be performed with a single highly purified DNA binding protein, mixes of highly purified proteins, partially purified protein preparations, or even nuclear or cell-free total protein extracts. In all instances, protein preparations and extracts should be devoid of nuclease activity that might digest the DNA probe.
10. Not all the material will be loaded in one well of the final gel electrophoresis under denaturing conditions, and thus smaller amounts can be used if the labeled material is limiting. However, notice that it is easier to dissolve the pellets in a larger volume and to load an aliquot only on gel. Another advantage of using a larger amount is that the remnants can be stored and used in another gel electrophoresis, if optimization of the running time or percentage acrylamide is required.
11. 5% or 6% polyacrylamide gels are commonly used for DNA fragments ranging from 100 bp to a few 100 bps. Notice that polyacrylamide concentrations below 5% are difficult to handle in the subsequent steps (autoradiography, recovery of DNA from the gel matrix). For short (<100 bp) to very short (few tens of bps) oligonucleotide duplexes, 8% or 10% gels are better suited. For large fragments (I kb up to a few kb), better use agarose gels.
12. Use a small amount of glass wool only (*see* Fig. 1c) and push it down gently to form a plug near the tip of the 1 mL pipette tip, but not at the very end as this will reduce the diameter of the plug and slow down the rate of the flow through. Don't compress the glass wool too hard as this will equally have a negative effect on the velocity and yield of the recovery.

13. Be careful. The liquid may come out quickly. Insert the pipette tip rapidly in the glass tube to avoid loss of radioactive material and contamination of gloves (wear a double pair) and working surface. Verify with the Geiger counter before proceeding further.

14. Yeast-tRNA is added to increase the total concentration of nucleic acids. This will improve the recovery of radiolabeled DNA since low concentrations of nucleic acids do not precipitate efficiently. tRNA is a small molecule that generally does not interfere with protein–DNA binding, as it is not labeled it will not be visible on the autoradiographs. DNA precipitation by dehydration requires about 0.3 M salt (the elution mix contains enough salt) and 2–3 volumes of ethanol per volume aqueous solution. Mix well by inverting the tube; otherwise, the aqueous solution will freeze at −20 °C and the DNA will not precipitate efficiently.

15. The binding buffer composition given here is commonly used, especially for transcription factors of the Lrp family with a relative high pI. Other proteins may require different salt concentrations and/or pH (solubility of the protein), concentrations of mono or divalent ions such as Ca^{2+}, Mg^{2+}, and the presence of a ligand. When the ligand is a charged molecule, it may be necessary to also include it in the gel composition and electrophoresis buffer to avoid dissociation of the complexes during the migration [14, 15]. Some other commonly used binding buffers are: 10 mM Tris-HCl pH 7.4, 250 mM KCl, 5 mM $MgCl_2$, 2.5 mM $CaCl_2$, 0.5 mM DTT, 10 mM L-arginine monochloride, 2.5 % glycerol (e.g., ArgR family transcription regulators with a wHTH DNA-binding motif); 25 mM Na_2HPO_4 pH 7.0, 150 mM NaCl, 0.1 mM EDTA, 2.0 mM $MgCl_2$, 1.0 mM DTT, 10% glycerol (for some OmpR-family response regulators), and 20 mM Tris pH 8, 150 mM NaCl, 1 mM TCEP (for some antitoxins of autoregulated TA systems).

16. Keeping the volume small and the penetration time short are important to obtain sharp bands on the autoradiograph. Fast penetration limits the time spent in the well where binding conditions are somewhat different from the ones in the test tube and limits the risk of dissociation.

17. Ideally complex formation is studied at the physiologic temperature. However, this may not always be possible in vitro due to, for instance, denaturation of short DNA duplexes at high temperature or the stability of the purified protein or protein–DNA complexes. Notice that in most instances binding affinity and stability of complexes will be improved at low

temperatures. The incubation time should at any rate be sufficiently long to reach the equilibrium of the binding reaction.

18. The migration should be sufficiently long to separate all molecular species present in the assay but not excessively long as this enhances the risk of complex dissociation, which will appear as smearing of the radiolabeled material between the position of free and bound DNA due to dissociation and concomitant acceleration of the DNA molecules at any time during the electrophoresis. This is certainly important for relatively unstable complexes with fast dissociation kinetics. If required, the electrophoresis can be performed at 4 °C, in an adapted electrophoresis tank or by placing the tank in a cold room.
19. In principle, it is also possible to first make an autoradiography to detect the position of the bands of interest, to cut out the cognate gel slices, and subsequently treat them with the chemical nuclease. However, this approach has the major disadvantage that dissociation of the complexes may occur during the entire exposure time of the X-ray-sensitive film. Therefore, it is not recommended.
20. The reaction time may have to be adapted in function of the length and the specific bp composition of the DNA probe. Statistically, a single cut per DNA molecule is ideal. Too long incubation times will result in too much cutting and very intense bands for short reaction products but very low intensity bands for longer ones. Inversely, too short reaction times will result in low intensity bands all along the gel but a very intense band of intact material.
21. Low pH results in spontaneous hydrolysis of the glycosidic bond at purines (but not pyrimidines), especially at high temperatures. The reaction conditions used here are such that statistically one purine residue is removed per DNA molecule [29]. This will generate a ladder of bands with increasing size upon subsequent cleavage of the phosphodiester bond at apurinic sites with piperidine (pH 11) at high temperature (90 °C) and gel electrophoresis under denaturing conditions.
22. It is important to remove all remnants of piperidine from the samples as it strongly affects the quality of the banding pattern upon gel electrophoresis under denaturing conditions.
23. Hydrazine disrupts the C5–C6 double bond of the pyrimidine ring. Subsequent treatment with piperidine results in cleavage of the phosphodiester bond at modified pyrimidine residues. Again, the reaction conditions are such that statistically one pyrimidine ring is modified per DNA molecule [29].
24. Depending on the position of the zone of interest with respect to the labeled end of the DNA probe, a 8% or 10% polyacrylamide gel might be better suited. Only the amount of

acrylamide/*bis*-acrylamide has to be adapted accordingly in the preparation of the denaturing stock solution and compensated for, by the volume of ddH_20. The concentrations of buffer and urea remain the same as for the 6% gel.

25. Always treat the same plate with "sticky" solution because once this has been done, it is difficult to remove it completely. Switching between treatments of plates with "repel" and "sticky" upon repeated use of the plates increases the risk that the gel will get disrupted when prying the plates apart at the end of the electrophoresis. Once the autoradiograph has been made and the gel is no longer required, the plate can be reutilized. Completely submerge the plate for at least 1 h in a plastic tray with tap water to rehydrate and soften the gel. Use an ice scraper to remove the gel from the glass plate (collect with radioactive waste). Then thoroughly clean the plate with detergent (use a brush), abundantly rinse with tap water, then demineralized water, wipe dry, and finally degrease with ethanol.
26. It is important to adjust all samples to the same number of counts in the same volume since more liquid in the tube will result in more quenching of the radiation. Furthermore, inevitably some material (even though this is only a very small fraction) will be lost during sample loading. Loading identical volumes on gel homogenizes this loss over the different samples.

Acknowledgments

Work in the Microbiology Research Group has been financed by the Research Foundation Flanders (FWO-Vlaanderen), the Institute for Innovation by Science and Technology (IWT), and the Research Council of the Vrije Universiteit Brussel (OZR-VUB). IB is a postdoctoral fellow of the Research Foundation Flanders (Fonds Wetenschappelijk Onderzoek-Vlaanderen).

References

1. Dey B, Thukral S, Krishnan S, Chakrobarty M, Gupta S, Manghani C, Rani V (2012) DNA-protein interactions: methods for detection and analysis. Mol Cell Biochem 365: 279–299
2. Revzin A (1989) Gel electrophoresis assays for DNA-protein interactions. BioTechniques 7: 346–355
3. Calladine CR, Collins CM, Drew HR, Mott MR (1991) A study of electrophoretic mobility of DNA in agarose and polyacrylamide gels. J Mol Biol 221:981–1005
4. Carey J (1991) Gel retardation. Methods Enzymol 208:103–117
5. Lane D, Prentki P, Chandler M (1992) Use of gel retardation to analyze protein-nucleic acid interactions. Microbiol Rev 56:509–528
6. Hellman LM, Fried MG (2007) Electrophoretic mobility shift assay (EMSA) for detecting

protein-nucleic acid interactions. Nat Protoc 2: 1849–1861

7. Brunelle A, Schleif RF (1987) Missing contact probing of DNA-protein interactions. Proc Natl Acad Sci USA 84:6673–6676
8. Siebenlist U, Gilbert W (1980) Contacts between *Escherichia coli* RNA polymerase and an early promoter of phage T7. Proc Natl Acad Sci USA 77:122–126
9. Wang H, Glansdorff N, Charlier D (1998) The arginine repressor of *Escherichia coli* K-12 makes direct contacts to minor and major groove determinants of the operators. J Mol Biol 277:805–824
10. Nguyen Le Minh P, Bervoets I, Maes D, Charlier D (2010) The protein-DNA contacts in RutR•*carAB* operator complexes. Nucleic Acids Res 38:6286–6300
11. Fried MG, Crothers DM (1981) Equilibria and kinetics of Lac repressor-operator interactions by polyacrylamide gel electrophoresis. Nucleic Acids Res 9:6505–6525
12. Garner MM, Revzin A (1981) A gel electrophoresis method for quantifying the binding of proteins to specific DNA regions: application to components of the *Escherichia coli* lactose operon system. Nucleic Acids Res 9: 3047–3060
13. Thompson JF, Landy A (1988) Empirical estimation of protein-induced DNA-bending angles: applications to lambda site-specific recombination complexes. Nucleic Acids Res 20:9678–9705
14. Szwajkajzer D, Dai L, Fukayama JW, Abramczyk B, Fairman R, Carey J (2001) Quantitative analysis of DNA binding by the *Escherichia coli* arginine repressor. J Mol Biol 312:949–962
15. Charlier D, Roovers M, Van Vliet F, Boyen A, Cunin R, Nakamura Y, Glansdorff N, Piérard A (1992) Arginine regulon of *Escherichia coli* K-12. A study of repressor-operator interactions and of in vitro binding affinities versus in vivo repression. J Mol Biol 226:367–386
16. Peeters E, Thia-Toong T-L, Gigot D, Maes D, Charlier D (2004) Ss-LrpB, a novel Lrp-like regulator of *Sulfolobus solfataricus* P2, binds cooperatively to three conserved targets in its own control region. Mol Microbiol 54: 321–336
17. Peeters E, van Oeffelen L, Nadal M, Forterre P, Charlier D (2013) A thermodynamic model of the cooperative interaction between the archaeal transcription factor Ss-LrpB and its tripartite operator DNA. Gene 524:330–340
18. Song N, Nguyen Duc T, van Oeffelen L, Muyldermans S, Peeters E, Charlier D (2013) Expanded target and cofactor repertoire for the transcriptional activator LysM from *Sulfolobus*. Nucleic Acids Res 41:2932–2949
19. van Oeffelen L, Peeters E, Nguyen Le Minh P, Charlier D (2014) The 'densitometric image analysis software' and its application to determine stepwise equilibrium constants from electrophoretic mobility shift assays. PLoS One 9: e85146
20. Nguyen Le Minh P, Velázquez-Ruiz C, Vandermeeren S, Abwoyo P, Bervoets I, Charlier D (2018) Differential protein-DNA contacts for activation and repression by ArgP, a LysR-type (LTTR) transcriptional regulator in *Escherichia coli*. Microbiol Res 206:141–158
21. Galas DJ, Schmitz A (1978) DNase footprinting: a simple method for the detection of protein-DNA binding specificity. Nucleic Acids Res 5:3157–3170
22. Tullius TD, Dombroski BA (1986) Hydroxyl radical "footprinting": high resolution information about DNA-protein contacts and application to λ repressor and Cro protein. Proc Natl Acad Sci USA 83:5469–5473
23. Marshall LE, Graham DR, Reich KA, Sigman DS (1981) Cleavage of deoxuribonucleic acid by the 1,10-phenanthroline-cuprous complex. Hydrogen peroxide requirement and primary and secondary structure specificity. Biochemistry 20:244–250
24. Kuwabara MD, Sigman DS (1987) Footprinting DNA-protein complexes *in situ* following gel retardation assays using 1,10-phenanthroline-copper ion: *Escherichia coli* RNA polymerase-*lac* promoter complexes. Biochemistry 26: 7234–7238
25. Sigman DS, Kuwabara MD, Chen CH, Bruice TW (1991) Nuclease activity of 1,10-phenanthroline-copper in study of protein-DNA interactions. Methods Enzymol 208:414–433
26. Papavassiliou AG (1995) Chemical nucleases as probes for studying DNA-protein interactions. Biochem J 305:345–357
27. Veal JM, Bull Rill RL (1988) Sequence specificity of DNA cleavage by *bis*(1,10-phenanthroline)-copper(I). Biochemistry 27: 1822–1827
28. Spassky A, Sigman DS (1985) Nuclease activity of the 1, 10-phenanthroline-copper ion. Conformational analysis and footprinting of the *lac* operon. Biochemistry 24:8050–8056
29. Maxam AM, Gilbert W (1980) Sequencing end-labeled DNA with base-specific chemical cleavages. Methods Enzymol 65:499–560
30. Carey J, Lewis DE, Lavoie TA, Yang J (1991) How does trp repressor bind to its operator? J Biol Chem 266:24509–24513

31. Xu Y, Sun Y, Huysveld N, Gigot D, Glansdorff N, Charlier D (2003) Regulation of arginine biosynthesis in the psychropiezophilic bacterium *Moritella profunda*: in vivo repressibility and in vitro repressor-operator contact probing. J Mol Biol 326:353–369

32. Vassart A, Van Wolferen M, Orell A, Hong Y, Peeters E, Albers S-V, Charlier D (2013) Sa-Lrp from *Sulfolobus acidocaldarius* is a versatile, glutamine-responsive, and architectural transcriptional regulator. MicrobiologyOpen 2(1):75–93

Chapter 12

Chemical Protection and Premodification-Binding Interference for the Identification of Phosphate and Base-Specific Contacts in Protein–DNA Complexes

Daniel Charlier

Abstract

The specificity and strength of protein–DNA complexes rely on tight interactions between side- and main chain atoms of amino acid residues and phosphates, sugars, and base-specific groups. Various (in-gel) footprinting methods (for more information, *see* Chapter 11) allow the identification of the global-binding region but do not provide details on the contribution to complex formation of individual sequence-specific constituents of the DNA-binding site. Here, we describe how various chemicals can be used to randomly and sparingly modify specific bases or phosphates and allow the identification of those residues that are specifically protected against modification upon protein binding (protection studies) or interfere with complex formation when modified or removed prior to protein binding (premodification-binding interference). Each one of these complementary approaches has its advantages and shortcomings and results have to be interpreted with caution, having in mind the precise chemistry of the modification. However, used in combination, these methods provide an accurate and high-resolution image of the protein–DNA contacts.

Key words Phosphate ethylation, Purine methylation, Depurination, Depyrimidation, Thymine oxidation, Uracil substitution, Missing contact probing, Gel retardation, Denaturing gel electrophoresis, Protein–DNA interactions

1 Introduction

Most prokaryotic sequence-specific DNA-binding proteins, especially transcription factors (TFs) with a helix-turn-helix (HTH) DNA-binding motif, bind as dimers or higher oligomeric forms to semi-palindromic targets, frequently 15–40 bp long [1, 2]. The binding specificity of TFs with an HTH motif is primarily ensured by hydrogen bonding between amino acid residues and major groove determinants of the operator, and hydrophobic interactions with the C5-methyl group of thymine [3–6]. Indeed, both the width of the major groove (about 22 Å) and the distribution of the principal interacting groups are better suited for a protein to

Eveline Peeters and Indra Bervoets (eds.), *Prokaryotic Gene Regulation: Methods and Protocols*,
Methods in Molecular Biology, vol. 2516, https://doi.org/10.1007/978-1-0716-2413-5_12,

unambiguously distinguish the four different bps from the major rather than from the minor groove side, by making at least two hydrogen bonds. Contacts with backbone constituents, especially the nonesterified oxygen atoms of phosphate groups, are generally considered nonspecific. However, the predominant B-form of DNA is not perfectly regular but shows local sequence-dependent structural heterogeneity and conformability due to rotational freedom around numerous single covalent bonds, differences in base pair stacking and groove compressibility. As a consequence, local variations in the position of phosphates and the conformability of the backbone may nevertheless contribute to the specificity of the interaction in a process also called indirect read-out [7].

Within the limits of a protein--binding site, not every position contributes equally to the specificity and strength of the interaction. Chemical protection and especially premodification-binding interference techniques allow the identification of those phosphates and bases that are tightly contacted and contribute most significantly to the energy of complex formation. In protection-type studies, the reactivity for the chemical probe of a specific residue in close contact with the protein will generally be reduced (though there may be exceptions, see below: Subheading 1.1 Purine methylation) upon protein binding and be visible as a band of reduced intensity after specific cleavage of the DNA backbone at modified positions and separation of the reaction products by gel electrophoresis under denaturing conditions (*see* Fig. 1). In the premodification-binding interference approach, the DNA is randomly but sparingly (statistically one modification per DNA molecule) modified prior to incubation with the protein (*see* **Note 1**). The heterogeneous mixture of DNA molecules, modified at different positions, is subsequently separated on basis of the affinity for the protein by gel electrophoresis in native conditions (EMSA, Electrophoretic Mobility Shift Assay). Free and bound DNA populations are recovered from gel, cleaved at the modified positions (*see* **Note 2**), and the reaction products separated by gel electrophoresis under denaturing conditions (*see* Fig. 2). DNA molecules modified at a position directly involved in complex formation are expected to have a reduced affinity for the protein and accordingly to be underrepresented in the bound form and overrepresented in the free form, whereas molecules modified at positions that are not relevant for protein binding should be evenly distributed over the free and bound forms. For the same amount of DNA loaded in all lanes of the denaturing gel, bands corresponding to positions that upon modification negatively affect complex formation will show a reduced intensity in the lane corresponding to the bound DNA form and an enhanced intensity in the free form compared to the input DNA (mixture of all modified DNA molecules not incubated with protein), whereas bands corresponding to positions that do not contribute significantly to complex formation should have the

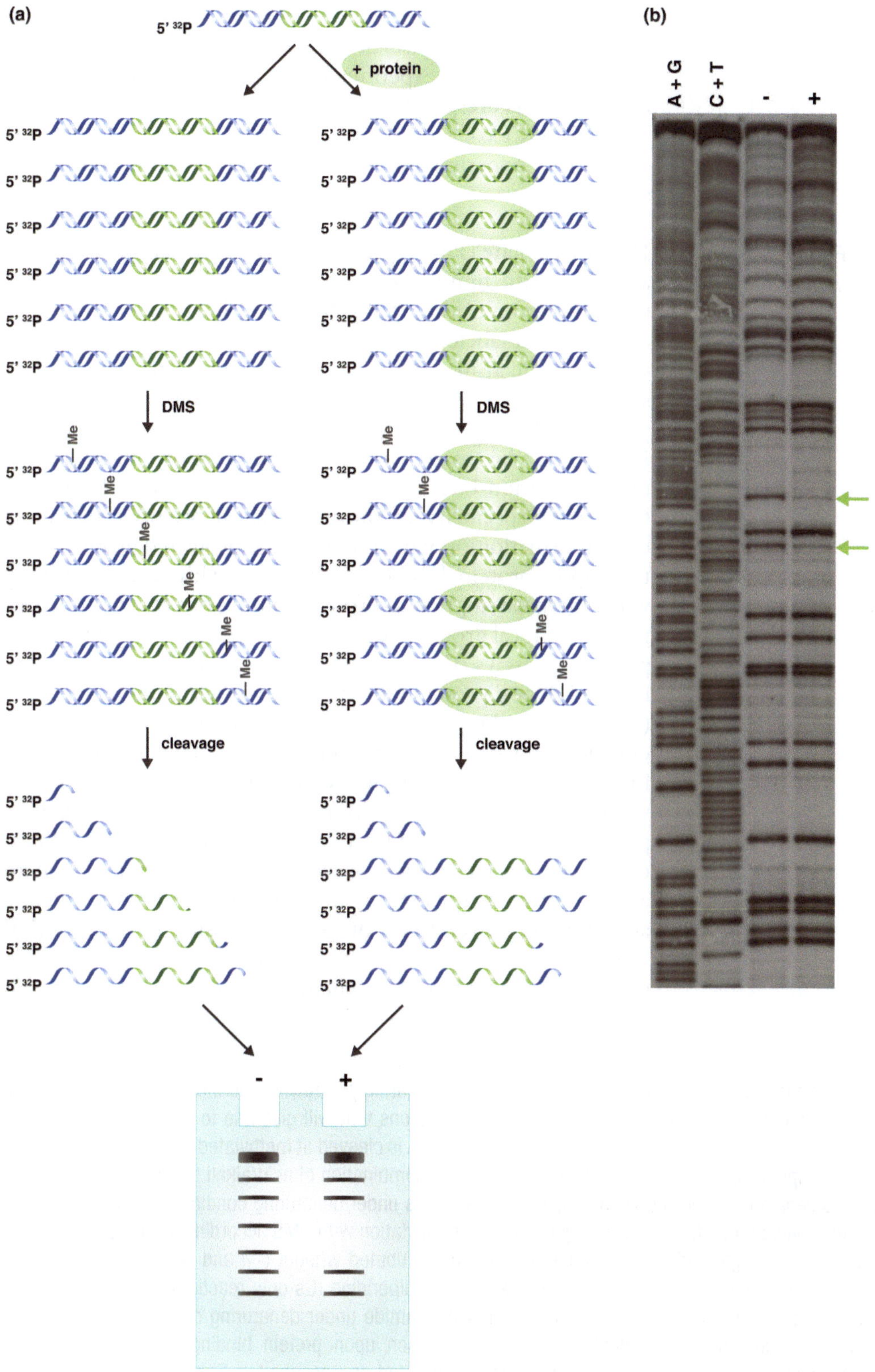

Fig. 1 Methylation protection assay. (**a**) Schematic representation of the methylation protection assay. [5′-^{32}P] single-end labeled operator DNA is incubated without and with (+) protein to allow complex formation and

same intensity in all three lanes (*see* Fig. 2). Several DNA modification reactions, their use in protection and/or premodification-binding interference experiments and guidelines for the interpretation of results are provided below (*see* Subheadings 1.1, 1.2, 1.3, 1.4, 1.5, and 1.6). All experiments require single-end labeled DNA molecules. Experimental details are provided in the Methods section.

1.1 Purine Methylation

Dimethyl sulfate (DMS) predominantly methylates the N3 atom of adenine on the minor groove side and the N7 atom of guanine on the major groove side of the DNA helix [8, 9]. Sparing methylation of single-end labeled DNA molecules and subsequent cleavage of the backbone at methylated residues results in the generation of a series of labeled DNA fragments with variable length that can be separated by gel electrophoresis under denaturing conditions. DMS can be used in both methylation protection and premodification-binding interference techniques. Furthermore, the groove-specificity of the methylation reactions provides not only information on which residues are protected against methylation upon protein binding or interfere with complex formation when methylated prior to protein binding but also furnishes information on local groove specificity of the interaction. In methylation protection assays, the reactivity of specific purine residues will be reduced if local protein binding reduces their accessibility for DMS (*see* Fig. 1). Occasionally, however, a purine may exhibit an enhanced reactivity upon protein binding. This may be due to an increased protein-binding-dependent local hydrophobicity in the vicinity of the base or to a local distortion of the DNA geometry, providing better accessibility to the methylating agent. Methylation of purine rings has several consequences: (i) it adds a methyl group and a positive charge and (ii) eliminates a potential hydrogen bond acceptor. Negative effects on protein binding in premethylation-binding interference assays may therefore be due to steric exclusion of protein binding or to a change in the resonance state of the

Fig. 1 (continued) subsequently treated with DMS to sparingly methylate purines (only methylation on the labeled strand is shown as these are the only modifications that will give rise to labeled cleavage products visible on the autoradiograph). After precipitation, the DNA is cleaved at methylated residues (at methylated Gs only with piperidine, or at methylated As and Gs with a combination of acid/alkali treatment) and the reaction products separated by polyacrylamide gel electrophoresis under denaturing conditions. (**b**) Autoradiograph of a protein-mediated protection assay against purine methylation with DMS according to the scheme presented in (**a**). [5′-^{32}P]-single-end labeled operator DNA was incubated without (−) and with (+) protein, sparingly methylated with DMS, and the backbone cleaved with piperidine (Gs only reaction). The reaction products were separated by gel electrophoresis on 8% polyacrylamide under denaturing conditions. Arrows indicate guanine residues that are protected against methylation upon protein binding. A+G and C+T are the corresponding chemical sequencing ladders [16] generated as described in Subheadings 3.4.2 and 3.4.3, respectively

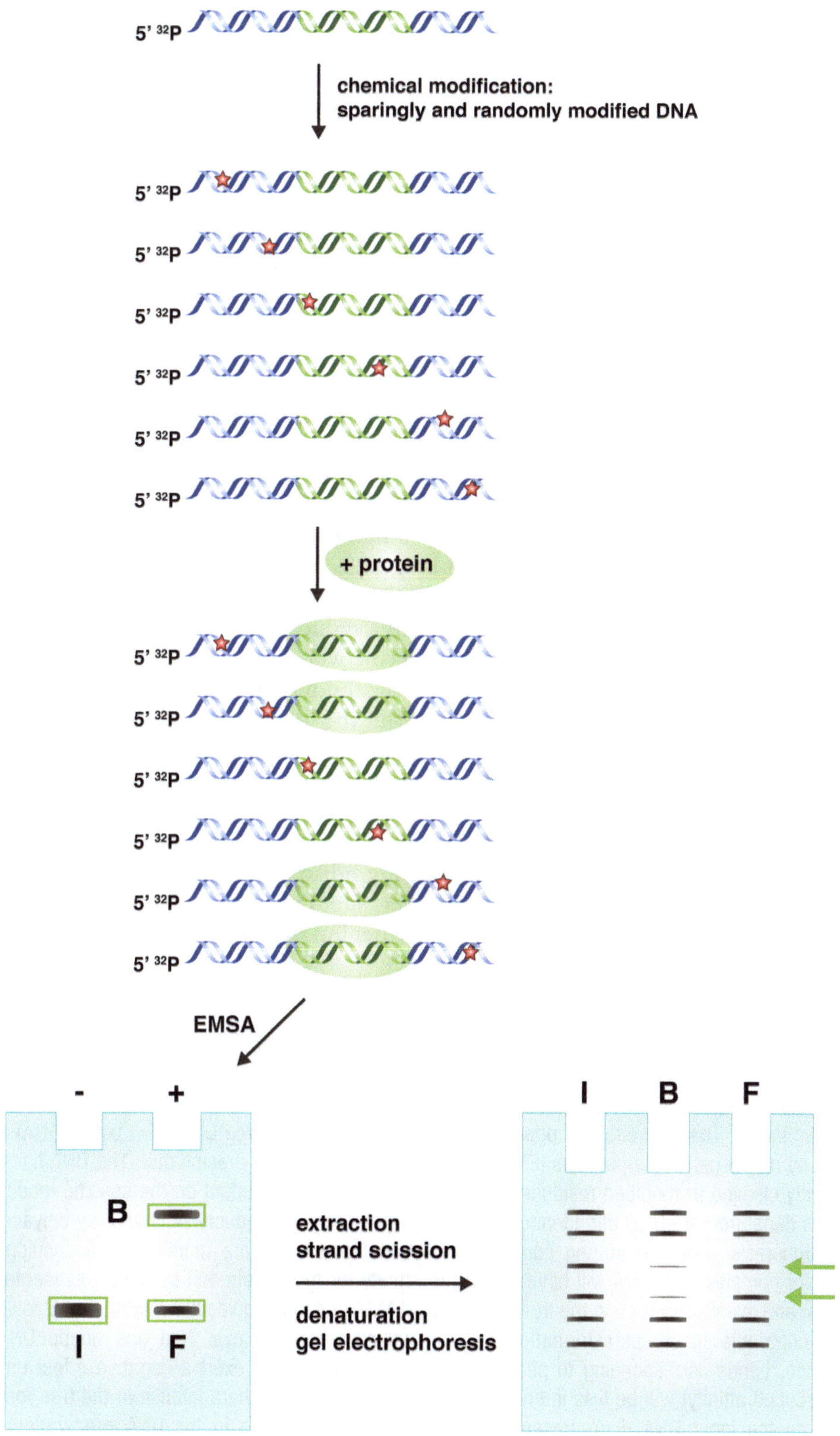

Fig. 2 Schematic representation of the premodification-binding interference assay. [5′-^{32}P] single-end labeled DNA is sparingly (statistically one modification per DNA molecule) and randomly modified by chemical

modified purine ring that may no longer be recognized by the protein. Hence, methylation protection and premethylation-binding interference will not provide exactly the same information but when both approaches point to the same residue, it is very likely that the latter directly interacts with the protein. Further confirmation can be sought by depurination-binding interference (*see* Subheading 1.2). For examples and comparisons of the different approaches targeting purines and cleavage at either G only or A and G (*see* Figs. 1b and 3a) and [10–15].

1.2 Depurination and Depyrimidation

The glycosidic bond at purine residues can be specifically hydrolyzed by incubation in acidic conditions (pH 4.0) at high temperature (80 °C) [16]. Pyrimidines can be specifically modified with hydrazine that results in base removal and opening of the deoxyribose ring [16, 17]. Sparingly modified DNA molecules can then be separated on basis of their affinity for the protein in an EMSA, free and bound forms recovered from gel, cleaved at abasic positions by piperidine treatment at high temperature (90 °C), and the reaction products separated by gel electrophoresis under denaturing conditions. Generally, the removal of a base that is directly contacted by the protein will result in a reduced affinity for the protein. Hence, the cognate DNA molecule will be underrepresented in the bound form of the EMSA and the residue subsequently identified as a band of reduced intensity on the denaturing gel. For examples of negative-binding interference effects, (*see* Fig. 3b, c) and [11, 14, 18]. Occasionally, however, a local inverse effect (higher binding affinity upon base removal) may be observed. This generally occurs at positions that are not directly contacted by the protein, such as linker regions connecting adjacent binding sites or at a position within a specific target site that deviates from the consensus sequence. The rationale behind these observations is the following.

Fig. 2 (continued) treatment (represented with a star on the DNA, only modifications on the labeled strand are shown since these are the only ones that will after strand scission result in fragments visible on the autoradiograph). Aliquots of the modified DNA are then incubated without (−) and with (+) protein to obtain about 50% binding. The different DNA populations (I = input DNA, F = free DNA, B = bound DNA) are then separated by native gel electrophoresis (EMSA), recovered from gel, and precipitated. The DNA backbone is subsequently cleaved at modified residues by chemical treatment (dependent on the specific modification), the strands denatured at 90 °C and identical amounts of the reaction products separated by polyacrylamide gel electrophoresis under denaturing conditions. DNA molecules that are modified at a position that is important for complex formation will have a reduced affinity for the protein and be underrepresented in the bound form and overrepresented in the free form on the EMSA, whereas molecules modified at a position that does not contribute to complex formation will be evenly distributed over free and bound DNA. As a consequence, bands corresponding to positions that upon modification exert a negative effect on protein binding (reduced affinity) will be less intense in the bound form (B) and more intense in the free form (F) as compared to the input DNA (I) (representing all possible modifications in the DNA preparation) on the autoradiograph of the denaturing gel (indicated with an arrow)

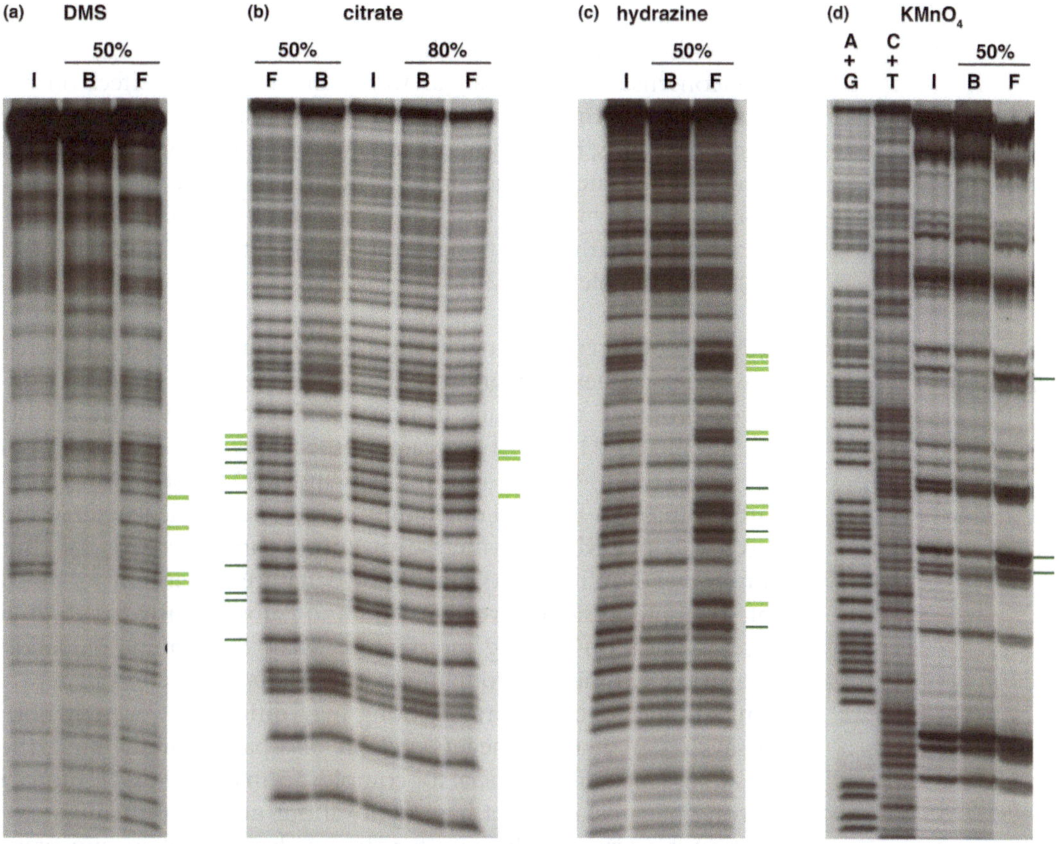

Fig. 3 Examples of chemical premodification-binding interference assays with different operators and DNA-binding transcription factors with a HTH DNA-binding motif. DNA was treated with (**a**) DMS to methylate purines, (**b**) citrate at low pH and high temperature for depurination, (**c**) hydrazine to modify specifically pyrimidines, (**d**) potassium permanganate to oxidize thymines. Sparingly modified DNA was incubated with protein (aiming at 50% or 80% binding, as indicated), and the different DNA populations separated on basis of their affinity for the protein in an EMSA. Input DNA (I, not incubated with protein), free (F), and bound (B) DNA populations were recovered from gel, precipitated, and the backbone cleaved at modified positions with piperidine. Finally, the reaction products were separated by gel electrophoresis on 8% polyacrylamide under denaturing conditions. A+G and C+T are the corresponding chemical sequencing ladders [16] generated as described in Subheadings 3.4.2 and 3.4.3, respectively. Small horizontal bars represent residues that upon modification or removal negatively interfere strongly (light green-colored symbol) or moderately (dark green-colored symbol) with complex formation. In (**b**) the 80% binding situation clearly reveals the three residues that are most critical for complex formation

Linker regions connecting two or more binding sites are generally not directly contacted by the protein but since they fix the distance between adjacent binding sites and frequently undergo deformations such as bending or wrapping, they may nevertheless contribute to the specificity and strength of complex formation. DNA deformations cost energy; base removal enhances local flexibility and hence reduces the energetic cost of DNA deformation. For examples, see [12, 19]. When a particular bp deviates from the consensus sequence, it may happen that not only a contact that is otherwise established with the consensus base is missing, but the nonconsensus residue may as well exert a negative effect on the interaction with a nearby residue (steric hindrance or repulsive forces). Hence, removal of the nonconsensus base will only benefit the binding energy since it adds no negative effect on complex formation (anyway the base can't be directly contacted) but eliminates the negative effect on the establishment of (a) nearby contact (s). For examples, see [20].

1.3 Thymine Oxidation

$KMnO_4$ preferentially attacks the C5–C6 double bond of unpaired thymine residues. Oxidation generates carboxylic acid and/or aldehyde products and results in ring opening but leaves the methyl group intact [21, 22]. This reaction specificity for unpaired Ts can be exploited to reveal local single-stranded DNA stretches generated by protein-induced unwinding as in, for example, bacterial and archaeal open RNA polymerase–promoter complexes [23, 24]. But linear ds DNA fragments can also be denatured, sparingly modified at thymine residues, reannealed, and subsequently used in a premodification-binding interference assay. As thymine oxidation leaves the C5-methyl group intact, the results gathered from depyrimidation and oxidation reactions may differ. However, if both assays reveal a strong negative effect of a particular thymine, the residue is most likely crucial for complex formation. Furthermore, a strong negative effect of base removal but not upon thymine oxidation is a strong indication for the involvement of an important local hydrophobic contact with the C5-methyl group of thymine in the major groove segment of the binding site, a hypothesis that can be further evaluated by uracil substitution of thymine residues (*see* Subheading 1.5). For examples of thymine oxidation-binding interference, (*see* Fig. 3d) and [11, 13, 25].

1.4 Cytosine Modification with Hydroxylamine Hydrochloride

At pH 6.0, $NH_2OH.HCl$ preferentially adds across the C5–C6 double bond of unpaired cytosine residues [21]. The fragilized cytosine ring undergoes further internal rearrangements, the reaction ends up in ring opening, and in a later step, the backbone can be cleaved with piperidine. As explained above for other chemical modifications, the sparingly modified DNA can be used in a premodification-binding interference experiment to reveal the importance of particular cytosine residues for protein binding. For an example, see [12].

1.5 Uracil Substitution

Crystallographic and other studies have shown that hydrophobic interactions with the C5-methyl group present on the major groove side of thymine residues are frequently involved in protein–DNA complex formation. Those critical methyl groups can be identified in premodification-binding interference assays with deoxyuride substituted DNA [26]. Uracil that lacks this C5-methyl group but is otherwise identical to thymine is not a normal constituent of DNA, but deoxyuridine can be occasionally and randomly incorporated upon PCR amplification in the presence of a mixture of dTTP and a lower concentration of dUTP. The mixture of random but sparingly substituted single-end-labeled DNA can then be used in a mobility shift assay to separate bound from free DNA molecules on the basis of their affinity for the protein. Input, free, and bound DNA are subsequently recovered from gel, treated with uracil-*N*-glycosylase to specifically hydrolyze the glycosidic bond at incorporated deoxyuridine residues, the phosphodiester bond cleaved at abasic positions with piperidine, and the reaction products separated by gel electrophoresis under denaturing conditions. As in the other premodification-binding interference techniques, molecules with a uracil substitution of thymine residues of which the methyl group contributes significantly to the energy of complex formation are expected to exhibit a reduced affinity for the protein and will be underrepresented in the bound form and overrepresented in the free form with respect to the input DNA. Hence, bands corresponding to such thymine residues will have a lower intensity in the lane corresponding to the bound form on the autoradiograph of the denaturing gel. For examples, see [11, 27].

1.6 Phosphate Ethylation

Phosphates can be randomly alkylated with *N*-ethyl-nitrosourea and the backbone specifically cleaved at ethylated phosphates by alkaline hydrolysis at high temperature that generates two kinds of reaction products, with either a 3'-OH or 3'-ethylphosphate group [9]. Local triester formation of the nonesterified oxygen atoms not only adds an ethyl group but also eliminates a negative charge. As a consequence, negative-binding interference effects of phosphate pre-ethylation on protein binding may be due to local steric hindrance or to the loss of an electrostatic interaction. Information gathered from a phosphate ethylation-binding interference assay is therefore complementary to footprinting with the hydroxyl radical [28] and in-gel footprinting with the copper-phenanthroline ion (*see* Chapter 11), which primarily attack the H1' position of the deoxyribose ring, another backbone constituent, and essentially reveals sugar moieties that are protected against the radical upon protein binding. For examples of phosphate ethylation-binding interference, see [11, 14, 29, 30].

2 Materials

2.1 Safety Measures

Many chemicals, including [γ-^{32}P]-ATP, *N*-ethyl-nitrosourea, DMS, $KMnO_4$, acrylamide/*bis*-acrylamide, ammoniumpersulfate, TEMED, EDTA, NaOH, HCl, hydrazine, and piperidine, are potentially carcinogenic, toxic, neurotoxic, irritant, and/or flammable. Avoid inhalation and all contact with the skin. Always wear gloves and protective clothing and work in a fume hood when necessary. Always consult the accompanying safety sheets prior to handling a chemical.

[^{32}P] is a high energy β-ray emitter with a half-life of 14 days. To limit exposure to radiation, all handlings with [^{32}P]-labeled material should be performed behind Plexiglas shields and samples transported and stored in Plexiglas boxes. Take into account that the radiation intensity is inversely proportional to the square of the distance from the source. Cover working surfaces with benchcoat (absorbing face on top), collect and store contaminated liquid and solid waste in separate authorized containers, regularly verify gloves, working areas, and equipment for potential spill and contamination with a Geiger counter and decontaminate as soon as possible. Working exclusively in a dedicated and specifically equipped room with dedicated racks, pipette sets, etc., reserved for manipulation of isotopes, is highly recommended. At any rate, authorization to manipulate nonencapsulated radionuclides should be obtained from the institutional health and safety office and its instructions/guidelines strictly applied. This generally includes wearing of a personal dosimeter to monitor cumulative exposure to radiation.

2.2 Preparation of [5′-^{32}P] Single-End Labeled DNA Fragments

2.2.1 Solutions, Enzymes, Chemicals

1. Freshly delivered [γ-^{32}P]-ATP (3000 Ci/mmol). Take into account that the half-life of [^{32}P] is 14 days. Store at −20 °C but make sure that the solution is completely thawed and homogenous before removing an aliquot.
2. T4 polynucleotide kinase (10 U/μl) (store at −20 °C, use on ice).
3. 10x commercial kinase buffer (store at −20 °C, thaw on ice).
4. Nuclease-free H_20.
5. Probe-specific oligonucleotides (dissolved in nuclease-free H_20) as primers for PCR amplification. Make stock solutions at 500 pmol/μL, conserved at −20 °C for long-term storage, and working solutions at 10 pmol/μL.
6. Template DNA (genomic or plasmid DNA, PCR fragment) (store at −20 °C in small aliquots to avoid multiple cycles of freeze and thaw).
7. 2x commercial PCR amplification mix or separate heat-resistant DNA polymerase and dNTPs.

2.2.2 Other Materials and Equipment

1. 1.5 mL sterile Eppendorf tubes
2. 0.25 mL sterile PCR tubes
3. Sterile pipette tips
4. Heating thermostatic water baths or heating blocks at 37 °C and 65 °C

5. Microcentrifuge
6. Thermocycler

2.3 Native Polyacrylamide Gel Electrophoresis for DNA Purification and EMSA

2.3.1 Solutions and Chemicals

1. 5x Electrophoresis buffer TBE: 445 mM Tris-base, 445 mM boric acid, 12.5 mM EDTA. TBE is commonly used but other low ionic strength buffers are equally possible to use for the separation of free from bound DNA (*see* **Note 3**).
2. 100 mg/mL ammoniumpersulfate stock solution dissolved in ddH_20. This can be stored up to a few weeks at 4 °C.
3. *N*,*N*,*N*′,*N*′-tetramethylenediamine (TEMED). Store in dark at 4 °C.
4. 30% Acrylamide/*N*, *N*′-methylene *bis*-acrylamide solution (ratio 29:1). Store in dark at 4 °C.
5. ddH_2O.
6. Repel solution (e.g., Blue Slick spray, Serva, to siliconize the glass plates).

2.3.2 Other Materials and Equipment

1. Pair of glass plates (19.5 × 16.0 cm and 19.5 × 19.0 cm). Other dimensions are possible but take into account that they should be compatible with the dimensions of the electrophoresis tank and sufficiently long to allow good separation of all bound DNA forms without loss of the free DNA by migration out of the gel.
2. Pair of spacers (0.1 cm thick) and cognate 14 well comb
3. Commercial rubber sealing system for glass plate assembly (adapted to the size of glass plates) or tape and clamps
4. Syringe (50 mL)
5. Büchner funnel and rubber tubing to connect to the vacuum pump
6. Filter paper for Büchner funnel
7. Vacuum pump
8. Power supply and electric cabling
9. Electrophoresis tank for vertical gel electrophoresis

2.4 Autoradiography

2.4.1 Solutions

1. Developer for photography (5x diluted in H_20). Store at room temperature but protected from light in a brown bottle. Can be used repeatedly but refresh when becoming brownish colored.
2. Fixer for photography (5x diluted in H_20). Store at room temperature but protected from light. Can be used repeatedly but refresh when becoming slightly pink colored.

2.4.2 Other Materials and Infrastructure

1. Radioisotope and chemiluminescent imaging film
2. Plastic foil (e.g., Saran Wrap or cling film)
3. Black plastic film or black plastic bags (sufficiently large to pack the gel + film and protect from light)
4. Dark room equipped with red light bulbs and a sink with running tap water and demineralized water
5. Two trays slightly larger than the dimensions of the autoradiograph for developer and fixer solutions

2.5 Recovery of DNA from Polyacrylamide Gel

2.5.1 Solutions and Chemicals

1. Elution buffer: 0.5 M ammonium acetate, 0.01 M magnesium acetate, 0.1 mM EDTA, 0.1% SDS. Sterilize by autoclaving, store at room temperature.
2. Ice-cold absolute ethanol. Store at −20 °C.
3. 80% ice-cold ethanol: 80% absolute ethanol, 20% nuclease-free H_2O. Store at −20 °C.

2.5.2 Other Materials and Equipment

1. Sterile glass tubes (L = 10 cm, d = 0.8 cm)
2. Sterile 1 mL pipette tips
3. 150 mm Pasteur pipettes
4. Drawn-out 150 mm Pasteur pipettes with thin tip to allow removal of a maximum of supernatant without touching the DNA pellet upon DNA precipitation in Eppendorf tubes
5. Sterile glass wool. Sterilize by autoclaving
6. Self-sealing tape
7. Nuclease-free scalpel
8. Water bath at 45 °C
9. Microcentrifuge
10. Vacuumcentrifuge
11. Vortex

2.6 Denaturing Polyacrylamide Gel Electrophoresis

2.6.1 Solutions and Chemicals

1. 5x Electrophoresis buffer TBE: 445 mM Tris-base, 445 mM boric acid, 12.5 mM EDTA.
2. 100 mg/mL ammoniumpersulfate stock solution dissolved in ddH_20. This can be stored up to a few weeks at 4 °C.
3. *N*,*N*,*N′*,*N′*-tetramethylenediamine (TEMED). Store in dark at 4 °C.
4. 40% acrylamide/*N*, *N′*-methylene *bis*-acrylamide solution (ratio 19:1). Store in dark at 4 °C.
5. Urea.
6. ddH_2O.
7. Repel solution (e.g., Blue Slick spray, Serva).

8. Sticky solution: 48.5 mL ethanol, 1.5 mL 10% acetic acid, 150 μL γ-methacryloxypropyltrimethoxysilane.
9. 2 L of 10% acetic acid (can be used for five gels, then refresh).

2.6.2 Other Materials and Equipment

1. A pair of heat-resistant sequencing-type glass plates (33.5 × 39.5 cm and 33.5 × 41.5 cm). Other dimensions are possible, but take into account that they should be compatible with the dimensions of the electrophoresis tank.
2. Pair of spacers (0.25 mm thick) and cognate 20 well comb.
3. Commercial rubber sealing system for plate assembly (adapted to the size of glass plates) or tape and clamps.
4. 50 mL syringe.
5. Büchner funnel and rubber tubing to connect to vacuum pump.
6. Filter paper for Büchner funnel.
7. Vacuum pump.
8. Power supply allowing electrophoresis at a constant power of at least 50 W (±1 500 V).
9. Electrophoresis tank for sequencing-type gel electrophoresis (ensuring homogenous heat distribution during migration).
10. Plastic tray slightly larger than dimensions of the glass plates.
11. Small electric heating system to blow hot air.
12. Heating magnetic stirrer

2.7 Methylation Protection

2.7.1 Products and Solutions

1. Purified [5′-^{32}P] single-end labeled DNA (about 100,000 cpm per lane) (*see* **Note 4**).
2. Purified protein.
3. Nonspecific competitor DNA: 0.5 mg/mL sonicated salmon or herring sperm DNA in nuclease-free H_2O (*see* **Note 5**).
4. Nuclease-free H_2O.
5. Dimethyl sulfate (DMS), analytical grade.
6. DMS reaction buffer: 50 mM sodium cacodylate pH 8.0, 10 mM $MgCl_2$, 1.0 mM EDTA.
7. DMS stop solution: 1.5 M sodium acetate pH 7.0, 1.0 M 2-mercaptoethanol.
8. 5 mg/mL yeast-tRNA dissolved in nuclease-free H_2O. Store stock solution at −20 °C and small working aliquots at 4 °C.
9. Ice-cold absolute ethanol. Store at −20 °C.
10. 80% ice-cold ethanol: 80% absolute ethanol (v/v), 20% (v/v) nuclease-free H_2O. Store at −20 °C.

11. Freshly prepared 1.0 M piperidine. Dilute a 10 M commercial stock solution 10× in nuclease-free H_2O. First add H_2O to the Eppendorf tube, then add 10 M piperidine and pipette a couple of times up and down. Close the tube and mix thoroughly by vortexing.
12. Formamide dye mix: 100 mL deionized formamide, 0.03 g xylene cyanol, 0.03 g bromophenol blue, 20 mM EDTA.
13. A denaturing polyacrylamide gel of suited concentration (*see* Subheading 2.6).

2.7.2 Other Materials and Equipment

All materials and equipment for the preparation of a denaturing polyacrylamide gel, gel electrophoresis under denaturing conditions, and visualization by autoradiography (*see* Subheadings 2.4 and 2.6).

2.8 Chemical Modification Reactions for Premodification-Binding Interference Assays

2.8.1 Products and Solutions Common to All DNA Modification Reactions

1. Purified [5′-^{32}P] single-end labeled DNA (*see* **Notes 4** and **6**).
2. Carrier DNA: 0.5 mg/mL sonicated salmon or herring sperm DNA in nuclease-free H_2O.
3. Nuclease-free H_2O.
4. 5 mg/mL yeast-tRNA dissolved in nuclease-free H_2O. Store stock solution at -20 °C and small working aliquots at 4 °C.
5. Ice-cold absolute ethanol. Store at −20 °C.
6. 80% ice-cold ethanol: 80% (v/v) absolute ethanol, 20% (v/v) nuclease-free H_2O. Store at −20 °C.

2.8.2 Methylation of Purines with DMS

1. Dimethyl sulfate (DMS), analytical grade
2. DMS reaction buffer: 50 mM sodium cacodylate pH 8.0, 10 mM $MgCl_2$, 1.0 mM EDTA
3. DMS stop buffer: 1.5 M Na-acetate pH 7.0, 1.0 M 2-mercaptoethanol

2.8.3 Depurination at Low pH and High Temperature

1. Citrate solution: 50 mM Na-citrate, 0.1 mM EDTA, pH 4.0 (adjusted with NaOH)
2. 0.3 M Na-acetate

2.8.4 Depyrimidation with Hydrazine

1. Hydrazine (NH_2-NH_2)
2. 0.3 M Na-acetate

2.8.5 Thymine Oxidation with $KMnO_4$

1. 30 mM Tris-HCl pH 8.0
2. 0.25 mM $KMnO_4$ freshly prepared in nuclease-free H_2O
3. Stop solution: 1.5 M Na-acetate pH 7.0, 1.0 M 2-mercaptoethanol
4. Renaturation buffer: 10 mM Tris-HCl pH 8.0, 1.0 mM EDTA, 30 mM NaCl

2.8.6 Cytosine Modification with $NH_2OH.HCl$

1. 30 mM Tris-HCl pH 8.0
2. 4.02 M $NH_2OH.HCl$ pH 6.0 (adjusted with diethylamine)
3. Stop solution: 0.3 M Na-acetate, 0.1 mM EDTA
4. Renaturation buffer: 10 mM Tris-HCl pH 8.0, 1.0 mM EDTA, 30 mM NaCl

2.8.7 Uracil Substitution

1. Deoxyuridine triphosphate (dUTP)
2. Set of dATP, dGTP, dCTP, and dTTP
3. Taq DNA polymerase
4. Template DNA (genomic, plasmid, or PCR fragment)
5. Pair of forward and reverse oligonucleotide primers for PCR amplification: one 5′-end labeled and one nonlabeled
6. PCR buffer: 10 mM Tris-HCl pH 8.3, 50 mM KCl, 3.5 mM $MgCl_2$, 0.01% (w/v) gelatin
7. Uracil-*N*-glycosylase

2.8.8 Phosphate Ethylation with N-Ethyl-Nitrosourea

1. Ethylation buffer: 50 mM Na-cacodylate pH 8.0, 10 mM $MgCl_2$, 1.0 mM EDTA.
2. Saturated solution of ethylnitrosourea in ethanol. Add ethylnitrosourea to absolute ethanol till the powder no longer dissolves after thorough mixing.
3. 2.0 M ammonium acetate.
4. 10 mM Na_3PO_4 pH 7.0.
5. 1.0 M NaOH.
6. 1.0 M HCl.

2.9 EMSA

2.9.1 Solutions and Products

1. [5′-^{32}P] single-end labeled DNA (*see* Subheading 2.2)
2. Nonlabeled, nonspecific competitor DNA (e.g., 0.5 mg/mL sonicated salmon or herring sperm DNA) (*see* **Note 5**)
3. DNA-binding protein, preferably purified to electrophoretic homogeneity
4. Protein-DNA binding buffer: 20 mM Tris-HCl pH 8.0, 0.4 mM EDTA, 0.1 mM DTT, 50 mM NaCl, 1 mM $MgCl_2$ (*see* **Note** 7)
5. 10× retardation loading dye mix: 25% ficoll, 0.1% xylene cyanol, 0.1% bromophenol blue
6. All solutions and products for native gel electrophoresis (*see* Subheading 2.3)
7. All solutions for autoradiography (*see* Subheading 2.4)

2.9.2 Other Materials and Equipment

1. All materials and equipment for native gel electrophoresis (*see* Subheading 2.3)
2. All materials and equipment for autoradiography (*see* Subheading 2.4)
3. Heating thermostatic water bath

3 Methods

3.1 Synthesis of [5′-^{32}P] Single-End Labeled DNA (see Note 4)

1. Mix in a 1.5 mL sterile Eppendorf tube 60 pmol of oligonucleotide with 3.0 μL of commercial 10× kinase buffer, 1.0 μL T4 polynucleotide kinase, 2.5 μL [γ-^{32}P]-ATP, and adjust the volume to 30 μL with nuclease-free H_20.
2. Mix gently by pipetting a couple of times up and down, spin briefly to collect all material at the bottom of the tube.
3. Incubate at 37 °C for 40 min.
4. Transfer the tube to 65 °C for 10 min to inactivate the enzyme and stop the reaction.
5. Let it cool down and use aliquots of the labeled oligonucleotide as primer in normal PCR amplification reactions with parameters of template denaturation, annealing, and elongation adapted to the nature of the template, the melting temperature of the primers, and the nature of the DNA polymerase.
6. Proceed to the purification of the amplicon of interest by gel electrophoresis and recovery from gel (*see* Subheading 3.2) or store the reaction mixes overnight at 4 °C or at −20 °C for longer periods.

3.2 Purification of [^{32}P]-labeled DNA

3.2.1 Aim

DNA protection and premodification-binding interference assays require highly purified DNA probes. Hence, native polyacrylamide gel electrophoresis is used to separate the labeled amplicon of interest from potential aberrant side products generated during the PCR amplification, an excess nonelongated labeled primer, and nonincorporated labeled [γ-^{32}P]-ATP.

3.2.2 Treatment of Glass Plates

1. Start by thoroughly cleaning the front and back glass plates of the gel casting system with detergent, rinse abundantly with tap water, then with demineralized water and wipe dry with absorbing paper.
2. Siliconize the inside face of the large front plate by spraying repel solution on the plate, spread equally all over the plate with absorbing paper, and let it evaporate a few minutes.
3. Assemble the back and front plates with the spacers with the help of a commercial rubber assembly system or seal the edges

and bottom with tape and clamps to avoid leaks. Deposit the assembly slightly tilted on a protected working surface (easily cleanable surface such as a Teflon plate or disposable benchcoat).

3.2.3 Preparation of a 6% Nondenaturing Polyacrylamide Gel

1. Mix in a 50 mL cylinder 10 mL of 5× TBE buffer with 10 mL of 30% acrylamide/*bis*-acrylamide solution, 0.5 mL ammoniumpersulfate solution and adjust to 50 mL with ddH_20.
2. Filter the solution on a Büchner funnel connected to a vacuum pump and degas for 10 min.
3. Remove a 10 mL aliquot and add 10 μL TEMED, mix and use a syringe to let 5 mL of the solution flow down along the spacers, on either side of the assembly.
4. Allow to polymerize (takes a few minutes only and can be verified by tilting the assembly). This reduces the risk of further leakage while pouring the bulk of the gel.
5. Add 20 μL TEMED to the remaining 40 mL of polyacrylamide solution and use a syringe to carefully fill up the plates till the top of the small back plate. Let the liquid flow down between the glass plates slowly but regularly to avoid trapping air bubbles in the gel solution.
6. Insert the comb and avoid trapping very small air bubbles beneath the teeth of the comb by starting with a tilted comb, insert one side, then progressively push down the remaining part of the comb.
7. Let polymerize for at least 30–45 min.
8. Remove sealing, comb, and potentially spilled polyacrylamide on the outer surface of the glass plates and mount the gel in the electrophoresis tank.
9. Add 1× TBE running buffer to the upper and lower chamber of the electrophoresis tank.
10. Use a syringe to rinse the wells and ensure that no polyacrylamide remains in the wells.
11. Switch on the current and prerun the gel for 30 min at 180 V (11 V/cm) to remove excess of nonpolymerized reagents.

3.2.4 Sample Preparation and Gel Electrophoresis

1. Mix 50 μL aliquots of the labeled PCR reaction mix (*see* Subheading 3.1) with 5 μL of 10× retardation-loading dye mix. Centrifuge briefly.
2. Switch off the current and use long, thin pipette tips to load the samples in the wells. Avoid blowing air bubbles in the samples.
3. Switch on the current and let it migrate at 180 V for 15 min to allow fast penetration of the material in the gel matrix.

4. Then reduce the current to 150 V and continue the migration to obtain a good separation, depending on the length of the labeled amplicon, using the dyes as a guide. On a 6% polyacrylamide gel bromophenol blue (the fastest migrating dye) co-migrates with an approximately 58 bp fragment (45 bp on 8%, 33 bp on 10%), and xylene cyanol with a 230 bp (160 bp on 8%, 117 bp on 10%).

3.2.5 Imaging by Autoradiography

1. Stop the electrophoresis.
2. Remove the assembly from the electrophoresis tank and wipe the glass plates dry. Make sure to remove all running buffer from the wells before disassembling the glass plates (aspire with a Pasteur pipette or shake heavily above the sink with the wells turned downward).
3. Put the assembly on a flat surface (behind a Plexiglas shield) with the large siliconized front plate on top.
4. Use a spacer or spatula as a lever to carefully pry the plates apart and remove the top plate. Avoid distorting the gel and remove the spacers.
5. Cover the gel still sticking to the small back plate with plastic foil (fold back on the edges to avoid disturbance upon later manipulations) to avoid any direct contact of the radioactive material in the gel with the X-ray-sensitive film.
6. Put a sheet of paper or cardboard on top of the plastic foil (to avoid that the latter would stick later to the front plate) and cover with the front plate to protect from radiation.
7. Move the assembly to the darkroom.
8. In the dark (red light) remove the front plate and the paper or cardboard.
9. Cut a piece of X-ray-sensitive film sufficiently large to cover the area of interest and deposit such that the film perfectly aligns with the bottom of the back plate and slightly sticks out on at least one side and the top of the back plate.
10. Place the front plate on top of the X-ray-sensitive film and fix the whole together with clamps.
11. Use the tip of scissors to scratch the contour of the small back plate on the film and indicate the orientation by making a small incision at one of the top corners of the film. This will allow a perfect realignment of the film and gel after development.
12. Expose for 10 min, then remove the X-ray-sensitive film from the assembly.
13. Pour a sufficient volume of 5× diluted developer to cover the entire surface of the film in a tray and develop the film for 2–3 min (gently shake the tray regularly).

14. Abundantly rinse the film with tap water.
15. Then move it to a second tray with 5× diluted fixer. Allow to stand with gentle shaking for a few minutes.
16. Rinse abundantly with tap water, then with demineralized water.
17. Suspend the film with a clamp and let air dry.
18. In the meantime, remove the running buffer from the electrophoresis tank. Be aware that the bottom chamber may be heavily contaminated with [γ-^{32}P]-ATP that may have migrated out of the gel, depending on the electrophoresis conditions. Verify with the Geiger counter. Rinse the electrophoresis apparatus abundantly with tap water and verify for potential remaining contamination with the Geiger counter.

3.2.6 Recovery of Labeled DNA from the Polyacrylamide Gel Matrix

1. Prepare one assembled elution set-up for every DNA band that has to be recovered. Start by inserting a small amount of sterile glass wool in a sterile 1 mL pipette tip and very gently push (don't press, *see* **Note 8**) the glass wool toward the bottom of the tip with a Pasteur pipette to form a plug that will retain the polyacrylamide particles but allows buffer to flow through. Then seal the tip of the 1 mL pipette tip with self-sealing tape and insert the lower part in a sterile glass tube.
2. Work behind a Plexiglas shield and deposit the autoradiography on a piece of benchcoat. Take the orientation into account and put the gel with the front plate on top of the autoradiography. Align the small back plate with the marked contour on the autoradiography and remove the front plate.
3. Cut and remove the plastic with a scalpel.
4. Use the scalpel to cut out a slice of acrylamide corresponding to the band of labeled DNA on the autoradiography. Keep the size of the gel slice as small as possible.
5. Use the scalpel to transfer the gel slice to the 1 mL pipette tip and use a Pasteur pipette to push it further down without touching the glass wool plug.
6. Crush the polyacrylamide between the wall of the 1 mL pipette tip and the Pasteur pipette by making a rotating movement (be careful not to touch or displace the plug; do it gently to avoid breaking the tip of the Pasteur pipette), till the slice has been transformed into small bead-like particles (this will improve the recovery by facilitating diffusion of the DNA) that stick to the Pasteur pipette.
7. Let 200 μL of elution mix flow down along the tip of the Pasteur pipette to collect the beads at the bottom of the 1 mL pipette tip, on top of the plug.

8. Repeat this operation to collect all acrylamide beads. Use a total of 400–600 μL elution mix per gel slice for elution, depending on the size of the gel slice. Make sure that all the acrylamide is submerged in the elution buffer. Use the Pasteur pipette to push all polyacrylamide in the liquid and to remove potential air bubbles in the solution.
9. Seal the top of the 1 mL pipette tip and the connection between the pipette tip and the glass tube with self-sealing tape.
10. Put the assembly in a floater and incubate the whole in a water bath at 45 °C, overnight.
11. Remove the tube from the water bath after overnight incubation.
12. Wipe dry, remove the self-sealing tape from the connection between the pipette tip and the glass tube. Verify that no water from the bath has penetrated the glass tube (otherwise replace with a clean tube).
13. Carefully remove the self-sealing tape at the tip of the 1 mL pipette tip and immediately insert it in the glass tube (*see* **Note 9**).
14. Subsequently, remove the self-sealing tape at the top of the 1 mL pipette tip and collect the liquid (containing labeled DNA) that will flow through the glass wool plug, whereas the polyacrylamide particles are retained.
15. When no more liquid comes out of the tip, add another 100 μL of elution mix to the 1 mL pipette tip to rinse the polyacrylamide and glass wool. Collect the flow-through in the same glass tube.
16. Transfer the collected liquid to a clean sterile 1.5 mL Eppendorf tube, add 3 μL yeast-tRNA and 1 mL ice-cold absolute ethanol (*see* **Note 10**).
17. Mix well by inverting the tube a couple of times and incubate at −20 °C for at least one hour.
18. Centrifuge for 15 min at maximum speed in a microcentrifuge to pellet the precipitated DNA. Immediately remove the supernatant with a drawn-out Pasteur pipette (use a different one for every DNA fragment).
19. Add 1 mL of 80% ice-cold ethanol to the pellet, spin 3 min at maximum speed, remove the supernatant.
20. Dry the pellet for a few minutes in the vacuum centrifuge.

3.3 Methylation Protection

3.3.1 Protein Binding and DNA Methylation

1. Mix in sterile 1.5 mL Eppendorf tubes 200 μL DMS reaction buffer with about 100 000 cpm of [5′-^{32}P] single-end labeled DNA, 1.0 μL nonspecific competitor DNA (*see* **Note 5**), and various amounts of protein (*see* **Note 7**). Always include a reference without protein.

2. Mix gently (don't vortex as this might denature the protein) and centrifuge briefly (few seconds) to collect all the material at the bottom of the tube.
3. Incubate the samples at room temperature (about 20 °C) for 25–30 min to allow complex formation to reach the equilibrium.
4. Add 1.0 μL DMS (work in the fume hood and behind a Plexiglas shield to protect from radiation) and let the reaction proceed for 1 min.
5. Immediately stop the reaction with 50 μL DMS stop solution.
6. Precipitate the DNA by adding 3.0 μL yeast-tRNA and 1 mL ice-cold absolute ethanol. Mix well by inverting the tube a couple of times and incubate for at least 1 h at −20 °C (*see* **Note 10**).
7. Centrifuge for 15 min at maximum speed in the microcentrifuge to pellet the DNA.
8. Immediately remove the supernatant with a drawn-out Pasteur pipette.
9. Add 1.0 mL 80% ice-cold ethanol, centrifuge for 3 min, and immediately remove the supernatant.
10. Briefly dry the pellet in the vacuum centrifuge.
11. The modified DNA can then be cleaved either at methylated Gs only (*see* Subheading 3.3.2) or at methylated Gs and As (*see* Subheading 3.3.3).

3.3.2 Strand Cleavage at Methylated Gs Only

1. Dissolve the dry pellets in 50 μL of 1 M piperidine (freshly diluted). Make sure all the DNA is dissolved (vortex thoroughly, *see* **Note 11**).
2. Centrifuge briefly to collect all material at the bottom of the tube.
3. Incubate the samples in a heating block at 90 °C for 30 min.
4. Cool down to room temperature and centrifuge briefly to collect the condensate formed on the lid.
5. Open the tubes and evaporate the piperidine in the vacuum centrifuge (*see* **Note 12**).
6. Dissolve the dry pellets in 40 μL nuclease-fee H_2O (vortex, centrifuge briefly).
7. Evaporate the H_2O in the vacuum centrifuge.
8. Repeat **steps 6** and 7 twice (*see* **Note 12**).
9. Store the dry samples at −20 °C or proceed further as indicated in Subheading 3.3.4 for sample preparation and separation of the reaction products.

3.3.3 Strand Cleavage at Methylated Gs and As

1. Dissolve the methylated DNA pellets in 25 μL of 0.1 M HCl. Make sure all the DNA is dissolved (vortex thoroughly, *see* **Note 11**).
2. Centrifuge briefly to collect all material at the bottom of the tube.
3. Incubate the samples on ice for 2 h.
4. Precipitate the DNA with yeast-tRNA and ethanol as described above (*see* Subheading 3.3.1, **steps 6–10**).
5. Dissolve the dry pellets in 43 μL of 10 mM Na_2PO_4 buffer pH 7.0.
6. Add 7.5 μL 1.0 M NaOH, mix, centrifuge briefly.
7. Incubate the tubes in a heating block at 90 °C for 30 min.
8. Cool down on ice and centrifuge briefly to collect all material at the bottom.
9. Neutralize with 7.5 μL 1.0 M HCl.
10. Precipitate the DNA with ethanol (*see* Subheading 3.3.1, **steps 6–10**).
11. Store the dry pellets at −20 °C or proceed further to sample preparation and gel electrophoresis under denaturing conditions (*see* Subheadings 3.3.4, 3.3.5, and 3.3.6).

3.3.4 Sample Preparation for Gel Electrophoresis Under Denaturing Conditions

1. Count the radioactivity in each tube with a Geiger counter (or by Cerenkov counting in a beta-counter).
2. Dissolve the pellets in an appropriate volume of formamide dye mix (vortex thoroughly). Take into account that no more than 7.0 μL per lane should be loaded on the final denaturing gel, that all lanes should contain the same number of counts (same number of DNA molecules) in the same volume (*see* **Note 13**), and that about 50,000 cpm are required for development of the autoradiography after overnight exposure.
3. Proceed to separation of the reaction products by gel electrophoresis under denaturing conditions as described (*see* Subheadings 3.3.6 and 3.3.7) or store the samples at −20 °C.

3.3.5 Preparation of a Denaturing Acrylamide/Bis-Acrylamide Stock Solution of Suited Concentration

1. For 500 mL of a 6% polyacrylamide gel, weigh 210 g urea in a 1 L glass beaker (*see* **Note 14**).
2. Add 100 mL 5× TBE electrophoresis buffer, 100 mL ddH_20, and 75 mL of 40% acylamide/*bis*-acrylamide solution (ratio 19/1).
3. Dissolve the urea on a heating magnetic stirrer (endothermic reaction).
4. Transfer to a 500 mL cylinder when all the urea is dissolved and adjust the volume to 500 mL with ddH_20.

5. Filter on a Büchner funnel and degas for 10 min.
6. Transfer the solution to a glass bottle and store at 4 °C in the dark (can be kept for a few months).

3.3.6 Preparation of a Denaturing Polyacrylamide Gel (see **Note 15**)

1. Start by thoroughly cleaning the front and back glass plates with detergent, rinse abundantly with tap water, then with demineralized water, wipe dry with absorbing paper and degrease the inside face with technical grade ethanol. Ensure that the plates are extremely clean since the slightest contamination will result in the formation of air bubbles in the gel.
2. Treat the inside face of the large front plate with "sticky" solution to ensure tight adhesion of the gel during the fixing and drying procedures subsequent to gel electrophoresis. Pour a few mL of sticky solution on the plate, spread all over the plate with absorbing paper and let it dry for at least a couple of hours (or overnight) (*see* **Note 16**).
3. Siliconize the inside face of the small back plate by spraying some repel solution on the plate, spread equally all over the plate with absorbing paper and let it evaporate a few minutes before assembling the plates with the spacers.
4. Assemble the small and large plates with 0.25 mm thick spacers (use a commercial rubber sealing system or seal the bottom and edges with tape and clamps) and deposit the assembly slightly tilted on the protected working surface (easily cleanable surface such as Teflon plate or disposable benchcoat).
5. Filter 60 mL of cold denaturing polyacrylamide stock solution (*see* Subheading 3.3.5) on a Büchner funnel and degas for 10 min.
6. Transfer the solution to a beaker.
7. Add 600 μL ammoniumpersulfate solution and 30 μL TEMED.
8. Mix by gentle swirling and use a 50 mL syringe to fill up the assembled glass plates. Let the solution flow down slowly but constantly (without interruption) from one side of the assembly and avoid trapping air bubbles.
9. Insert the comb (avoid trapping small air bubbles underneath the teeth of the comb by starting at one end and having some solution on top of the border of the small back plate).
10. Let polymerize at least 45 min and proceed to Subheading 3.3.7 or store the gel at −4 °C (*see* **Note 15**).

3.3.7 Gel Electrophoresis Under Denaturing Conditions and Visualization by Autoradiography

1. Remove the sealing and comb from the gel assembly. Remove all remaining acrylamide above the wells on the front plate and use a capillary (made from the tip of a Pasteur pipette) to clean the wells, if required. Clean the outer surface of the glass plates.

2. Install the assembly in an appropriate electrophoresis tank that allows for an homogenous distribution of the heat developed during the electrophoresis (metal back plate or buffer layer behind the back plate).
3. Fill the upper and lower chamber with 1× TBE buffer and use a syringe to rinse and remove air bubbles from the wells.
4. Connect to the power supply and perform a prerun of 30 min at constant power of 60 W.
5. Immediately prior to loading, heat the samples for 3 min at 90 °C in a heating block to completely denature the DNA strands and remove any secondary structures in single-stranded DNA to ensure that separation of the reaction products will only rely on length.
6. Stop the prerun, use a syringe to rinse the wells, and load the samples.
7. Start the electrophoresis (constant power of 60 W). Take into account that bromophenol blue and xylene cyanol co-migrate with approximately 26 nt and 106 nt long fragments, respectively, on a 6% polyacrylamide gel (19 nt and 75 nt on 8%; 12 nt and 55 nt on 10%) to calculate the required migration distance for an optimal resolution of the bands in the zone of interest.
8. Stop the electrophoresis.
9. Remove the gel assembly from the electrophoresis tank and deposit it with the small back plate (siliconized) on top on a flat surface.
10. Use a small wedge to pry open the plates and remove the back plate and spacers.
11. Incubate the front plate with the gel sticking to it for 10 min in a plastic tray with 2 L of 10% acetic acid and shake regularly (work in a chemical fume hood).
12. Remove the plate from the tray, let a maximum of liquid drain off.
13. Wipe the back side of the plate dry with absorbing paper.
14. Put on a flat surface and remove liquid from the gel by rolling an absorbing paper roll over the gel (don't be afraid, the gel should not detach or get damaged because it sticks very well to the plate if the "sticky" treatment has been performed correctly).
15. Use a small electric heating device blowing hot air to completely dry the gel that should become even thinner and very hard (blow from different sides to ensure homogenous drying). This will inhibit diffusion during exposure of the X-ray-sensitive film and generate sharp bands on the autoradiograph.

16. Let the plate cool down.
17. Cover the gel with plastic foil (put it very flat, without wrinkles, fold back and attach with tape on the back face of the plate) to avoid any direct contact of the radioactive material in the gel with the X-ray-sensitive film.
18. Cover with the back plate to protect from radiation and move to the dark room.
19. Remove the back plate, place a piece of X-ray-sensitive film on top of the gel, cover again with the back plate, fix the assembly together with clips and wrap the whole in black plastic. Store in the dark.
20. Proceed further as described in Subheading 3.2.5, **steps 12–17**, but use a longer, appropriate exposure time. As a guide, take into account that about 50,000 cpm per lane are required to have sufficiently strong signals after overnight exposure. If the band intensity is not appropriate, a new exposure can be made for an adapted period of time. If the signals are very low, an intensifying screen placed on top of the X-ray-sensitive film and exposure at −80 °C can be used. This will increase the signal intensity by approximately a factor two.

3.4 Premodification-Binding Interference Assays

3.4.1 Preparation of Sparingly Modified DNA

Always use at least 300,000 cpm of [5′-^{32}P] single-end labeled DNA to prepare the different types of sparingly modified DNA described below (*see* **Note 6**). The reaction conditions described below normally ensure the generation of statistically one modified residue per DNA molecule (*see* **Note 1**).

3.4.2 Depurination

1. Add 2.5 μL carrier DNA to the labeled DNA and adjust to a total volume of 10 μL with nuclease-free $H_2 0$.
2. Add 10 μL citrate solution, homogenize, centrifuge briefly to assemble all droplets at the bottom of the tube.
3. Incubate at 80 °C for 4 min.
4. Immediately cool down on ice and centrifuge briefly to collect the condensate from the cover.
5. Add 200 μL 0.3 M Na-acetate, 3.0 μL yeast-tRNA, and 1 mL ice-cold absolute ethanol. Mix well by inverting the tube a couple of times and put at −20 °C for at least 1 h to precipitate the DNA.
6. Centrifuge for 15 min at maximum speed in the microcentrifuge.
7. Immediately remove the supernatant with a drawn-out Pasteur pipette.

8. Rinse the pellet with 1.0 mL 80% ice-cold ethanol, spin down for 3 min, immediately remove the supernatant.
9. Briefly dry the pellet in the vacuum centrifuge.

3.4.3 Depyrimidation

1. Add 2.5 μL carrier DNA to the labeled DNA and adjust to a total volume of 10 μL with nuclease-free H_20.
2. Add 30 μL hydrazine, homogenize, and centrifuge briefly to collect all the material at the bottom of the tube.
3. Incubate at 25 °C for 10 min.
4. Stop the reaction, precipitate the modified DNA, and dry the pellet as described above for depurination (*see* Subheading 3.4.2, **steps 4–9**).

3.4.4 Thymine-Specific Oxidation with $KMnO_4$

1. Add 10 μL of 30 mM Tris-HCl buffer (pH 8.0) to the labeled DNA.
2. Incubate at 95 °C for 2 min (strand separation).
3. Immediately put on ice, centrifuge briefly to collect the condensate.
4. Add 40 μL of freshly prepared 0.25 mM $KMnO_4$ and incubate at 20 °C for 10 min.
5. Stop the reaction with 100 μL stop solution.
6. Add 500 μL ice-cold absolute ethanol, mix well, and incubate at least 1 h at −20 °C.
7. Centrifuge for 15 min and immediately remove the supernatant with a drawn-out Pasteur pipette.
8. Rinse the pellet with 500 μL 80% ice-cold ethanol, centrifuge for 3 min, immediately remove the supernatant.
9. Briefly dry the pellet in the vacuum centrifuge.
10. Dissolve the DNA pellet in 10 μL renaturation buffer.
11. Incubate for 5 min at 60 °C in a water bath.
12. Switch off the heating of the water bath and let it cool down slowly to room temperature for strand annealing.

3.4.5 Cytosine-Specific Modification with $NH_2OH.HCl$

1. Add 10 μL of 30 mM Tris-HCl buffer pH 8.0 to the labeled DNA.
2. Incubate at 95 °C for 2 min (strand separation).
3. Immediately cool down on ice and centrifuge briefly to collect the condensate.
4. Add 2.5 μL carrier DNA and 20 μL freshly prepared 4.02 M $NH_2OH.HCl$ (pH 6.0), mix and centrifuge briefly.
5. Incubate for 10 min at 20 °C.
6. Stop the reaction with 200 μL ice-cold stop solution.

7. Add 500 μL ice-cold absolute ethanol, mix well and incubate at −20 °C for at least 1 h.
8. Centrifuge for 15 min and immediately remove the supernatant with a drawn-out Pasteur pipette.
9. Rinse the pellet with 500 μL of 80% ice-cold ethanol, centrifuge for 3 min and immediately remove the supernatant.
10. Briefly dry the pellet in the vacuum centrifuge.
11. Dissolve the DNA pellet in 10 μL renaturation buffer, centrifuge briefly.
12. Incubate for 5 min at 60 °C in a water bath.
13. Switch off the heating of the water bath and let it cool down slowly to room temperature for strand annealing.

3.4.6 Uracil Substitution

Sparingly uracil-substituted DNA can most easily be prepared by PCR amplification with, as substrates the four dNTPs supplemented with a smaller amount of dUTP and as primers two oligonucleotides, one of which is labeled at the 5′-end with [γ-^{32}P]-ATP and T4 polynucleotide kinase (*see* Subheading 3.1).

1. Mix in a PCR tube 20 pmol of labeled Fw and 20 pmol non-labeled Rv primer (and/or the inverse combination) with 0.2 pmol template DNA and 100 μL PCR buffer supplemented with 200 μM of each of the four dNTPs, 50 μM dUTP, and 5 units Taq DNA polymerase. Adjust the total volume to 110 μL.
2. After initial template denaturation at 94 °C for 4 min, perform eight cycles of denaturation (94 °C, 30 s), annealing (5–10 °C below the Tm of the primers, 30 s) and elongation (1 min, 72 °C), and a finishing step of 10 min elongation at 72 °C.
3. Purify the labeled amplicon by native 6% polyacrylamide gel electrophoresis and recover the DNA from gel as described (*see* Subheading 3.2).

3.4.7 Phosphate Ethylation with N-Ethyl-Nitrosourea

1. Add 100 μL ethylation buffer and 100 μL of a saturated solution of *N*-ethyl-nitrosourea (in ethanol) to the labeled DNA. Mix, centrifuge briefly.
2. Incubate at 50 °C for 60 min.
3. Cool down on ice and centrifuge briefly to collect the condensate.
4. Immediately stop the reaction with 50 μL 2 M ammonium acetate, 2.5 μL carrier DNA, 3.0 μL yeast-tRNA.
5. Add 1.0 mL ice-cold absolute ethanol, mix and incubate at −20 °C for at least 1 h.

6. Centrifuge for 15 min, immediately remove the supernatant, rinse the pellet with ice-cold 80% ethanol and dry the pellet as described (*see* Subheading 3.4.2, **steps 6–9**).

3.5 EMSAs with Sparingly Modified DNA

The procedure to separate sparingly modified DNA molecules on basis of their affinity for the protein in an EMSA is the same for all modification reactions described above. However, before doing this, the binding (buffer composition, temperature, concentration of protein and nonspecific competitor DNA, ligand) and electrophoresis conditions (polyacrylamide concentration, electrophoresis buffer, migration time) should be optimized and the apparent K_D of the interaction determined with nonmodified DNA in EMSA (*see* **Note** 7).

Use at least 100,000–150,000 cpm of modified DNA per lane (*see* **Note 6**) in the EMSA and use protein concentrations aiming at about 50% of the DNA being bound and always include a control without protein; the DNA recovered from this lane (input DNA) represents the mixture of all molecules modified at a different position present in the preparation and will provide the reference intensity for all bands corresponding to the products of the cleavage reaction after separation by gel electrophoresis under denaturing conditions. Furthermore, it may be interesting to include a protein concentration resulting in more than 50% binding (about 80%). The comparison allows for a better differentiation between weak, moderate, and strong negative effects on protein binding of premodification at specific residues. Indeed, at a higher protein concentration, complex formation is favored and only those molecules with a strongly reduced affinity for the protein will still be largely underrepresented in the bound form. For an example (*see* Fig. 3b).

1. Prepare and install a 6% polyacrylamide in the electrophoresis tank as described (*see* Subheadings 3.2.2, 3.2.3, and 3.2.4).
2. Combine in different sterile 1.5 mL Eppendorf tubes 10 μL of 2× binding buffer (*see* **Note** 7) with at least 100,000 cpm of labeled modified DNA, 1.0 μL of carrier DNA as nonspecific competitor (*see* **Note 5**), an appropriate amount of protein to reach 50% binding (and possibly more, see above), and adjust the volume to 20 μL with nuclease-free H_20.
3. Mix gently (don't vortex), centrifuge briefly, and incubate at an appropriate temperature (*see* **Note** 7) for 25–30 min to reach the equilibrium of complex formation.
4. Start the prerun of the polyacrylamide gel (180 V; 11 V/cm).
5. Add 2.0 μL of retardation dye mix to the samples and mix gently (don't vortex).
6. Load the samples on the native polyacrylamide gel to separate free from bound DNA.

7. Start the migration at 180 V (11 V/cm) till the material has penetrated the gel, then reduce the current to 130 V (8 V/cm), to avoid too much heating of the gel. Use an adapted migration time as determined from the optimization in EMSA with non-modified DNA to separate free from bound DNA molecules (see above, **Note 7**).
8. Stop the electrophoresis and reveal the position of input, free, and bound DNA on the gel by making an autoradiography as described (*see* Subheading 3.2.5), but extend the exposure time to 2–3 h.
9. Recover the different DNA populations from gel as described (*see* Subheading 3.2.6).

3.6 Cleavage of Sparingly Modified DNA After Recovery from Gel

The method for strand scission of depurinated and depyrimidated DNA and molecules specifically modified at thymine ($KMnO_4$) or cytosine ($NH_2OH.HCl$) residues is identical and performed with piperidine at high temperature.

3.6.1 Reactions Common to Several Modifications

1. Dissolve the different dry DNA pellets (input, free, bound DNA populations) in 100 μL freshly diluted 1 M piperidine (vortex and make sure all the DNA is dissolved) (*see* **Note 11**).
2. Collect all the material at the bottom of the tubes (centrifuge briefly).
3. Incubate for 30 min at 90 °C in a heating block.
4. Cool down, collect the condensate from the lid (centrifuge briefly).
5. Open the tubes and evaporate the piperidine in the vacuum centrifuge (*see* **Note 12**).
6. Dissolve the pellets in 40 μL nuclease-free H_2O, centrifuge briefly.
7. Evaporate the H_2O in the vacuum centrifuge.
8. Repeat **steps 6** and 7 twice to ensure that all remnants of piperidine have been removed (*see* **Note 12**).
9. Proceed to sample preparation and separation of the reaction products by gel electrophoresis under denaturing conditions (*see* Subheading 3.7) or store the dry pellets at −20 °C.

3.6.2 Strand Scission of Methylated DNA

Strand scission at methylated purines (Gs only or As and Gs) is performed as described (Methylation Protection, Subheadings 3.3.2 and 3.3.3).

3.6.3 Strand Scission at Ethylated Phosphates

1. Dissolve the DNA pellets in 43 μL of 10 mM Na_3PO_4 (pH 7.0), centrifuge briefly.
2. Add 7.5 μL of 1.0 M NaOH, mix, centrifuge briefly, and incubate at 90 °C for 30 min.

3. Cool down on ice, centrifuge briefly, and neutralize with 7.5 μL of 1.0 M HCl.
4. Add 200 μL of 0.3 M Na-acetate and 1.0 mL ice-cold absolute ethanol, mix and let the DNA precipitate for at least 1 h at −20 °C.
5. Centrifuge for 15 min at maximum speed in a microcentrifuge and immediately remove the supernatant with a drawn-out Pasteur pipette.
6. Add 1.0 mL 80% ice-cold ethanol, centrifuge for 3 min, and immediately remove the supernatant.
7. Briefly dry the pellets in the vacuum centrifuge.
8. Proceed to sample preparation and separation of the reaction products by gel electrophoresis under denaturing conditions (*see* Subheading 3.7) or store the dry pellets at −20 °C.

3.6.4 Strand Scission of Uracil-Substituted DNA

1. Dissolve the dry pellets in 60 μL of PCR buffer.
2. Add 1.0 U uracil-*N*-glycosylase and incubate for 1 h at 37 °C.
3. Precipitate the DNA and cleave the DNA backbone at abasic sites by piperidine treatment as described (*see* Subheading 3.6.1).

3.7 Separation of the Reaction Products by Gel Electrophoresis Under Denaturing Conditions

1. Count all samples with a Geiger counter (or by Cerenkov counting in a beta-counter).
2. Dissolve the pellets in an appropriate volume of formamide dye mix (vortex thoroughly). Take into account that no more than 7.0 μL per lane should be loaded on the final denaturing gel and that all lanes should contain the same number of counts (same amount of DNA) in the same volume (*see* **Notes 6** and **13**).
3. Separate the reaction products by gel electrophoresis under denaturing conditions and reveal the results by autoradiography as described (*see* Subheadings 3.3.4, 3.3.5, 3.3.6, and 3.3.7). To calculate the migration distance, take into account that bromophenol blue and xylene cyanol co-migrate with approximately 26 nt and 106 nt long fragments, respectively, on a denaturing 6% polyacrylamide gel (19 nt and 75 nt on 8%; 12 nt and 55 nt on 10%).

3.8 Potential Additional Investigations

The specificity and contribution to the energy of complex formation of bases and base-specific groups identified to be of particular interest in premodification-binding interference assays can be further investigated quantitatively by measuring the K_D of binding to synthetic oligonucleotide duplexes and variants thereof carrying a single classical Watson–Crick bp substitution, an abasic site, or a nonnatural base, such as uracil (U), inosine (I), 2-aminopurine (2-AP), or C5-methyl cytosine in EMSAs.

Saturation mutagenesis, the substitution at every position of a binding site (or a half site for a palindromic target) of a bp by all three other possible ones, allows the establishment of an energy normalized logo that can subsequently be used to screen the genome sequence for potential similar sites and estimate their physiologic relevance [27]. This approach is particularly interesting when only one binding site has been experimentally characterized. Furthermore, the comparison of the effects of all three different bp substitutions at one position allows to formulate hypotheses on the involvement of base-specific atoms and exocyclic groups in the minor and major groove segments of the binding site. Along the same lines, measurement of the K_D of the interaction with homogenous populations of substituted or abasic duplexes (rather than heterogeneous mixes of sparingly but randomly modified DNA as in premodification-binding interference) allows for a better quantitative determination of the effect on protein binding of a particular base or even a base-specific group. Thus, substitution of one specific T by U in synthetic DNA allows the determination of the importance for complex formation of one specific hydrophobic C5 methyl group in the major groove of the binding site. Similarly, the importance of hydrogen bonding to the C6 carbonyl atom of G (major groove side of the duplex) can be estimated with the use of 2-AP and I substituted duplexes. 2-AP lacks this C6 carbonyl (major groove), whereas I lacks the exocyclic 2-amino group of G on the minor groove side but is otherwise identical to G on the major groove side. A drop in affinity of the 2-AP but not I substituted duplex points to the involvement of hydrogen bonding of the local C6 carbonyl of G. For an example, see [27]. Still, better binding to a C-I rather than the normal C-G bearing duplex in region where the minor groove is facing the protein is generally a strong indication for protein-induced bending accompanied by local minor groove compression, a phenomenon that frequently occurs near the center of palindromic binding sites. I lacks the exocyclic amino group of G in the minor groove and hence allows for an easier (less energy requiring) minor groove compression. For examples, see [11, 27].

4 Notes

1. DNA modification reactions have to be very partial in order to generate randomly modified molecules with statistically one modification per DNA molecule only. Indeed, if more than one modification is present, the more distally located

modification(s) with respect to the labeled end might affect complex formation without being detected since only the fragment bearing the labeled end will be visible on the autoradiograph. Therefore, and since modification and subsequent cleavage of the nonlabeled strand leaves the labeled one intact (statistically 50% of the molecules), autoradiographs of premodification-binding interference experiments should always exhibit a very intense band corresponding to the full-length probe (*see* Figs. 1b and 3). If this is not observed, and if in contrast short fragments are abundantly generated, this is an indication of excess modifications and the reaction conditions should be optimized.

2. In contrast to modification reactions, subsequent cleavage of the backbone should go to completion to reveal all positions that have been modified in the DNA preparation.
3. Some alternative running buffers for EMSA:

   0.5× TBE: 45 mM Tris, 45 mM boric acid, 1.25 mM EDTA

   TE: 10 mM Tris, 1.0 mM EDTA

   TAE: 40 mM Tris-acetate, 1 mM EDTA, pH 7.9

   Tris-glycine: 50 mM Tris, 0.4 M glycine, 2.5 mM EDTA

   Always use the same buffer in the preparation of the gel solution and as running buffer in the electrophoresis apparatus. Notice that the pH of the gel and running buffer may have a significant impact on the migration velocity and resolving power of complexes, also dependent on the pI of the protein. As an example, see [30].
4. DNA molecules used in protection and premodification-binding interference experiments are ideally no longer than 150 bp, with the binding site near the middle of the fragment. This allows the generation in parallel of DNA molecules with either the coding or noncoding strand labeled with the same pair of primers (switching the labeled one) in the same conditions for PCR amplification and purification. Furthermore, all further steps in the protection and premodification-binding interference assays (% of native and denaturing polyacrylamide gels, migration times) can then also be performed for both strands in identical conditions and in parallel, thus saving time and materials.
5. Nonspecific competitor DNA is used to reduce the contribution of nonsequence specific interactions in complex formation. The amount has to be determined experimentally and depends on the ratio of the binding affinities of the protein for specific/nonspecific sites. Too much nonspecific

competitor DNA might abolish complex formation if that ratio is small. Different types of nonlabeled nonspecific competitor can be used in protein–DNA binding assays, including commercially available sonicated salmon or herring sperm DNA, sonicated calf thymus DNA, synthetic polymers, such as poly (dI:dC), or an unrelated DNA fragment. However, notice that the use of naturally occurring DNA better reflects the physiologic situation. Furthermore, synthetic polymers do not have exactly the same structure as naturally occurring DNA and hence are not truly equivalent. At any rate, do not use genomic DNA from a species that is too closely related and might bear binding sites resembling the specific target.

6. Make sure to modify a sufficient amount of DNA. Take into account that in premodification-binding interference assays, at least 100,000–150,000 cpm per lane are needed in the subsequent EMSA that should include a reference lane without protein (input DNA) and that it may be desirable to use more than one protein concentration aiming at about 50% binding and more (approximately 80%). It is also interesting to have excess material after the final cleavage reactions. This allows the use of a larger volume of formamide dye mix to resuspend the pellets (makes it easier). Furthermore, if not all the material is loaded on the denaturing gel, the remnants can be stored frozen and used later in another gel electrophoresis under denaturing conditions if migration conditions have to be optimized.

7. Binding conditions and K_D of the interaction can best be determined prior to proceeding to methylation protection or premodification-binding interference assays to ensure that suited amounts of protein are used to guarantee the protection of a significant fraction of the DNA molecules in protection type experiments and about 50% binding in premodification-binding interference. The binding buffer used here is commonly used for TFs of the Lrp family. However, other buffers with a different concentration/composition of mono- and divalent ions and/or pH might be better suited for other DNA-binding proteins. Furthermore, some DNA-binding proteins require a specific ligand. Optimization of the binding conditions can be done by EMSA as described under Subheading 3.5 but with nonmodified DNA and about 10,000–15,000 cpm per lane (allows development of the autoradiography after overnight exposure). Complex formation is ideally performed at a physiologically relevant temperature. However, this is not always possible because short linear DNA duplexes denature at high temperatures (relevant for proteins of hyperthermophiles) or complexes might be too

unstable in the in vitro conditions. Hence, complex formation is frequently performed at lower temperature since protein–DNA interactions generally comprise an important ionic component. Ionic interactions are exotherm, and hence stabilized at low temperature. More information on EMSA can be found in the Chapter 11.

8. Use a small amount of glass wool only and push it down gently to form a plug near the tip of the 1 mL pipette tip, but not at the very end as this will reduce the diameter of the plug and slow down the rate of the flow through. Don't compress the glass wool too hard as this will equally have a negative effect on the velocity and yield of the recovery.
9. Be careful. Liquid may come out quickly. Insert the pipette tip rapidly in the glass tube to avoid loss of radioactive material and contamination of gloves (wear a double pair) and working surface. Verify with the Geiger counter before proceeding further.
10. Yeast-tRNA is added to increase the total concentration of nucleic acids. This will improve the recovery of radiolabeled DNA since low concentrations of nucleic acids do not precipitate efficiently. tRNA is a small molecule that generally does not interfere with protein–DNA binding and as it is not labeled it will not be visible on the autoradiographs. DNA precipitation by dehydration requires about 0.3 M salt (the elution mix contains enough salt) and 2–3 volumes of ethanol per volume aqueous solution. Mix well by inverting the tube, otherwise the aqueous solution will freeze at −20 °C and the DNA will not precipitate efficiently.
11. It is important that all DNA is dissolved. A vortex can be used since short DNA fragments are very resistant to mechanical shearing.
12. Piperidine strongly affects DNA migration. Therefore, all remnants should be carefully removed before loading the samples on the denaturing polyacrylamide gel.
13. Quenching of the radiation is dependent on the sample volume. Therefore, all samples loaded on the gel should contain the same number of counts in the same volume (adjusted with formamide dye mix). Loading the same volume in all wells of the denaturing gel will also ensure a more homogenous loss of a small fraction in the loading process. To calculate the exposure time of the autoradiography, take into account that 50,000 cpm per lane allows development after overnight exposure.

14. 6% gels are commonly used but depending on the position of the zone of interest with respect to the labeled end of the DNA probe, a 8% or 10% polyacrylamide gel might be better suited. Only the amount of acrylamide/*bis*-acrylamide has to be adapted accordingly in the preparation of the denaturing stock solution and compensated for by the volume of ddH_20. The concentrations of buffer and urea remain the same as for the 6% gel.
15. Denaturing polyacrylamide gels can be prepared in advance on forehand and stored overnight (or maximally a few days) at 4 °C. Cover the wells with wet paper and wrap the gel in a plastic foil to avoid drying.
16. Always treat the same plate of the pair with "sticky" solution because once this has been done it is difficult to remove it completely. Switching between treatments of plates with "repel" and "sticky" upon repeated use of the plates increases the risk that the gel will get disrupted when prying the plates apart at the end of the electrophoresis. Once the autoradiograph has been made and the gel is no longer required, the plate can be reutilized. Completely submerge the plate for at least one hour in a plastic tray with tap water to rehydrate and soften the gel. Use an ice scraper to remove the gel from the glass plate (collect with radioactive waste). Then thoroughly clean the plate with detergent (use a brush), abundantly rinse with tap water, then demineralized water, wipe dry, finally degrease with ethanol.

Acknowledgments

Work in the Microbiology Research Group has been financed by the Research Foundation Flanders (FWO-Vlaanderen), the Institute for Innovation by Science and Technology (IWT), and the Research Council of the Vrije Universiteit Brussel (OZR-VUB).

References

1. Nelson HC (1995) Structure and function of DNA-binding proteins. Curr Opin Genet Dev 5:180–189
2. Aravind L, Anantharaman V, Balaji S, Babu M, Iyer LM (2005) The many faces of the helix-turn-helix domain: transcription regulation and beyond. FEMS Microbiol Rev 29: 231–262
3. Choo Y, Klug A (1997) Physical basis of a protein-DNA recognition code. Curr Opin Struct Biol 7:117–125
4. Nadassy K, Wodak SJ, Janin J (1999) Physical features of protein-nucleic acid recognition sites. Biochemistry 38:1999–2017
5. Cheng AC, Cheng WW, Fuhrmann CN, Frankel AD (2003) Recognition of nucleic acid

bases and base-pairs by hydrogen bonding to amino acid side-chains. J Mol Biol 327: 781–796

6. Pan Y, Tsai C-J, Ma B, Nussinov R (2009) How do transcription factors select specific binding sites in the genome? Nat Struct Mol Biol 16:1118–1120
7. Rice PA, Yang S, Mizuuchi K, Nash HA (1996) Crystal structure of an IHF-DNA complex: a protein-induced U-turn. Cell 87:1295–1306
8. Johnsrud L (1978) Contacts between *Escherichia coli* RNA polymerase and the *lac* operon promoter. Proc Natl Acad Sci U S A 75: 5314–5318
9. Siebenlist U, Gilbert W (1980) Contacts between *Escherichia coli* RNA polymerase and an early promoter of phage T7. Proc Natl Acad Sci U S A 77:122–126
10. Charlier D, Roovers M, Van Vliet F, Boyen A, Cunin R, Nakamura Y, Glansdorff N, Piérard A (1992) Arginine regulon of *Escherichia coli* K-12. A study of repressor-operator interactions and *in vitro* binding affinities versus *in vivo* repression. J Mol Biol 226(2):367–386
11. Wang H, Glansdorff N, Charlier D (1998) The arginine repressor of *Escherichia coli* K-12 makes direct contacts to minor and major groove determinants of the operators. J Mol Biol 277:805–824
12. Enoru-Eta J, Gigot D, Glansdorff N, Charlier D (2002) High resolution contact probing of the Lrp-like DNA-binding protein Ss-Lrp from the hyperthermoacidophilic crenarchaeote *Sulfolobus solfataricus* P2. Mol Microbiol 45: 1541–1555
13. Xu Y, Sun Y, Huysveld N, Gigot D, Glansdorff N, Charlier D (2003) Regulation of arginine biosynthesis in the piezopsychrophilic bacterium *Moritella profunda*: *in vivo* repressibility and *in vitro* repressor-operator contact probing. J Mol Biol 326:353–369
14. Nguyen Le Minh P, Bervoets I, Maes D, Charlier D (2010) The protein-DNA contacts in RutR•*carAB* operator complexes. Nucleic Acids Res 38:6286–6300
15. Ali MM, Provoost A, Mijnendonckx C, Van Houdt R, Charlier D (2020) DNA-binding and transcription activation by unphosphorylated response regulator AgrR from *Cupriavidus metallidurans* involved in silver resistance. Front Microbiol 11:1635. https://doi.org/10.3389/fmicb.2020.01635
16. Maxam A, Gilbert W (1980) Sequencing end-labelled DNA with base-specific chemical cleavages. Methods Enzymol 65:499–560
17. Brunelle A, Schleif RF (1987) Missing contact probing of DNA-protein interactions. Proc Natl Acad Sci U S A 84:6673–6676
18. Nguyen Le Minh P, Velázquez Ruiz C, Vandermeeren S, Abwoyo P, Bervoets I, Charlier D (2018) Differential protein-DNA contacts for activation and repression by ArgP, a LysR-type (LTTR) transcriptional regulator in *Escherichia coli*. Microbiol Res 206:141–158
19. Peeters E, Thia-Toong T-L, Gigot D, Maes D, Charlier D (2004) Ss-LrpB, a novel Lrp-like regulator of *Sulfolobus solfataricus* P2, binds cooperatively to three conserved targets in the own control region. Mol Microbiol 54: 321–336
20. Caldara M, Nguyen Le Minh P, Bostoen S, Massant J, Charlier D (2007) ArgR-dependent repression of arginine and histidine transport genes in *Escherichia coli* K-12. J Mol Biol 373: 251–267
21. Rubin CM, Schmid CW (1980) Pyrimidine-specific chemical reactions useful for DNA sequencing. Nucleic Acids Res 8:4613–4619
22. Truss M, Chalepakis G, Beato M (1990) Contacts between steroid hormone receptors and thymines in DNA: an interference method. Proc Natl Acad Sci U S A 87:7180–7184
23. Sasse-Dwight S, Gralla D (1989) $KMnO_4$ as a probe for *lac* promoter DNA melting and mechanism *in vivo*. Proc Natl Acad Sci U S A 264:8074–8081
24. Hausner W, Thomm M (1999) Events during initiation of archaeal transcription: open complex formation and DNA-protein interactions. J Bacteriol 183:3025–3031
25. Vassart A, Van Wolferen M, Orell A, Hong Y, Peeters E, Charlier D (2013) Sa-Lrp from *Sulfolobus acidocaldarius* is a versatile, glutamine-responsive, and architectural transcriptional regulator. MicrobiologyOpen 2:75–93
26. Pu WT, Struhl K (1992) Uracil interference, a rapid and general method for defining protein-DNA interactions involving the 5-methyl group of thymines: the GCN4-DNA complex. Nucleic Acids Res 20:771–775
27. Peeters E, Wartel C, Maes D, Charlier D (2007) Analysis of the DNA-binding sequence specificity of the archaeal transcriptional regulator Ss-LrpB from *Sulfolobus solfataricus* by systematic mutagenesis and high resolution contact probing. Nucleic Acids Res 35: 623–633
28. Tullius TD, Dombroski BA (1986) Hydroxyl radical "footprinting": high-resolution information about DNA-protein contacts and

application to λ repressor and Cro protein. Proc Natl Acad Sci U S A 83:5469–5473

29. Phillips SE, Manfield I, Parsons I, Davidson BE, Rafferty JB, Somers WS, Margarita D, Cohen GN, Saint-Girons I, Stockley PG (1989) Cooperative tandem binding of Met J repressor of *Escherichia coli*. Nature 341: 711–715
30. Manfield IW, Reynolds LA, Gittins J, Kneale GG (2000) The DNA-binding domain of the gene regulatory protein ArcA extends beyond the minimal zinc-finger region conserved between GATA proteins. Biochem Biophys Acta 1493:325–332

Chapter 13

Analysis of Protein–DNA Interactions Using Isothermal Titration Calorimetry: Successes and Failures

San Hadži and Jurij Lah

Abstract

Isothermal titration calorimetry (ITC) is a golden standard for the characterization of protein–DNA binding affinities and allows direct assessment of the accompanying thermodynamic driving forces. Their interpretation can give insight into role of electrostatics, specificity of the DNA recognition, contribution of protein folding upon DNA binding and help to distinguish between minor and major groove binders. The main advantages of ITC are that the binding is measured in solution, and it requires no labeling of the samples, however, the method is not well suited for high-performance studies. Here we describe the sample preparation, a procedure to perform a typical ITC experiment, data analysis, and lastly discuss how to interpret the obtained thermodynamic parameters. In conclusion, we show examples of several unsuccessful ITC experiments and identify the underlying reasons for failed experiments. In most cases with a proper adjustment of the experimental setup, it was possible to obtain data appropriate for further analysis.

Key words ITC, Protein, DNA recognition, Repressor, Thermodynamics, DNA specificity, Calorimetry

1 Introduction

Biological processes are essentially linked to molecular recognition. Biological macromolecules interact with ions, small molecules, and macromolecules with varying affinities and specificity. Specific interactions between proteins and DNA are crucial for many biological processes, such as DNA replication and transcription. The structures of numerous protein–DNA complexes have provided a detailed picture of protein–DNA binding and revealed a diverse pattern of interactions that proteins can use to recognize DNA. However, a deeper understanding of the molecular basis of protein–DNA recognition requires characterization of the thermodynamic driving forces that control protein–DNA complex formation [1]. The only technique that allows us to obtain the fundamental thermodynamic driving forces at constant (physiological) temperature, T, and constant pressure, p, within a single experiment is

Eveline Peeters and Indra Bervoets (eds.), *Prokaryotic Gene Regulation: Methods and Protocols*, Methods in Molecular Biology, vol. 2516, https://doi.org/10.1007/978-1-0716-2413-5_13,

isothermal titration calorimetry (ITC) that measures the thermal effects that accompany the titration of protein solution into DNA solution or vice versa. In other words, the ITC measures the heat generated or absorbed by the protein–DNA association. Thermodynamic analysis relates these heat effects to the standard thermodynamic parameters that accompany protein–DNA binding (driving forces): the enthalpy, ΔH_b, which represents the difference in potential energy of the interactions, the entropy, ΔS_b, which reflects the difference in molecular degrees of freedom between the bound and unbound states, and the Gibbs free energy, ΔG_b, which determines the stability of the formed protein–DNA complex. In addition, experiments performed at different temperatures yield the corresponding heat capacity of binding, $\Delta C_{p,b}$, which mainly reflects the changes in protein and DNA hydration upon binding. The physical meaning of these parameters depends essentially on the adequacy of the model used to describe the calorimetric isotherms. Only the fitting of the appropriate model provides thermodynamic driving forces that contribute to a better characterization of the molecular mechanisms of protein–DNA binding.

In this chapter, we describe typical ITC experiments used in the studies of protein–DNA interactions. Preparation of samples, execution of the experiment, and the analysis of results are presented, followed by a description of the interpretation of obtained thermodynamics parameters in the context of protein–DNA interactions. Several types of more advanced experiments are presented, and we conclude by showing a series of unsuccessful experiments and discuss the reasons for the obtained results. However, with an adequate optimization of the experimental setup, these experiments could be amended to obtain useful results.

2 Materials

2.1 Preparation of Duplex DNA

Short DNA oligos (<50 base pairs) can be purchased from commercial suppliers (*see* **Note 1** regarding purity of oligos). Lyophilized oligos are then dissolved in the ITC buffer at concentration around 200–600 μM, depending on the concentrations used in ITC experiments. To prepare a DNA duplex, both strands need to be mixed at equimolar ratio and annealed. Absorbance at 260 nm and the appropriate extinction coefficients (*see* **Note 2**) are used to calculate concentration of single strand oligo stock solutions. Mind that due to the high absorbance of DNA solutions, an appropriate dilution needs to be prepared such that the measured values of absorbance are below 1, that is, within the region where the linear dependence between concentration and absorbance still holds. Based on the calculated concentration of single strands, the appropriate volumes are mixed together to obtain equimolar solution of both strands. This mix is then annealed by incubating at 90 °C for 10 min followed by cooling to the room temperature over a course of few hours. A pot of boiling water or a thermoblock can be used for this purpose. In ideal scenario, it is not necessary to purify DNA duplex any further after annealing, however, *see* **Note 3**.

2.2 Proteins

For the ITC experiments, use pure proteins that were subjected to an additional polishing step via size exclusion chromatography. Any additional information regarding protein–DNA interaction will be useful for the design and analysis of ITC results. For example, performing an electrophoretic mobility shift assay will provide an estimate of the protein–DNA affinity which is necessary to appropriately design the ITC experiment (*see* Subheading 3.2). Also, the knowledge of oligomerization state (monomer, dimer) of the studied protein is needed in order to meaningfully interpret the ITC-binding isotherms.

2.3 Buffers

Buffers need to be prepared using the ultrapure MQ water and filtered through the standard 0.22 μm filters. The choice of an appropriate ITC buffer mainly depends on the requirements of studied protein and DNA system. The aim is to use a buffer where samples remain stable during the experiment and neither the reactants nor products aggregate. If the presence of reducing agents is absolutely necessary, TCEP (tris(2-carboxyethyl) phosphine) is preferred over DTT (1,4-dithiothreitol) or mercaptoethanol, which can cause drifts in the ITC baseline. Second, it is a common practice to use buffers with low ionization enthalpy, such as phosphate buffer. In such cases, the presence of potential protonation equilibria (exchange of proteins during protein–DNA binding) will not significantly affect the measured heat effects (*see* **Note 4**). It is also important to emphasize that protein–DNA binding is usually strongly dependent on ionic strength; therefore, using buffers with overall low ionic strength (e.g., 2 mM phosphate, 13 mM NaCl) will lead to the higher apparent affinities but can also promote nonspecific binding (*see* Subheading 3.5.2). Thus, the salt concentration can be optimized if the experiments lead to unsatisfactory results (*see* Subheading 3.7.1). As an example, a typical ITC buffer would consist of 10 mM phosphate pH 7, 150 mM NaCl.

3 Methods

3.1 Sample Dialysis

Both protein and DNA need to be extensively dialyzed against the buffer prior to the experiment. A buffer mismatch between the two samples results in the dilution heat effects during the titration, which can in extreme cases override the heat signal related to the protein–DNA interaction (*see* Subheading 3.7.4). We typically prepare 1 L of buffer and dialyze the samples three times (2 h, 5 h, overnight). The dialysis cassettes with appropriate cut-off values need to be washed and prepared according to the instructions. If proteins are prone to degradation, dialysis should be performed at 4 °C, and addition of 1 mM EDTA can be used to inhibit metalloproteases.

3.2 Preparation of Solutions for the ITC Experiment

After dialysis, it is necessary to determine concentrations of solutions and prepare appropriate dilutions that will be used for the actual experiment. First, visually check for the presence of aggregates that can occur during dialysis, then the samples should be centrifuged (10 min) or filtered. Determine the dialyzed sample concentrations spectrophotometrically (*see* **Note 2**). In the next step, DNA and protein need to be prepared at the appropriate concentration for the ITC experiment. We usually place DNA solution in the cell and titrate in the protein solution, since the concentration of protein can be estimated more accurately. Concentration errors in the cell and syringe have different effects on the isotherm (*see* **Note 5**).

The required concentrations of samples depend on the protein–DNA interaction affinity and stoichiometry. For a strong protein–DNA interaction, where one protein molecule binds to the DNA with the affinity in the $K_b = 10^8\ M^{-1}$ range, the aim is to over titrate protein into DNA till the final stoichiometry is around $r = 2.5$ and concentration of DNA between 2 and 4 μM can be used. For a weaker interaction with $K_b = 10^6\ M^{-1}$, one should aim for higher 10–15 μM concentration of DNA in the cell. Thus, depending on the protein–DNA affinity, different DNA concentrations and stoichiometry may be used (*see* **Note 6**). As the rule of thumb, the concentration of the proteins is around ten times higher than that of the DNA in the cell. More exactly, the protein concentration can be calculated as:

$$c_{syr} = V_{cell} R\, c_{cell} / V_{syr} \tag{1}$$

where V_{cell} and V_{syr} are the volume of the ITC cell and syringe, r is the desired final stoichiometry, and c_{cell} is the concentration of DNA in the cell. For the Microcal VP-ITC instrument, V_{cell} and V_{syr} are about 1386 and 298 μL, respectively, while for the Microcal iTC200 and PEAQ models, these are 204 μL and 40 μL. Note that higher volumes of solutions than the nominal ones are needed to fill the cell and syringe (*see* **Note 7**).

3.3 Performing an ITC Experiment

Solutions should be briefly centrifuged before the experiment to remove any precipitates or fibers. Next, immediately prior to the experiments the solutions need to be degassed. Samples are placed in a degassing chamber, where the pressure is lowered to remove dissolved air, which could form air bubbles during the experiment. Typically, we place the samples into the degassing station

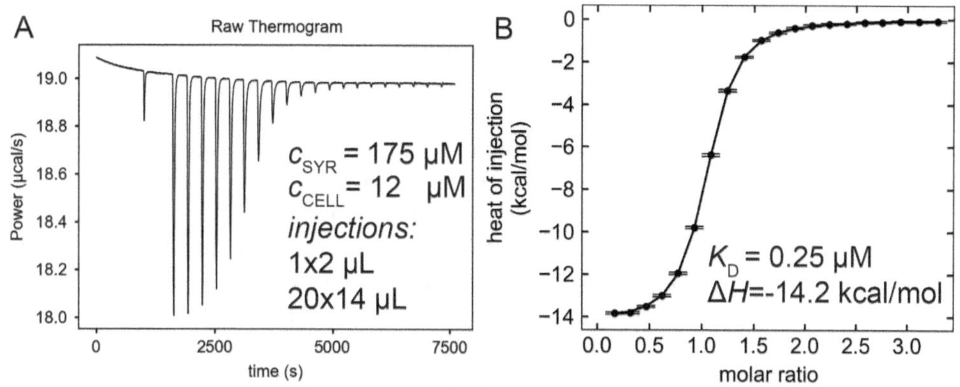

Fig. 1 A typical protein–DNA experiment and expected results. (**a**) Parameters used for a titration of Nanobody 55 to the noncanonical AGCGA-quadruplex. The resulting thermogram shows instrument power as a function of time. (**b**) The binding isotherm shows that one molecule of Nanobody 55 binds to the quadruplex. The single site-binding model was used to describe the isotherm and estimate binding parameters

(TA instruments) for 10 minutes and stir the solution using small magnetic stirrers. After degassing, the cell and the syringe are filled with the solution, and special care should be taken not to introduce any air bubbles during the filling stage (*see* **Note 7**).

In the next step, the experimental parameters need to be set in the instrument's user interface. Generally, macromolecular reactions have relatively small heat effects compared to the other chemical reactions; therefore, instruments should be set in a high feedback mode with a low reference power (e.g., 10–20 µcal/s or around 5 µcal/s for iTC200 or similar). Instruments are usually set to fast equilibration scheme and an initially delay is used to obtain a fully equilibrated flat baseline (although shorter equilibrations are sometimes suggested, we typically use initial delay of 3000 s for VP-ITC and 300 s for iTC200 instruments). Finally, the procedure for the titration experiment needs to be defined. One should aim for about 20 points to define the isotherm. Fewer points will mean that the volumes of added protein will be larger and the measured signals will be stronger and therefore measured more accurately. In principle, it is better to have a bit fewer points which are measured more accurately than many points which are poorly determined (this is discussed in reference [2]). A typical titration would thus have 20 injections with 14 µL and 5 min delay time between injections (*see* Fig. 1a). For technical reasons, the measured heat of the first injection is discarded, and therefore, a small injection volume is used, typically 2 µL for VP-ITC instrument. Obviously, many different titration schemes can be used, depending on the specific requirements and type of the instrument. One should be careful not to set the injection volume too low and should ideally aim for strong signals around 0.5 µcal/s (VP-ITC) or 0.1 µcal/s (ITC200 and PEAQ); however, good isotherms are obtained even with lower signals. After the experiment is finished, remove the solution from the cell using a filling syringe and visually check for presence of aggregates.

3.4 Analysis of Results

3.4.1 Integration of Thermograms

During titration, the instrument records spikes in the power, which compensates the differences in the temperature gradient between the sample and the reference cell. This difference becomes more pronounced when, due to binding of protein to DNA, the heat is exchanged in the sample cell. Such a recording is also referred to as a thermogram (*see* Fig. 1a). The first step in the analysis is to integrate the power signal with respect to time to obtain total heat exchanged upon addition of protein from the syringe. Next, the exchanged heat (in cal) is normalized with respect to the amount of added protein to obtain ΔH (in cal/mol) per injection. The programs accompanying ITC instruments can automatically perform integration of the thermogram. However, in our experience, this integration is not ideal and it is usually necessary to correct the integration manually, especially in cases where the measured signals are weak. Much better integration results can be obtained using the freely available NITPIC program [3] which preforms singular value decomposition to distinguish signal peaks from the baseline.

3.4.2 Analysis of Binding Isotherms

The result of the integration is a binding isotherm, which is the measured ΔH per mole of protein as a function of molar ratio r, which is the molar ratio between protein (P) and DNA (D) (*see* Fig. 1b). Some information can be already obtained from the visual inspection of the isotherm. Plateau at low r values followed by a significant ΔH magnitude drop or a sigmoidal ΔH vs. r shape are characteristic for a high degree of protein binding at low r values. In other words, almost all the added protein becomes bound in the first injections corresponding to the plateau region. Such calorimetric isotherms may be observed due to binding to a single site, statistical binding to equivalent independent binding sites or in the case of maximally cooperative binding to multiple sites (all-or-none). On the other hand, isotherms with small but significant slope at low r are usually characteristic for low degree of protein binding at low r values. Isotherms with more than one shoulder (plateau) are characteristic of more complex sequential binding to different binding sites (*see* Subheading 3.5.3 and Fig. 2c).

Such a qualitative characterization leads us to the selection of the appropriate model for quantitative analysis of the isotherms. For model analysis, the measured enthalpy changes at a given p and T induced by the binding of protein (P) to DNA (D) can be expressed as [4]:

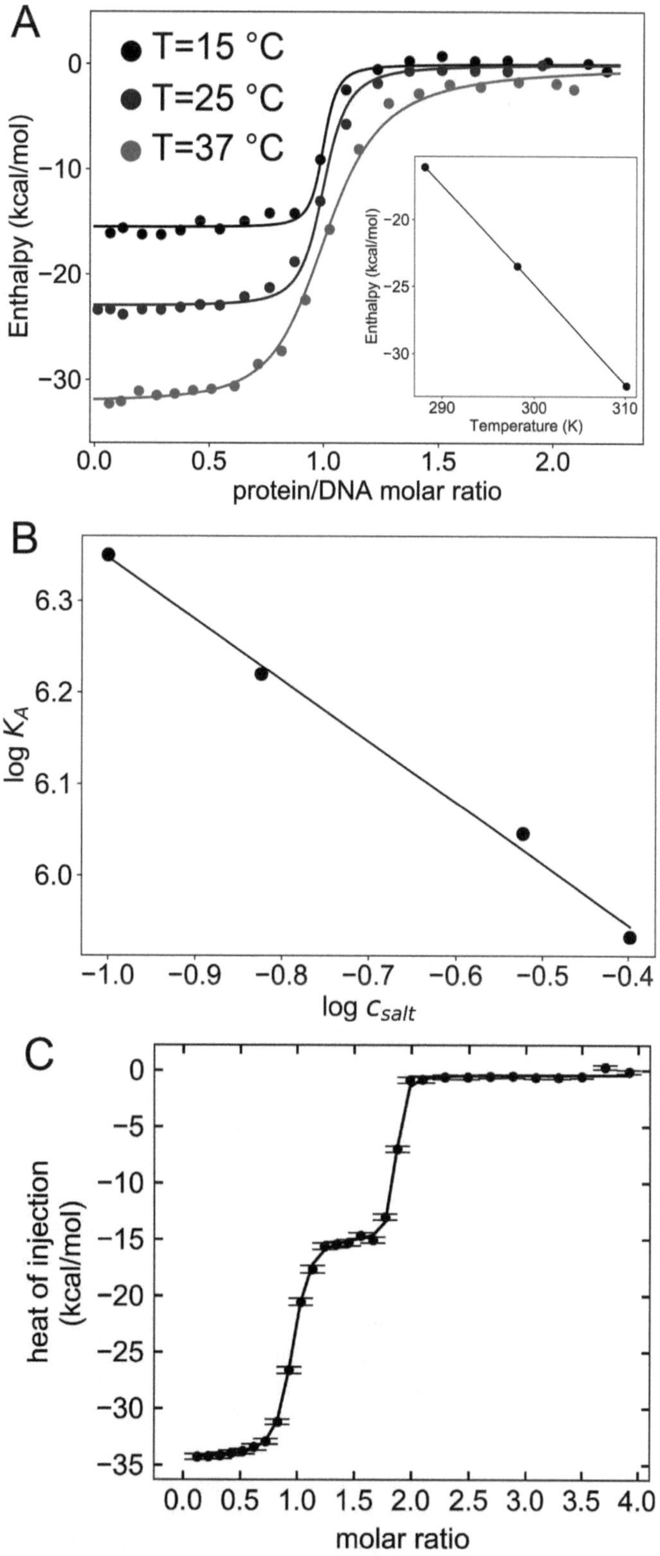

Fig. 2 Advanced ITC experiments. (**a**) Changing the temperature of titration can be a way of optimizing the titration (*see* Subheadings 3.7.3, 3.7.6, and 3.7.7) or

$$\Delta H = \sum_i \Delta H_i \left(\frac{\partial n_i}{\partial n_P} \right)_{p,T} \tag{2}$$

where ΔH_i is the enthalpy of formation of the complex i from P and D, and $(\partial n_i/\partial n_P)_{p,\ T}$ is the corresponding partial derivative at given p and T in which n_i is the amount of the complex i, and n_P is the total amount of added P. The simplest and widely used approach to analyze ITC data resulting from protein–DNA binding is to treat protein-binding sites on DNA as equivalent independent sites. By applying this model and assuming that ΔH_i is independent of concentration (and thus equal to its value in the standard state), Eq. 2 transforms into:

$$\Delta H = \Delta H_b \left(\frac{\partial \nu}{\partial r} \right)_{p,T} \tag{3}$$

where ν represents the average number of occupied sites and ΔH_b the standard enthalpy of P binding to the binding site on D. ν can be defined as:

$$\nu = \frac{[\mathrm{P}]_b}{[\mathrm{D}]_t} = \frac{nK_b[\mathrm{P}]}{1 + K_b[\mathrm{P}]} \tag{4}$$

where $[\mathrm{P}]_b$ is equilibrium concentration of the bound P, [P] the corresponding concentration of unbound P, $[\mathrm{D}]_t$ is the total concentration of D, K_b is the apparent microscopic equilibrium constant (given at the molar scale) characterizing binding of P to a

Fig. 2 (continued) used to determine the heat capacity increment. This example shows a titration of HigA2 repressor into its operator fragment performed at three temperatures. Note that the combination of negative Δc_P (typical for all protein–DNA interactions) and negative ΔH (typical for major groove binders) leads to more negative ΔH and lower affinity at higher temperatures. The inset shows a simple way to determine Δc_P, which equals to the slope in the ΔH (determined from the individual fits of each isotherm) vs. T plot. (**b**) Experiments at different concentration of salt reveal the role of electrostatics in protein–DNA binding. This example shows a titration of HigA2 repressor into its operator performed at different concentrations of salt. A basic analysis requires fitting each isotherm separately to determine the affinity constant, followed by a log-log plot of affinity and salt concentration. The slope is proportional to the number of protein–DNA ionic contacts [11]. (**c**) Nonbinary reactions can be performed by using a mixture of two different molecules in the cell and titrating in a third molecule. This example shows a titration of co-repressor MazF into the 1:1 mixture of repressor MazE and its corresponding DNA operator. Upon addition of co-repressor, different oligomeric complexes interconvert, which cannot be studied using binary titrations. However, analysis of a set of binary and ternary titrations using a global model allows to determine the association parameters for all the underlying reactions. (See another example in Ref. [12])

binding site on D, and n is the total number of binding sites. In Eq. 4 [P] can be expressed as:

$$[\mathrm{P}] = [\mathrm{P}]_t - [\mathrm{D}]_t\nu = [\mathrm{D}]_t(r - \nu) \tag{5}$$

where $[\mathrm{P}]_t$ represents the total concentration of P in the measuring cell. It follows that for known K_b and n values the equilibrium populations of all species in solution and the model function (*see* Eq. 3) for description of calorimetric isotherm are completely defined at any $[\mathrm{P}]_t$, $[\mathrm{D}]_t$ or r through Eqs. 4 and 5. Fitting of the model function (equivalent independent binding sites) to the experimental ITC curve is usually based on the nonlinear Levenberg–Marquardt χ^2 regression procedure. This can be performed using ORIGIN program that is supplied with the VP-ITC and iTC200 instruments. A more advanced and very versatile freeware program is SADPHAT, which resolves many drawbacks of the ORIGIN software [5]. For more complex models, a specialized, home-written code is necessary to analyze isotherms.

3.5 Advanced Experiments

3.5.1 Temperature-Dependent Experiments

The aim of temperature-dependent experiments is to determine the heat capacity contribution ($\Delta C_{P,b}$) of protein–DNA interaction, an important parameter that offers insight into the binding specificity and is also related to dehydration of protein and DNA surfaces. Typically, after successfully obtaining an isotherm at 25 °C, a set of titrations is performed at 5, 15, and 35 °C (*see* Fig. 2a). It should be verified that the protein does not unfold at higher temperatures, for example, by monitoring temperature melting using a spectroscopic technique. One should keep in mind that macromolecular binding is accompanied by a negative heat capacity change, meaning that the strength of ITC power signal (as well as the enthalpy) will change with temperature; heat effects will generally increase with temperature (*see* **Note 8**). The temperature-dependent changes of enthalpy and affinity can therefore be used to optimize the experiments, see the examples discussed in Subheading 3.7.6.

After the data is obtained, the simplest approach is to analyze each isotherm separately with an appropriate model and plot the resulting ΔH vs. T (*see* Fig. 2a, inset). The slope of the regression line corresponds to the heat capacity contribution, which is assumed to be temperature-independent in the studied range:

$$\Delta H_b(T) = \Delta H_b(T_0) + \Delta C_{P,b}(T - T_0) \tag{6}$$

Temperature T_0 is the reference temperature (usually 25 °C) and $\Delta H(T_0)$ is the enthalpy at the reference temperature. A better approach to determine $\Delta C_{P,b}$ is to perform a global analysis and fit the whole set of curves simultaneously. A global fit provides more reliable estimation of $\Delta C_{P,b}$, since it extracts more information from each isotherm [6, 7]. Finally, the temperature-dependent experiments may reveal whether the DNA-binding protein folds upon binding to DNA, which is typically observed as the curvature in the ΔH vs. T plot (see example of a homeodomain refolding upon DNA binding, [8–10]). In these cases, ΔH_b values need to be corrected for the refolding contribution to obtain an accurate estimate of $\Delta C_{P,b}$ or only data in the temperature range where protein is fully folded is used for the analysis.

3.5.2 Experiments at Different Ionic Strengths

The aim of these experiments is to determine the value of the electrostatic free energy contribution and the number of released ions, which offers more insight into the driving forces governing the protein–DNA interactions. Typically, after successfully obtaining an isotherm using a buffer with a given ionic strength (e.g., at 150 mM salt concentration), one performs additional experiments in buffers where the concentration of salt is changed. Since the association constant depends on the logarithm of salt concentration, one should sample a sufficiently large range of salt concentrations, for example, from 15 to 300 mM. Special attention should be paid to the presence of aggregates at lower ionic strengths, where protein binding may become unspecific due to decreased screening of the electrostatic interactions. After the data is obtained, the simplest analysis is to plot the logarithm of association constant against the logarithm of salt concentration (*see* Fig. 2b). According to the counter-ion condensation theory, the slope of the plot is proportional to the number of released ions upon interaction, which is related to the number of ionic contacts between protein and DNA [11]. It is expected that the binding enthalpy remains unchanged at different salt concentrations, since changes in affinity are of entropic origin.

3.5.3 Nonbinary Reactions

The aim of these experiments is to gain insight into more complex mechanisms of protein interaction with DNA. Very often the transcription regulation involves other proteins, such as co-repressors or co-activators or different cooperative mechanisms that govern DNA binding. In these cases, one usually performs binary titrations (repressor into DNA), followed by a second titration that is a titration of co-repressor into the repressor–DNA mixture (*see* **Note 8**). Alternatively, repressor–DNA mixture can be prepared separately, put in the cell, and then titration is performed at the desired repressor–DNA ratio. In this manner, we studied many systems from toxin-antitoxin modules (see for example [12]). Figure 2c shows an experiment where MaxF toxin from *E. coli* (conditional co-repressor) was titrated into the mixture of MazE antitoxin (repressor) and the operator DNA. The shape of the isotherms indicates formation of different macromolecular complexes. With help of binary ITC titrations (MazE+MazF and MazE+DNA), it was possible to detangle a rather complex mechanism of toxin-antitoxin regulation that involved formation of different toxin-antitoxin-DNA species.

3.6 Interpretation of Thermodynamics Parameters

The obtained thermodynamic parameters can be used to predict the behavior of the systems beyond the conditions used in the experiments. Moreover, the interpretation of these parameters can reveal information regarding the role of electrostatic interactions, number of ionic contacts, protein–DNA specificity, differences between minor and major groove binders, and the degree of coupled folding upon DNA binding. The interaction affinity K_b (inverse of the dissociation constant $K_b = 1/K_d$) is related to the free energy of interaction (ΔG), which can be further separated into the enthalpy (ΔH) and entropy contributions (ΔS):

$$\Delta G_b = -RT \ln (K_b) = \Delta H_b - T\Delta S_b \tag{7}$$

Generally, the enthalpy contribution is related to formation of van der Waals interactions, hydrogen bonds, and is affected by dehydration of protein and DNA surface and as such is regarded to represent the sequence-specific contribution to binding. The entropy contribution reflects the favorable release of ions and specifically bound waters from the DNA, dehydration of surfaces, protein rigidification, and loss of rotational and translational degrees of freedom upon complex formation. Formation of protein–DNA complexes is strongly stabilized by the ionic contacts between the positively changed protein sidechains and the negatively charged DNA backbone. The role of electrostatic contribution can be investigated with the experiments at different salt concentrations, [salt], given in mol/L (*see* Subheading 3.5.2):

$$\Delta G_b = \Delta G_{NE} - n_S RT \ln \text{[salt]} \tag{8}$$

where ΔG_{NE} represents the nonelectrostatic free energy (obtained by the extrapolation of ΔG_b to 1 M salt concentration), and n_S is the number of released ions upon binding, which is proportional to number of ionic contacts [11]. The electrostatic part of free energy at physiological salt concentration, which is considered to be purely entropic, is calculated as: $\Delta G_E = \Delta G_b - \Delta G_{NE}$. It has been observed from many protein–DNA complexes that the nonelectrostatic component is related to the sequence-specific binding, even though the majority of the free energy comes from the electrostatic component [11].

Values of thermodynamic parameters can distinguish between binding and specific or nonspecific DNA sequences. For many protein DNA complexes, it had been found that specific binding leads to a much larger heat capacity increment compared to the nonspecific one. For example, binding of lac repressor to the specific sequence is accompanied by the $\Delta C_{P,b} = -1200$ cal/molK, whereas binding to the nonspecific one by only $\Delta C_{P,b} = -200$ cal/molK [13]. The value of heat capacity can also be directly correlated to the buried solvent accessible area; however, these values should be interpreted with care since the surface area parametrizations developed for the protein surface cannot be simply transposed to the DNA surface [14]. Different empirical parametrizations relating to heat capacity and the surface area changes have been introduced to describe dehydration of DNA [15].

Another example of how thermodynamics parameters offer insight into DNA binding is related to different thermodynamics signatures of minor and major groove binders. It has been observed that proteins that bind to major groove are associated with negative ΔH increment, while for those that bind to the minor groove, their ΔH is positive [1, 11]. This is due to the release of tightly bound waters in the minor groove (spine of hydration), which are accompanied by positive entropic and enthalpic signatures. Therefore, binding to the minor groove is usually an entropy-driven process.

3.7 Unexpected Results

Here we briefly present results from the unsuccessful experiments and discuss the probable causes. In many cases, modifications of the experimental setup resulted in a successful experiment. We do not consider situations where the unexpected thermograms are a result of instrument malfunction. These situations are generally covered in the instrument's manual. Performing any of the standard reactions (e.g., water–water titration, titration of Ca^{2+} to EDTA, see Ref. [16]) will quickly reveal if the problem is related to the studied system or to the instrument.

3.7.1 Aggregation

Figure 3a–c shows the examples of ITC thermograms that are affected by the different degrees of aggregation. Severe aggregation was observed upon titrating HigA2 repressor into its operator DNA (*see* Fig. 3a). This is manifested as very strong power spikes of about 1–2 μcal/s (aggregation is usually strongly exothermic reaction), where presumably all DNA in the solution are aggregated together with the antitoxin. The aggregation was caused by low salt concentration used in the experiment that promoted nonspecific protein–DNA interactions. The problem was solved by preforming titrations at higher salt concentration. Figure 3b shows a thermogram with less strong aggregation. After a very strong signal accompanying the first injection, the following injections produced only minor signals, most likely because majority of molecules precipitated out of solution. The last example on Fig. 3c shows a thermogram where only minor aggregation occurs at the end of titration. This is observed as a noisy baseline and visual inspection of the cell contents after the titration confirmed presence of precipitates. In this case, the problem was solved by reversing the titration, that is, switching the solutions in cell and syringe.

3.7.2 Air Bubbles

Air bubbles can be formed due to the dissolution of air from the solution during the experiments or they can be accidently introduced when filling the cell or syringe. Usually, the bubble remains in the cell for some time and at a certain stage pops out of the cell causing a strong power spike in the thermogram, as shown in Fig. 3d. In some cases, the baseline becomes shifted because the absence of an air bubble in the cell changes its heat capacity.

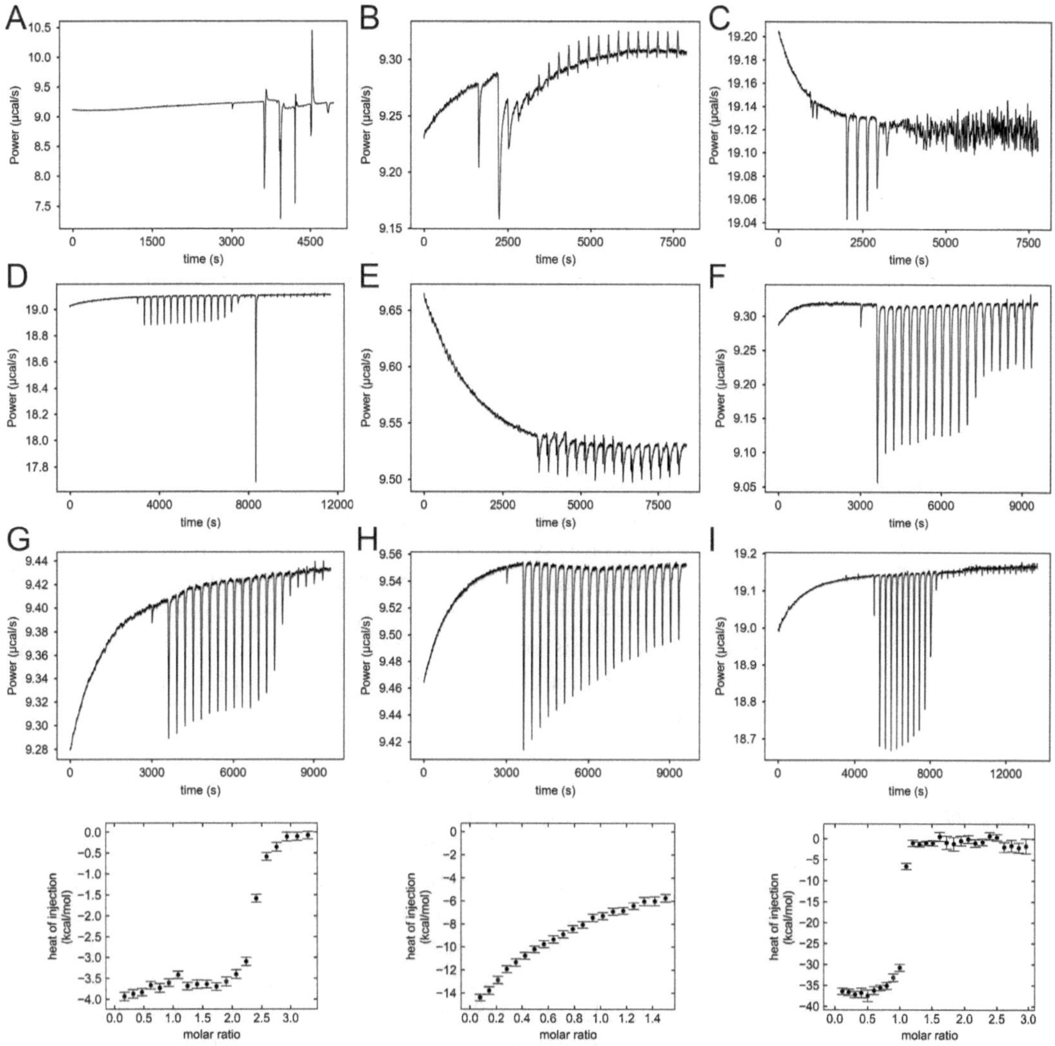

Fig. 3 The wall of "infamous" ITC experiments. (**a–c**) shows ITC thermograms affected by extreme, strong, and mild aggregation, respectively. (**d**) A strong power spike caused by the release of the air bubble from the cell. (**e**) No heat signal even though protein binds to DNA. This situation can occur in some systems where at the studied temperature ΔH_b is close to zero. (**f**) Strong dilution effects at the end of thermogram indicate that samples were not sufficiently dialyzed or a bad batch of protein used. (**g**) Incorrect stoichiometry ($n = 2.3$, see the isotherm in the panel below) was due to the error in the DNA concentration. (**h**) Low affinity of the protein–DNA interaction leads to monotonic isotherms that contain little information. (**i**) Too high affinities result in steep isotherms, which cannot be used to determine the binding affinity

3.7.3 No Signal

No signals or a very small one usually indicates that there is no binding or that binding is too weak in the studied concentration ranges (*see* **Note 5**). In such cases, one would only observe very small power signals (0.02 μcal/s) which are caused by the heat of dilution (dilution of protein into the buffer–DNA mixture). Such dilution enthalpy effects are typically less than 1 kcal/mol.

However, in some rare cases, the protein may bind the DNA, but at the studied temperature, the enthalpy of the reaction is exactly zero, resulting in absence of heat effects. Equation 6 and the inset in Fig. 2a show that due to the nonzero heat capacity, the enthalpy changes may pass through zero at some temperature. The thermogram in Fig. 3e illustrates this situation for binding of IdaS repressor to its operator DNA fragment at 15 °C, where enthalpy of reaction is zero resulting in no binding-induced heat effect although the protein does bind to DNA. The result of such experiment can be confusing, especially when it is known from other experiments that the two molecules bind at these concentrations. One only needs to repeat the experiment at different temperature to obtain a measurable binding-induced heat effects.

3.7.4 High Dilution Effects

High dilution effects are usually caused by insufficient dialysis of the samples and can be observed as persistently strong signals even near the end of titrations (*see* Fig. 3f). The expected heat signal due to dilution effects should be lower than 0.05 μcal/s. Strong signals near the end of titration may also indicate that the protein still binds to the DNA: this may happen when affinity is low. Additional dialysis or a new batch of protein or DNA samples typically resolves the problem.

3.7.5 Unexpected Stoichiometry

In many cases, binding stoichiometry (n) is known from the complementary techniques. It may thus come as a surprise when the stoichiometry, observed as the position of the inflection point of the isotherm, differs from what is expected. The example in Fig. 3g shows a titration of IdaS into its operator sequence where the isotherm shown below suggests the 2.3 molecules bind to the operator sequence. However, it is known from other experiments that 2 molecules of IdaS bind to the operator fragment. Such a discrepancy is commonly observed and is related to the uncertainty in determination of DNA duplex concentration. The simplest approach is to ignore small differences between the observed and expected stoichiometry and adjust the DNA concentration manually to obtain physically reasonable value of n ($n = 1, 2$, etc.). Some programs adjust concentrations of species to operate with the physically realistic values of stoichiometry (e.g., SADPHAT software corrects the concentrations by defining the corresponding "incompetent fraction"), while others do not (e.g., Origin). Alternatively, systematic discrepancy in the stoichiometry may also reveal some unexpected process and should be investigated further. The best approach to find out whether the shift in the stoichiometry is due to concentration error is to perform a reverse titration, that is, to titrate the DNA in the protein. If the concentration of DNA is corrected based on the first, direct titration (to obtain physically reasonable value of n), the binding enthalpies should be the same irrespective of direction of titration, confirming the presumed stoichiometry and indicating an error in the DNA concentration.

3.7.6 Affinity Is Too Low

When protein–DNA affinity is too low ($K_b < 10^4$ M^{-1}), the isotherm loses sigmoidal shape and looks monotonic, which can be observed from the nonspecific protein–DNA binding. A typical isotherm for low affinity binding would have strong heat signals that change little during the titration and persist till the end of titration since the DNA molecule is still not saturated. The thermogram in Fig. 3h shows low affinity binding of a truncated version of HigA2 antitoxin to its cognate DNA. Such monotonic isotherms are difficult to analyze and accurate determination of stoichiometry, enthalpy, and interaction affinity is problematic. In some cases, even monotonic isotherms with affinities around $K_b = 10^3$–10^4 M^{-1} can be adequately analyzed, particularly using a global approach, or using the stoichiometry as a fixed parameter, or by supplying additional information from the reverse titrations (see reference [18]). Generally, the problem may be solved by increasing protein and DNA concentrations to obtain more informative sigmoidal isotherms and by increasing the final ratio r (*see* **Note 5**). In the case of the truncated HigA2, this was achieved by increasing the concentration of DNA in the cell form 2 μM to 20 μM, enabling accurate estimation of thermodynamic parameters. However, one may run into problems with aggregation and high consumption of the sample. In such case, one may also try to increase the affinity by lowering the temperature or by lowering ionic strength (see Subheadings 3.5.1 and 3.5.2). For illustration, if we consider that the binding enthalpy is −20 kcal/mol, a decrease of temperature from 25 to 10 °C can increase the apparent affinity constant by around three to five times. Similarly, lowering the ionic strength from 250 mM to 100 mM and assuming that about 4 ionic contacts are made upon interaction can increase the interaction affinity by an order of magnitude.

3.7.7 Affinity Is Too High

The problem of very high affinity ($K_b > 10^9$ M^{-1}) is more often encountered in protein–protein and protein–ligand reactions. The resulting isotherms are too steep, and the affinity cannot be determined accurately. However, such isotherms provide an excellent estimate of enthalpy and stoichiometry. An example of such experiments is shown in Fig. 3i for the binding of two proteins. To decrease steepness of such isotherms, it is common to use competition titrations (see [18]) but one may also try reducing the cell concentrations as low as possible, increase the temperature if the reaction is exothermic, or increase salt concentration to reduce the apparent binding constant. Generally, VP-ITC is better suited to study high affinity interactions, since higher volume of the cell permits using lower sample concentrations while still obtaining a measurable signal.

4 Notes

1. In our experience, it is sufficient to order oligos that have been purified using the reverse-cartage step that removes DNA truncamers. In our study, using oligos with higher purity (PAGE or HPLC purification) gave identical results. A recent study has found that PAGE purification produces only marginal increase in oligo purity [18].
2. To calculate concentration, it is necessary to use appropriate extinction coefficients of single DNA strands (to prepare an equimolar mix) and, in the second stage, of duplex DNA. Note that due to hypochromic effect, the extinction coefficient of DNA in duplex is smaller compared to one in the single strand. Extinction coefficient of short oligos can be estimated using the nearest-neighbor method by Cantor et al. with the revised parameters [20, 21]. This method is also implemented on several online servers or websites of commercial DNA suppliers (e.g., IDT DNA).
3. Ideally, if the solution contains an equimolar mix of both strands, then a pure DNA duplex should be obtained after annealing. However, this may not be the case because small amounts of either strand can be present in solution due to inexact estimation of concentration. In principle, the presence of a small amount of single-stranded DNA should not affect the ITC experiment; however, it can increase the error in duplex concentration. This is because the measured duplex absorbance will contain an additional contribution of the residual single strand, leading to the overestimation of duplex concentration. We suggest that after annealing, the purity of duplex DNA is verified using a native-PAGE. If the solution of duplexes is carefully prepared, we found that the resulting ITC isotherms are identical to those using FPLC-purified duplexes.
4. If a buffer with high ionization enthalpy is used (such as Tris), the enthalpy of reaction can be overestimated in the cases where exchange of protons occurs during binding. The ionization enthalpies of different buffers are listed in reference [22]. Thus, if buffers with high ionization enthalpy are used, it is recommended that a control experiment is performed with any other buffer to verify if the obtained binding enthalpy is the same. This indicates no exchange of protons during protein–DNA binding.
5. Effects of concentration errors are discussed in the reference [23]. As a general rule, the sample whose concentration can be determined more reliably should be used in the syringe. The reason is that the error in the syringe concentration affects both

the value of ΔH and the curvature of the isotherm, therefore, also the determination of affinity and stoichiometry. However, error in the cell concentration affects only the determination of stoichiometry. In our experience, concentrations of proteins, provided they have sufficient number of aromatic residues, can be determined more accurately compared to DNA and are therefore placed in the syringe.

6. Choosing an appropriate concentration depending on the protein–DNA affinity. The aim is to (1) measure the injection heat with sufficient accuracy, (2) to obtain isotherms with sufficient curvature, and (3) to perform titration till the reaction is complete (saturation). The sigmoidal shape of the isotherm depends on the product between the protein–DNA association (binding) constant (K_b) and the concentrations of reactant in the cell (c) such that the value of this product should be in the range between 1 and 1000: $1 < K_b\,c < 1000$. Thus, for the initial experiments, it is required to approximately estimate the magnitude of association constant and choose the appropriate concentration of reactant in the cell. Thus, for very low affinity ($K_b = 10^5$ M^{-1}), concentration in the cell should be around 100 μM. In contrast for high affinities ($K_b = 10^9$ M^{-1}), low concentrations should be used, around 1 μM.
7. Although the nominal cell volume is 1386 μL, one needs at least 1800 μL to fill the cell. One approach to fill the cell is to inject the sample with 2–3 fast pushes of the syringe to expel any bubbles residing on the wall of the cell. Similarly, while the nominal volume of the titrator syringe is 287 μL, one requires a minimum of 430 μL to fill it. For small cell instruments (iTC200 and similar), about 300 μL are needed to fill the cell and about 50 μL for the syringe.
8. When reaction is exothermic (negative ΔH_b), increasing temperature will lead to stronger signals while the association constant will decrease. In contrast, signal strength will decrease and association constant will increase at lower temperatures, therefore one may need to use higher concentrations of samples for experiments performed at 5–15 °C. The reverse holds for the endothermic reactions (positive ΔH_b).
9. In this setup, the solution in cell (after the titration) becomes slightly diluted and the concentrations of species in the cell need to be adjusted by factor 0.894. Therefore, concentration of the DNA in the cell for the second titration is only about 90% of that used for the first titration.

References

1. Privalov PL, Crane-Robinson C (2018) Forces maintaining the DNA double helix and its complexes with transcription factors. Progress in Biophysics and Molecular Biology 135: 30–48
2. Tellinghuisen J (2012) Designing isothermal titration calorimetry experiments for the study of 1:1 binding: Problems with the "standard protocol". Analytical Biochemistry 424(2): 211–220
3. Scheuermann TH, Brautigam CA (2015) High-precision, automated integration of multiple isothermal titration calorimetric thermograms: New features of NITPIC. Methods 76: 87–98
4. Lah J, Drobnak I, Dolinar M, Vesnaver G (2008) What drives the binding of minor groove-directed ligands to DNA hairpins? Nucleic Acids Research 36(3):897–904
5. Houtman JC, Brown PH, Bowden B, Yamaguchi H, Appella E, Samelson LE, Schuck P (2007) Studying multisite binary and ternary protein interactions by global analysis of isothermal titration calorimetry data in SEDPHAT: Application to adaptor protein complexes in cell signaling. Protein Science 16(1):30–42
6. Drobnak I, De Jonge N, Haesaerts S, Vesnaver G, Loris R, Lah J (2013) Energetic basis of uncoupling folding from binding for an intrinsically disordered protein. Journal of the American Chemical Society 135(4): 1288–1294
7. Drobnak I, Vesnaver G, Lah J (2010) Model-based thermodynamic analysis of reversible unfolding processes. The Journal of Physical Chemistry B 114(26):8713–8722
8. Dragan AI, Li Z, Makeyeva EN, Milgotina EI, Liu Y, Crane-Robinson C, Privalov PL (2006) Forces driving the binding of homeodomains to DNA. Biochemistry 45(1):141–151
9. Freiburger L, Auclair K, Mittermaier A (2015) Global ITC fitting methods in studies of protein allostery. Methods 76:149–161
10. Privalov PL, Dragan AI, Crane-Robinson C, Breslauer KJ, Remeta DP, Minetti CA (2007) What drives proteins into the major or minor grooves of DNA? Journal of Molecular Biology 365(1):1–9
11. Privalov PL, Dragan AI, Crane-Robinson C (2011) Interpreting protein/DNA interactions: Distinguishing specific from nonspecific and electrostatic from nonelectrostatic components. Nucleic Acids Research 39(7): 2483–2491
12. Vandervelde A, Drobnak I, Hadži S, Sterckx YG, Welte T, De Greve H, Charlier D, Efremov R, Loris R, Lah J (2017) Molecular mechanism governing ratio-dependent transcription regulation in the ccdAB operon. Nucleic Acids Research 45(6):2937–2950
13. Frank DE, Saecker RM, Bond JP, Capp MW, Tsodikov OV, Melcher SE, Levandoski MM, Record MT Jr (1997) Thermodynamics of the interactions of lac repressor with variants of the symmetric lac operator: Effects of converting a consensus site to a non-specific site. Journal of Molecular Biology 267(5):1186–1206
14. Hadži S, Lah J (2021) Origin of heat capacity increment in DNA folding: The hydration effect. Biochimica et Biophysica Acta - General Subjects 1865(1):129774
15. Dragan AI, Read CM, Crane-Robinson C (2019) Hydration differences between the major and minor grooves of DNA revealed from heat capacity measurements. European Biophysics Journal 48(2):131–138
16. Medoš Ž, Čobanov I, Bešter-Rogač M et al (2020) Usually overlooked problems related with measurements of high-heat effects using power compensation isothermal titration calorimetry. Journal of Thermal Analysis and Calorimetry
17. Bončina M, Lah J, Reščič J, Vlachy V (2010) Thermodynamics of the lysozyme-salt interaction from calorimetric titrations. The Journal of Physical Chemistry B 114:4313–4319
18. Velazquez-Campoy A, Freire E (2006) Isothermal titration calorimetry to determine association constants for high-affinity ligands. Nature Protocols 1(1):186–191
19. Pinto A, Chen SX, Zhang DY (2018) Simultaneous and stoichiometric purification of hundreds of oligonucleotides. Nature Communications 9:2467
20. Cantor CR, Warshow MM, Shapiro H (1970) Oligonucleotide interactions. III. Circular dichroism studies of the confromation of deox- yoligonucleotides. Biopolymers 9: 1059–1077
21. Cavaluzzi MJ, Borer PN (2004) Revised UV extinction coefficients for nucleoside-5-′-monophosphates and unpaired DNA and RNA. Nucleic Acids Research 32(1):e13

22. Goldberg RN, Kishore N, Lennen RM (2002) Thermodynamic quantities for the ionization reactions of buffers. Journal of Physical and Chemical Reference Data 31:231–370
23. Kantonen SA, Henriksen NM, Gilson MK (2017) Evaluation and minimization of uncertainty in ITC binding measurements: Heat error, concentration error, saturation, and stoichiometry. Biochimica et Biophysica Acta - General Subjects 1861(2):485–498

Chapter 14

Determination of RNA Structure with In Vitro SHAPE Experiments

Rani Baes, Daniel Charlier, and Eveline Peeters

Abstract

The structure of an RNA molecule is often critical for its correct functioning, post-transcriptional regulation, and/or translation. Predicting RNA secondary structures with in silico tools is relatively straightforward with the large array of software and webservers available. However, for long RNAs and RNA at high temperatures, in silico predictions are less accurate and require experimental validation. To this end, a variety of structural probing reagents are commonly used, both for in vitro and in vivo mapping of RNA structure. Selective 2′-hydroxyl acylation analyzed by primer extension (SHAPE) experiments make use of a nonbase-specific modifying reagent, acylating all conformationally flexible (mainly single-stranded or unpaired) nucleotides and have been employed both for in vitro and in vivo modification of RNA. Here, we describe a protocol for an in vitro SHAPE experiment, starting from in vitro transcribed RNA using a T7 polymerase system. This RNA is folded and subsequently modified in vitro with the SHAPE-reagent N-methyl isatoic anhydride (NMIA). Primer extension employing a radioactive ^{32}P-labeled primer that binds the RNA downstream of the structure of interest will generate cDNA until the reverse transcriptase enzyme is halted by the introduced SHAPE modifications. Denaturing acrylamide gel electrophoresis of the pool of ^{32}P-labeled cDNAs and the corresponding sequencing ladders, followed by autoradiography, will expose these stops in reverse transcription (RT) and will therefore enable to identify single-stranded nucleotides in the RNA of interest. These RT stops and NMIA-modification efficiencies can be quantified with ImageJ software and can be used to validate or increase the accuracy of RNA secondary structure predictions.

Key words RNA structure, SHAPE, NMIA, in vitro transcription, RNA gel electrophoresis, Primer extension, Sequencing ladders, Denaturing acrylamide gel electrophoresis, ^{32}P-labeling, Autoradiography

1 Introduction

It has been known for over 30 years that the structure of nucleic acids can affect DNA replication, gene expression, and post-transcriptional regulation and that this is often very dynamic. Besides the typical structure of DNA, the RNA structure is also often critical for the correct functioning of an RNA species (e.g., many noncoding RNA species such as tRNA, rRNA, ribozymes,

Eveline Peeters and Indra Bervoets (eds.), *Prokaryotic Gene Regulation: Methods and Protocols*, Methods in Molecular Biology, vol. 2516, https://doi.org/10.1007/978-1-0716-2413-5_14,

and small RNAs), its correct post-transcriptional regulation and translation (e.g., mRNA riboswitches, RNA thermometers).

The two-dimensional RNA secondary structure encompasses the representation of the RNA due to Watson–Crick (WC) base pairing of the nucleotides (nts) (A-T and G-C pairing and the G-U wobble pair). However, pairing between RNA nts is not limited to these WC interactions and could also include any hydrogen-bonding interactions between sugar- and Hoogsteen-edge of nts and helical stacking, giving rise to a particular three-dimensional RNA tertiary structure [1]. Whereas prediction of these non-WC interactions is still challenging (mainly because of lack of experimental data) [1, 2], a variety of software tools and online webservers are available for the prediction of secondary RNA structures, of which the most known ones include RNAfold web server (Vienna RNA Websuite) [3], Mfold [4], and RNAstructure [5]. These software tools enable to determine the most probable RNA secondary structure(s), for example, based on minimum-free energy models derived from thermodynamic parameters. These parameters are all based on an extensive set of early experiments that were performed in the late 1990s and early 2000s, from which a set of probabilities have been derived for nts to be part of a particular substructure (GU pairs, hairpin loops, bulges, etc.) [6]. However, these experiments were typically performed in "standard conditions", and therefore predictions of RNA secondary structure are often only reliable in a temperature range around 37 °C. Furthermore, predictions are less accurate with increasing length of the RNA sequence. Whereas these software tools are undoubtedly very useful, experimental validation of their predictions is advised or can even be necessary, for example, when interested in RNA structure at multiple temperatures or at temperatures above 50 °C. Furthermore, experimental data can also be added as an input in many RNA structure predictor tools in order to increase the accuracy of the predictions [2].

A wide array of structure probing reagents used for the identification of unpaired nts or mapping of RNA structure is currently available, both for in vivo and in vitro mapping. On the one hand, base-specific modifying reagents such as dimethyl sulfate (DMS) and 1-cyclohexyl-3-(2-morpholinoethyl)carbodiimide metho-*p*-toluenesulfonate (CMCT) have been used extensively in many eukaryotic and bacterial model organisms (e.g., human cells, *Mus musculus*, *Saccharomyces cerevisiae*, *Arabidopsis thaliana*, *Escherichia coli,* and *Salmonella enterica*) [7]. On the other hand, the focus has shifted to reagents with "nonbase-specific" modifying reagents, such as backbone-cleaving reagents or selective 2-′-hydroxyl acylation analyzed by primer extension (SHAPE) reagents that modify conformationally flexible 2′hydroxyl groups of the RNA. Upon employing both types of reagents, "RNA structuromes" have been established for many model organisms by coupling the probing reagent to next-generation sequencing [7, 8].

Arguably one of the most straightforward techniques to determine or validate RNA structure of a well-defined set of RNA species is the "selective 2′-hydroxyl acylation analyzed by primer extension" (SHAPE) technique, which makes use of a SHAPE reagent for the covalent modification of all conformationally flexible nts (which mainly occurs when the nt is not engaged in base pairing) in a particular structure by acylation of the 2′-hydroxyl group of the ribose. A wide array of SHAPE reagents have currently been described, all with their own characteristics in terms of stability (half-life), working concentration, commercial availability, and ease of chemical synthesis [9]. This SHAPE technique can in principle be used in vivo by modifying total, cellular RNA of a (prokaryotic) cell culture and subsequently extracting the modified RNA, thus identifying single-stranded RNA nts in a biological context. However, not all SHAPE reagents are well-suited for an in vivo application since hydrolyzation of the reagent can pose a serious problem under certain growth conditions (e.g., low pH and high temperature) [10, 11]. Moreover, the modification rate of the RNA of interest can be low in the case of a less abundant mRNA, leading to a low specific output signal. Accordingly, a large experimental optimization might be required when establishing an in vivo SHAPE experiment, especially in nonmodel organisms.

Therefore, depending on the research question, in vitro SHAPE (*see* Fig. 1) can be a good alternative, since RNA modification takes place in a well-defined environment and allows making use of easily available SHAPE reagents such as N-methyl isatoic anhydride (NMIA) [8, 12–15]. To this end, the nt sequence corresponding to the RNA of interest can conveniently be cloned in a high copy number plasmid under control of a T7 promoter (*see* Fig. 2) and can then be in vitro transcribed using a T7 polymerase system. This RNA is folded in vitro in a buffer and a temperature of choice and unbound nts are modified with a SHAPE reagent. These acylations can be detected with a primer extension approach, by reverse transcribing the RNA to cDNA starting from a ^{32}P-5′end-labeled primer hybridizing downstream of the region of interest. Modifications of unpaired nts will manifest themselves in stops in the reverse transcription (RT) reaction, therefore yielding cDNAs of different lengths of which the last incorporated nt is the one preceding the modified, unpaired nt. These ^{32}P-labeled cDNAs are separated by denaturing acrylamide gel electrophoresis and visualized by autoradiography. By incorporating a sequence ladder on the gel, the RT stops and last incorporated nt in the cDNA can be identified, and so the single-stranded RNA nts can be inferred. These RT stops and corresponding NMIA modification efficiencies of a particular flexible nt can be quantified with densitometry, for example, using ImageJ software [12].

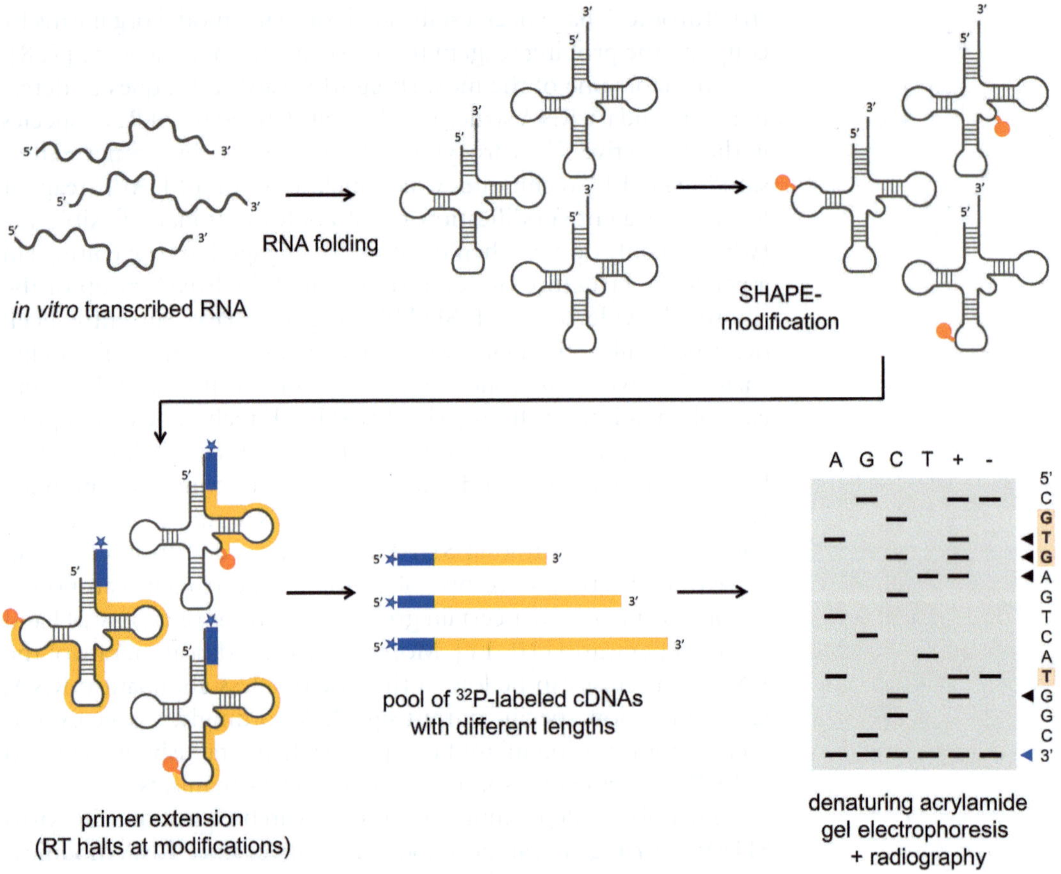

Fig. 1 Workflow of in vitro SHAPE for the mapping of unpaired nucleotides of an in vitro transcribed, folded, and SHAPE-modified (orange label, +) RNA species. Primer extension using a radioactive (^{32}P-) labeled primer (blue with star), allows for reverse transcribing the modified RNA to the complement DNA strand (cDNA) (yellow) up until the nucleotide preceding the modification. By running the corresponding pool of ^{32}P-labeled cDNAs with different lengths on a denaturing acrylamide gel and exposing the latter to autoradiography film, the abundance of a particular length of cDNA can be compared between the SHAPE-modified sample (+) and mock control sample (−) (arrows). The modified, unpaired nucleotide in the RNA of interest is at position $n + 1$ in the 3′→5′ direction (orange background)

2 Materials

2.1 Precautionary Measures

2.1.1 Working RNase-Free

While working with RNA, it is especially important to work sterile and in an RNase-free environment to avoid degradation of the synthesized RNA. This includes wearing gloves, working in a clean laminar flow/biosafety cabinet (this might not always be possible, e.g., when working with radioactively labeled material), decontaminating the surfaces and all material used with 70% ethanol and an RNase decontamination solution (including tip boxes, micropipettes, racks, etc.) and making use of sterile and nuclease-free tubes and filter tips.

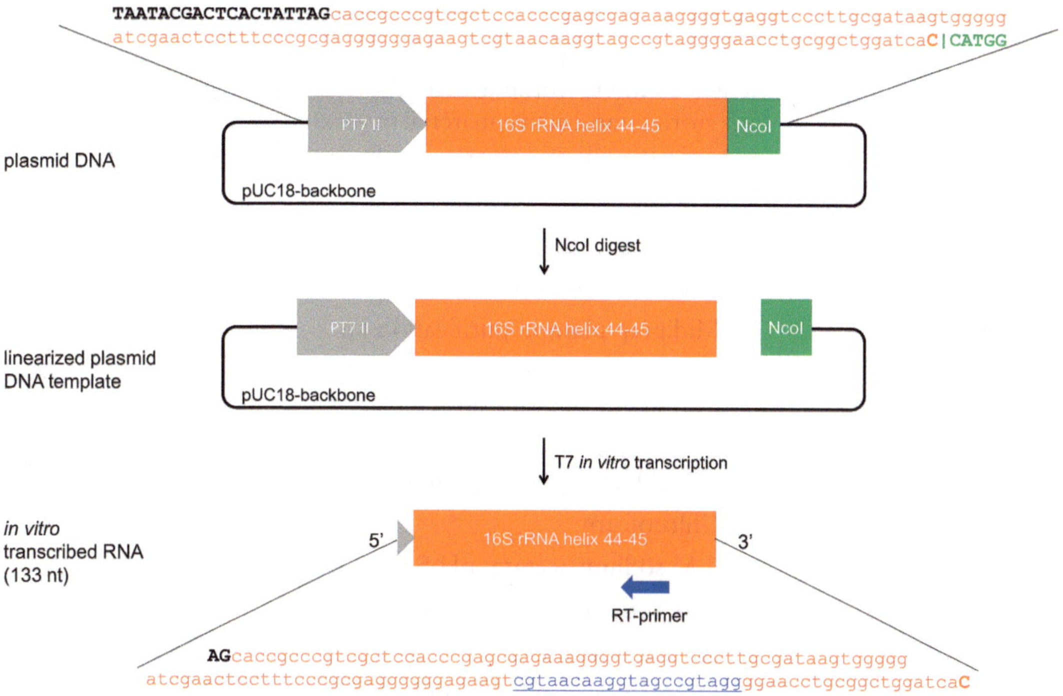

Fig. 2 Example of a plasmid DNA template and in vitro transcribed RNA encoding the last two helices (44 and 45) of *Sulfolobus acidocaldarius* 16S rRNA (S*aci_1300*). DNA encoding the RNA of interest (orange) was cloned under control of the T7 class II promoter (grey) and upstream of the NcoI-restriction site (green) in the high-copy number plasmid pUC18. Transcription is initiated at the last "AG" from PT7 II. The addition of these two extra nts in the final transcript does not affect RNA secondary structure according to the RNA structure software [5]. Note that the addition of "GG" or "GGG" (originating from PT7 III) would have induced changes in the original secondary structure and was therefore not chosen as the promoter sequence. The NcoI-linearized plasmid was used as a template for T7 in vitro transcription, yielding a 133-nt long RNA. The 20-nt long RT primer (blue; 5′-CCTACGGCTACCTTGTTACG-3′) was designed at the 3′ end of the transcript in correspondence with the RT-primer used in [10] and would yield a full-length cDNA of 114 nts during primer extension (in absence of SHAPE modifications)

2.1.2 Working with Ethidium Bromide

The procedure of RNA gel electrophoresis makes use of ethidium bromide (EtBr) for the visualization of nucleic acids. EtBr is a toxic and mutagenic compound. Handle with care, in a dedicated working place. Wear a lab coat and gloves when handling, and discard all possibly contaminated material in a dedicated EtBr waste (e.g., gloves, tips, gels, used running buffer, and paper). Discard gloves when exiting the EtBr area.

2.1.3 Working with Radioisotope ^{32}P

The radioisotope of phosphorus ^{32}P is used for the radioactive labeling of cDNA and undergoes beta decay with a half-life of 14 days. Due to risk of radiation, ^{32}P-ATP can only be ordered with the appropriate licenses, and authorization for isotope handling should only be handled in specialized and well-equipped rooms. Furthermore, all handlings should be performed in a way that limits exposure to radiation, including regularly checking the working surface and gloves for radioactive contamination with a Geiger counter, working behind plexiglass screens and transporting/storing radioactive material in plexiglass or lead boxes. Discard contaminated (liquid) waste in appropriate authorized containers. Wear a personalized dosimeter.

2.2 Preparing the in vitro Transcription Template

1. *Escherichia coli* strain harboring the plasmid template for in vitro transcription.
2. Midiprep plasmid purification kit.
3. Restriction enzyme, corresponding to the introduced restriction site in the plasmid template.
4. Phenol:Chloroform:Isoamyl Alcohol 25:24:1, saturated with 10 mM Tris, pH 8.0, 1 mM EDTA (stored at 4 °C).
5. Chloroform.
6. 3 M sodium acetate, pH 5.2 (stored at 4 °C)
7. 100% ethanol (stored at 4 °C)
8. Nuclease-free water.
9. DNA agarose gel electrophoresis system.
10. Fume hood.
11. Gloves.
12. Vortex.
13. Microcentrifuge.
14. −20 °C incubator (freezer)
15. Flat-tip gel-loading pipet tips (e.g., 1–200 μL flat 0.2 mm thick gel-loading pipet tips).

2.3 T7 In Vitro Transcription

1. MEGAscript™ T7 transcription kit, including dNTP solutions, 10× reaction buffer, T7 enzyme mix, Turbo DNase.
2. Nuclease-free water.
3. Purified linearized plasmid template.
4. Water bath at 37 °C.
5. Vortex.
6. Microcentrifuge.
7. Nanodrop spectrophotometer.
8. Crushed ice.

2.4 RNA Purification

1. MEGAclear™ transcription clean-up kit, including binding solution, wash solution, filter cartridge, collection tube, elution tube.
2. Nuclease-free water.

3. 100% ethanol (molecular biology grade).
4. Heat block at 95 °C.
5. Nanodrop spectrophotometer.
6. Crushed ice.

2.5 Validation of RNA Integrity with RNA Gel Electrophoresis

1. Agarose.
2. DEPC-treated water. Add 1 mL diethyl pyrocarbonate (DEPC) to 1 L double-distilled water, shake vigorously, and incubate overnight at 37 °C. Autoclave.
3. 10× FA gel buffer: 20 mM MOPS, 5 mM sodium acetate, 1 mM EDTA, pH 7.0 in DEPC-treated water. Stored in the dark (e.g., wrapped in aluminum foil).
4. 1× FA gel running buffer: 100 mL 10× FA gel buffer, 20 mL 37% formaldehyde, up to 1 L with DEPC-treated water.
5. 37% formaldehyde.
6. 10 mg/mL ethidium bromide (EtBr) stock solution.
7. 5× RNA loading buffer: 16 μL saturated aqueous bromophenol blue solution, 80 μL 500 mM EDTA, pH 8.0, 720 μL 37% formaldehyde, 2 mL 100% glycerol, 3.084 mL formamide, 4 mL 10× FA gel buffer, 100 μL nuclease-free water. Stored at 4 °C.
8. DNA molecular weight ladder.
9. Dedicated EtBr working place.
10. Dedicated gel tray and electrophoresis tank for RNA work.
11. UV transilluminator.
12. Microwave.

2.6 RNA Folding and RNA Modification with the NMIA SHAPE Reagent

1. Nuclease-free water.
2. 5 M NaOH. Add 10 g of NaOH to 50 mL nuclease-free water. Autoclave.
3. 1 M HEPES, pH 8 (adjust with 5 M NaOH). HEPES cannot be autoclaved for removal of RNase contamination after bringing the solution to pH. Therefore, prepare the solution twice in parallel. Twice add 2.383 g HEPES to 5 mL nuclease-free water. Use one solution to determine with a probe-type pH meter how much volume 5 M NaOH should be added to the solution to bring the pH to 8. Add this volume to the other solution and verify pH with sterile pH sticks. Fill up the total volume to 10 mL with nuclease-free water. Filter sterilize and protect from light by wrapping in aluminum foil.
4. 5 M NaCl. Add 1.461 g NaCl in 5 mL nuclease-free water. Autoclave.

5. 1.12 M $MgCl_2.6H_2O$. Add 2.14 g $MgCl_2.6H_2O$ in 9.36 mL nuclease-free water. Autoclave.
6. 3.3× RNA folding mix: 333 mM HEPES, pH 8.0, 20 mM $MgCl_2$, 333 mM NaCl in nuclease-free water. Mix 333 μL of 1 M HEPES, pH 8, 33.3 μL 1.12 M $MgCl_2.6H_2O$, 66.6 μL 5 M NaCl, and 567 μL nuclease-free water in a sterile, RNase-free 1.5 mL tube.
7. N-Methylisatoic anhydride (NMIA) SHAPE reagent (commercially available). All SHAPE reagents are relatively unstable and should be handled with care. They should be stored anhydrous in a desiccator at 4 °C.
8. Anhydrous DMSO 3 mL aliquots.
9. 100% ethanol (stored at 4 °C).
10. 100 mM EDTA, pH 8 with 5 M NaOH. Dissolve 0.292 g EDTA in 5 mL nuclease-free water. Bring to pH 8 with 5 M NaOH. Fill up the total volume to 10 mL with nuclease-free water. Autoclave.
11. 20 mg/mL RNA-grade glycogen.
12. Two heat blocks or thermocyclers suitable for PCR tubes, one at 95 °C one at temperature of interest (e.g., 37 °C).
13. Vacuum pomp for pulling the desiccator of NMIA vacuum.
14. −80 °C incubator.
15. Nanodrop spectrophotometer.
16. Microcentrifuge for PCR tubes.
17. Microcentrifuge for 1.5 mL tubes.
18. Crushed ice.
19. Very fine, RNase-free, pipet tip.
20. Sterile, RNase-free PCR tubes.
21. Sterile, RNase-free 1.5 mL tubes.

2.7 Primer Extension

2.7.1 [γ-^{32}P] Labeling of the Reverse Transcription Primer

1. Fresh ATP, [γ-^{32}P]-3000 Ci/mmol 10 mCi/mL, 250 μCi.
2. Nuclease-free water.
3. 10 pmol/μL RT primer.
4. T4 polynucleotide kinase (10 U/μL).
5. 10× T4 reaction buffer A (forward reaction).
6. Water bath at 37 °C.
7. Heat block at 65 °C.

2.7.2 Reverse Transcription of the Modified RNA

1. Nuclease-free water.
2. 2 pmol/μL [γ-^{32}P]-labeled RT primer.

3. GoScript™ reverse transcription system, including GoScript Reverse Transcriptase, GoScript 5× Reaction buffer.
4. 25 mM $MgCl_2$, RNA grade.
5. dNTP mix, RNA grade (10 mM each dATP, dGTP, dCTP, and dTTP).
6. 100 mM RNaseOUT™ Recombinant Ribonuclease Inhibitor.
7. 0.1 M DTT.
8. Deionized formamide. Add 3 g Serdolit MB-1 to 200 mL formamide, stir for 10 min, and filter-sterilize.
9. 2× formamide dye. Mix 0.03 g xylene cyanol, 0.03 g bromophenol blue, 96 mL deionized formamide, 4 mL 0.5 M EDTA, pH 8.
10. RNase-free 1.5 mL tubes.
11. RNase-free PCR tubes.
12. Heat block.
13. Thermocycler.
14. Crushed ice.

2.7.3 Preparing the Dideoxy Chain-Termination Sequencing Ladder

1. Thermo sequenase cycle sequencing kit, including ddA termination mix, ddT termination mix, ddG termination mix, ddC termination mix, thermo sequenase reaction buffer, thermo sequenase DNA polymerase, stop solution dye.
2. 2 pmol/μL [γ-^{32}P]-labeled RT primer.
3. Midiprep plasmid with the cloned in vitro transcription construct.
4. Nuclease-free water.
5. Thermocycler.

2.8 Denaturing Acrylamide Gel Electrophoresis

2.8.1 Preparing the Denaturing Acrylamide Gel

1. Urea.
2. 5× TEB: 446 mM Tris, 445 mM boric acid, 10 mM EDTA.
3. 40% acrylamide/bisacrylamide (19:1) (stored at 4 °C). Wear gloves when handling.
4. Double-distilled water.
5. 100% ethanol.
6. 10% acetic acid.
7. γ-methacryloxypropyltrimethoxysilane.
8. 6% polyacrylamide stock solution (500 mL). Wear gloves and mix 210 g urea, 100 mL 5× TEB, 75 mL 40% acrylamide/bisacrylamide (19:1) and 100 mL double-distilled water in a beaker. Heat the solution to 150 °C on a magnetic stirrer until all urea is dissolved. Adjust the volume to 500 mL with double-

distilled water. Filter the solution on a Buchner filter and degas for 10 min. Transfer the solution to a flask and store at 4 °C (up to a couple of months).

9. Sticky solution (50 mL). Mix 48.5 mL 100% ethanol, 1.5 mL 10% acetic acid, and 150 μL γ-methacryloxypropyltrimethoxysilane in a flask. Store the solution at 4 °C (up to a couple of months).
10. Detergent.
11. Repellant (e.g., Blue slick).
12. 70% ethanol (technical).
13. 0.1 g/mL ammonium peroxodisulfate (APS) (stored at 4 °C).
14. *N*,*N*,*N*′,*N*′-tetramethylenediamine (TEMED) (stored in dark, at 4 °C).
15. Heating, magnetic stirrer.
16. Buchner filter and degassing system (vacuum pomp).
17. A set of heat-resistant, glass plates with dimensions of 33.4 × 42.4 × 0.5 cm (large plate) and 33.4 × 39.3 × 0.5 cm (small plate), depending on the type of electrophoresis system.
18. Warm room at 30 °C or incubator large enough to fit the largest glass plate.
19. Two 0.25 mm spacers, long enough to fit on the largest side of the large plate.
20. 20 well comb, fitting perfectly in between the assembly of small and large glass plate with spacers.
21. Rubber sealing system.
22. 50 mL syringe.
23. Clamps to temporarily secure the glass-plate sandwich and to exert pressure on the comb when the gel is polymerizing.

2.8.2 Acrylamide Gel Electrophoresis

1. Vaseline.
2. 5× TEB: 446 mM Tris, 445 mM boric acid, 10 mM EDTA.
3. 1× TEB. Dilute 5× TEB in ddH_2O (final volume of at least 2 L).
4. 10% acetic acid (at least 2 L). Can be reused up to five times.
5. Large plastic tray, big enough to fit the large plate.
6. Flat-tip gel-loading pipet tips (e.g., 1–200 μL flat 0.2 mm thick gel-loading pipet tips).
7. Vertical electrophoresis system with power supply that can go up to 1500 V (usually for a sequencing-type gel electrophoresis, e.g., Hoefer SQ3 Sequencer).

8. 50 mL or 100 mL syringe.
9. Heat block.
10. Small heating system used as hot air blower.
11. Plastic foil (e.g., Saran Wrap or cling film).

2.9 Autoradiography and Interpretation and Quantification of the SHAPE Results

1. Autoradiography film (35 cm × 43 cm).
2. Developer solution for photography (1:5 dilution with water). Protect from light (e.g., brown bottle). Can be reused until color turns brown.
3. Fixator solution for photography (1:5 dilution with water). Protect from light (e.g., brown bottle). Can be reused until a precipitate forms and color turns orange.
4. Dedicated "dark room" with red lights.
5. Black plastic bags, big enough to fit the glass plate/gel/film assembly and protect the film from light.
6. Clamps to secure the glass-plate sandwich with film and close off the black bags.
7. Two large plastic trays for the developer and fixator solutions; big enough to fit the large plate with gel.
8. Office scanner.
9. ImageJ software (imagej.nih.gov/ij/download.html).

3 Methods

Whether it is the first time that a SHAPE experiment is set up, or whether new modification conditions are explored, it is often useful to incorporate a "positive control" in the experiment besides the RNA of interest. This can be an RNA of which the secondary structure is known from previous experimental work or a structure that is conserved in other species (e.g., tRNA and last helical stretches of 16S rRNA). Incorporating this control allows to validate the SHAPE outcome by suggesting the formation of a stable RNA secondary structure in the folding step and proving the modification efficiency of the SHAPE reagent in the conditions tested. The protocols below can be applied and performed in parallel for the RNA of interest and the positive control.

3.1 Designing the DNA Template and Reverse Transcription Primer for In Vitro Transcription

As a first step, a DNA template and reverse transcriptase (RT) primer need to be designed for each RNA species under study. An example of a DNA template for 16S rRNA is given in Fig. 2. The template DNA for the in vitro transcription reaction, encoding the RNA of interest can be PCR-amplified from genomic gDNA using a forward primer that incorporates a suiting T7 promoter (*see* **Note 1**). In some cases, it can be efficient to attach specific linkers to the 5′ and 3′ end as well, DNA sequences encoding stable RNA stem-loop structures, which will force stable folding of the RNA of interest.

Whereas the T7 in vitro transcription with the MEGAscript Kit should work with pure PCR fragments encoding the DNA for the RNA of interest under control of a suiting T7 promoter as a template, performing run-off in vitro transcription from the DNA template cloned in a high-copy number plasmid backbone (e.g., pUC18) resulted in higher RNA yields. This can be most likely explained by the presence of contaminants of the PCR reaction in the preparation of the purified template. In this case, it is necessary to include a transcription terminator site or restriction site downstream of the region of interest (or 3′ end spacer) (*see* **Note 2**).

Lastly, the primer which will be used for the RT reaction should be designed according to the classical "rules" of primer design. However, it should be noted that depending on the electrophoresis system and acrylamide percentage of the gel, very short and very long cDNA fragments might be difficult to distinguish. Therefore, the primer should be located about 5–10 nts upstream of the region of interest and can be used for the RT reaction of an RNA fragment of about 90–150 nts (*see* **Note 3**). When designing the DNA template, it can be opted to attach a specific binding site for the primer at the 3′ end of the construct. This allows making use of one-labeled primer and can be beneficial when many primer extension reactions are carried out in parallel, however, this is not required.

3.2 Preparing the In Vitro Transcription Template

1. Extract the plasmid with the cloned in vitro transcription construct of the RNA of interest from an *E. coli* culture with a plasmid midiprep kit according to the manufacturer's instructions (*see* **Note 4**).
2. Linearize the plasmid with the restriction enzyme of which the corresponding restriction site was added downstream of the construct, according to manufacturer's instructions. Perform multiple reactions in parallel to get sufficient template DNA for the in vitro transcription reaction (0.2–1.5 pmol of digested template DNA will be required). Verify efficient restriction by agarose gel electrophoresis (*see* **Note 5**).
3. After linearization, extract the digested plasmid with phenol-chloroform extraction and ethanol precipitation as follows. Work in the fume hood and wear gloves.
4. Pool all restriction reactions to a particular total volume in a 1.5 mL tube (=Tube 1).
5. Extract DNA with a 1:1 phenol:chloroform mixture. Add an equal volume of cold Phenol:Chloroform:Isoamyl Alcohol 25:

24:1, saturated with 10 mM Tris, pH 8.0, 1 mM EDTA to the linearized plasmid DNA in Tube 1. Vortex for 5 s and centrifuge for 5 min in a microcentrifuge at maximum speed. Transfer the upper (aqueous) phase, containing the DNA, to a new tube (=Tube 2). The approximate volume corresponds to the initial volume of the pooled reactions.

6. Extract twice with an equal volume of chloroform to remove residual phenol. Add an equal volume of chloroform to Tube 2. Vortex for 5 s and centrifuge for 3 min at maximum speed. Transfer the upper (aqueous) phase, containing the DNA, in a new tube (=Tube 3). Repeat a second time and transfer the upper (aqueous) phase in a new tube (=Tube 4).
7. Leave Tube 4 open for ±10 min under the fume hood to evaporate the remaining chloroform but avoid overdrying the pellet.
8. Precipitate the DNA with ethanol. Add one-tenth volume of cold 3 M sodium acetate, pH 5.2, and two volumes of cold 100% ethanol to Tube 4 and invert a couple of times to mix well. Incubate at −20 °C for at least 60 min (you can even leave this overnight). Pellet the DNA in a microcentrifuge for 15 min at top speed. Remove the largest volume of supernatant and carefully remove all remainder with a very fine pipet tip.
9. Rinse the pellet by adding 500 μL of cold 70% ethanol and centrifuging for 15 min at maximal speed. Remove the largest volume of supernatant and carefully remove all remainder with a very fine pipet tip.
10. Air dry the pellet in Tube 4 for about 10 min under the fume hood to evaporate the remaining ethanol but avoid overdrying the pellet, since this complicates redissolving the DNA.
11. Add 10–100 μL nuclease-free water to the DNA pellet in Tube 4 while ensuring not to touch the pellet. Vortex for a couple of seconds to dissolve the pellet and give it a quick spin in the microcentrifuge.
12. Store the purified linearized plasmid at −20 °C.

3.3 T7 in vitro Transcription

This protocol is based on the manufacturer's protocol of MEGAscript™ T7 Transcription Kit. However, since in most cases the RNA of interest for SHAPE is rather short (<150 nt), the reaction conditions mentioned below are optimized for small transcripts (<500 nt) (*see* **Note 6**).

1. Preheat the water bath at 37 °C.
2. Continue working in an RNase-free environment for the preparation of the components of the in vitro transcription reaction.

3. Vortex and briefly spin down the template DNA (purified linearized plasmid with the correct insertion of the DNA encoding the RNA of interest). Keep at room temperature.
4. Thaw and vortex the four dNTP solutions (ATP, CTP, GTP, and UTP) until completely in solution. Make a mix of the four dNTP solutions by adding equal volumes of each (2 μL of each dNTP for each in vitro transcription reaction). Vortex and spin down briefly. Keep the mix at room temperature and store the separate dNTP solutions at −20 °C.
5. Thaw and vortex the 10× reaction buffer until completely in solution. Make an aliquot of the volume needed (2 μL per in vitro transcription reaction) and keep this at room temperature. Store the remainder of reaction buffer at −20 °C.
6. Prepare an aliquot of the volume of T7 enzyme mix needed (2 μL per in vitro transcription reaction). Keep on ice and put it at room temperature just before assembling the reactions. Store the remainder of T7 enzyme mix at −20 °C.
7. In an RNase-free environment, assemble the in vitro transcription reaction at room temperature in the following order. Do not assemble on ice.

   x μL nuclease-free water (up to a total volume of 20 μL).

   8 μL dNTP mix.

   2 μL 10× reaction buffer.

   0.2–1.5 pmol of linear plasmid DNA template (*see* **Note 7**).

   2 μL T7 Enzyme Mix.
8. Mix well by pipetting and quick spin in a microcentrifuge.
9. Incubate the in vitro transcription reaction in a 37 °C water bath for 6–16 h (*see* **Note 8**).
10. After incubation, quick spin the reaction in a microcentrifuge and add 1 μL of Turbo DNase to the 20 μL reaction to get rid of the template DNA. Incubate in the 37 °C water bath for 15 min.
11. Quick spin in a microcentrifuge and keep the reaction on ice from now on.
12. Measure the RNA concentration with a nanodrop spectrophotometer (*see* **Note 9**).
13. *! Pause Point!* Store the in vitro transcription reactions at −80 °C (up to a couple of days) or proceed immediately with RNA purification.

3.4 RNA Purification

Unincorporated nts, salts, and proteins will be removed from the RNA in the reaction using the MEGAclear™ Kit. Although it is mentioned in this kit's instructions that it can only be used for the purification of ssRNA molecules larger than 100 nts, RNA molecules with a length of 90 nts were successfully purified in our experiment.

1. In case the RNA is frozen, thaw on ice. Otherwise, continue working on ice.
2. Preheat an aliquot of nuclease-free water (110 μL per sample) to 95 °C in a heat block until required for the elution step (*see* **steps 9** and **10**).
3. Add nuclease-free water (at room temperature) to the in vitro transcription reaction up to a total volume of 100 μL (normally 80 μL water to the 20 μL reaction) and mix by pipetting.
4. Add 350 μL of the binding solution to the sample and mix by pipetting.
5. Add 250 μL of 100% ethanol to the sample and mix by pipetting.
6. Prepare the column by inserting a filter cartridge into a collection tube. Pipet the sample (±700 μL) onto the filter cartridge and centrifuge 1 min at 10,000–15,000 × *g* at room temperature. Discard the flow-through and place the filter cartridge back into the collection tube.
7. Wash the filter by adding 500 μL wash solution (*see* **Note 10**) to the column and centrifuge 1 min at 10,000–15,000 × *g* at room temperature. Discard the flowthrough.
8. Wash the filter by adding a second aliquot of 500 μL wash solution to the column and centrifuge 1 min at 10,000–15,000 x *g* at room temperature. Discard the flow-through and centrifuge for an additional 30 s to remove the last traces of wash solution.
9. Place the filter cartridge with the bound RNA into a new elution tube and elute the RNA by adding 50 μL of the pre-heated nuclease-free water to the filter and centrifuging for 1 min at 10,000–15,000 × *g* at room temperature.
10. Increase the RNA yield by adding a second aliquot of 50 μL pre-heated nuclease-free water and centrifuging for 1 min at 10,000–15,000 x *g* at room temperature.
11. Keep the 100 μL of purified RNA on ice from now on.
12. Measure the RNA concentration and determine the purity of the undiluted sample with a nanodrop spectrophotometer (*see* **Note 11**).
13. Prepare an aliquot of the purified RNA for RNA gel electrophoresis (6–12 μL).

14. *! Pause Point!* Store the remainder of purified RNA at −80 °C or proceed immediately with RNA folding, SHAPE modification and primer extension.

3.5 Validation of RNA Integrity with RNA Gel Electrophoresis

Whereas the spectrophotometric measurements by nanodrop already give an idea of the RNA quality, an additional validation of the RNA integrity by RNA gel electrophoresis is often useful to ensure that the RNA was not degraded during manipulation or after freeze thaw in case the purified RNA was stored for a long time at −80 °C.

1. Prepare the gel mixture for a 1% agarose gel by mixing 1.0 g agarose with 10 mL 10× FA gel buffer and 90 mL DEPC-treated water in a sterile flask. Heat the mixture in a microwave until the agarose is completely dissolved and cool down the agarose solution to 65 °C (*see* **Note 12**).
2. After cooling of the gel mixture, add 1.8 mL of 37% formaldehyde to the agarose solution in a fume hood. Wear gloves.
3. In the EtBr laboratory, clean a dedicated gel tray and electrophoresis tank for RNA work with 70% ethanol and RNase decontamination solution. Pour approximately 100 mL of the agarose solution onto the gel tray. Add 1 μL of a 10 mg/mL EtBr stock solution to the poured agarose solution and mix by moving the pipette tip throughout the solution. Insert the comb and let the gel solidify for at least 30 min.
4. Remove the comb and place the gel in an electrophoresis tank filled with 1× FA gel running buffer. Equilibrate the gel by running for at least 30 min at 100 V.
5. Prepare the RNA sample by mixing 6–12 μL of the RNA solution with 3 μL of 5× RNA loading buffer and complete with nuclease-free water until a total volume of 15 μL. Denature the RNA by incubating for 5 min at 65 °C in a heat block and immediately put on ice afterward for a couple of minutes.
6. Load the denatured samples on the pre-ran gel (*see* **Note 13**) and run the gel electrophoresis at ±100 V for about 45 min. Visualize under UV light.

Pure and intact ssRNA should result in one clear band visible on gel (*see* Fig. 3). Bands (or smears) at a lower molecular weight indicate that the RNA was degraded during the process of in vitro transcription or purification. In this case, the source of RNase contamination needs to be identified and the in vitro transcription reaction should be repeated with more care to work RNase-free. Bands at a higher molecular weight indicate that there was still undigested template plasmid left in the in vitro transcription reaction (no run-off transcription). However, this should not complicate the SHAPE modification and primer extension if the RNA band of interest is overrepresented.

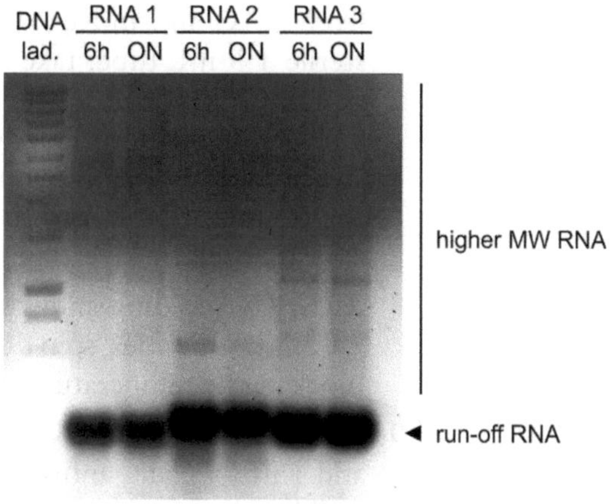

Fig. 3 Example of an RNA gel electrophoresis analysis of three different RNA species with length 90–133 nts, constructed by run-off T7 in vitro transcription (either 6 h or overnight incubation (ON)) followed by purification. The DNA ladder (DNA lad.) is solely for following the electrophoresis run and cannot be used for alignment purposes. Whereas one clear RNA band is visible in all lanes, some higher molecular weight (MW) RNA bands are present, most likely originating from transcription of undigested plasmid left in the template plasmid preparation

3.6 RNA Folding and RNA Modification with the NMIA SHAPE Reagent

1. Preheat two heat blocks or thermocyclers suitable for PCR tubes at 95 °C and a temperature of interest (e.g., 37 °C).
2. Prepare the RNA sample by adding 5 pmol of purified, in vitro transcribed RNA in a total volume of 12 μL nuclease-free water in a sterile PCR tube. This amount of RNA is sufficient for one NMIA-modification reaction (+) and one negative mock control (−) (*see* **Note 14**).
3. Denature the RNA by incubating for 2 min at 95 °C and then put on ice for 2 min.
4. When the RNA sample has cooled down, add 6 μL of 3.3× RNA folding mix and mix by pipetting (*see* **Note 15**). Keep on ice.
5. Prepare two new PCR tubes and label them with "+" and "−" for the NMIA-modification reaction and one negative mock control, respectively. Add 9 μL of the RNA sample in folding mix to both tubes. Keep on ice.
6. Incubate the "+" and "−" tubes at the temperature of interest (e.g., 37 °C) for 20 min.
7. In the meanwhile, prepare the 10× NMIA stock solution in DMSO (130 mM). Make the stock solution as fresh as possible. In order to avoid hydrolysis of the NMIA reagent, work as fast as possible and close the NMIA container as quickly as possible. Take out the desiccator with the container of NMIA powder

from 4 °C and move it toward the vacuum pomp. Using a sterile 1.5 mL tube, take some NMIA powder from the container. Close the tube immediately. Close the NMIA container and desiccator immediately and pull vacuum. Using a second sterile tube, weigh off 69 mg of NMIA. In the laminar flow, add the 69 mg of NMIA to the vial with 3 mL anhydrous DMSO and mix well by inverting.

8. After RNA folding and when condensation is observed on the "+" and "−" tubes, spin down briefly and place back in the heat block at the temperature of interest.
9. Add 1 μL of the 10× NMIA stock solution to the "+" tube and add 1 μL of plain, anhydrous DMSO to the "−" tube and mix well by pipetting.
10. Incubate the reactions at the temperature of interest for five NMIA hydrolysis half-lives (e.g., 45 min at 37 °C) (*see* **Note 16**).
11. After RNA modification, spin down the "+" and "−" tubes briefly and transfer the 10 μL reactions to new, RNase-free 1.5 mL tubes.
12. Precipitate the modified RNA by ethanol precipitation. To each tube, add 350 μL 100% ethanol, 90 μL nuclease-free water, 4 μL 5 M NaCl, 2 μL 100 mM EDTA pH 8, and 1 μL 20 mg/mL glycogen. Incubate the tubes at −80 °C for at least 1 h. Pellet the RNA in a microcentrifuge at 4 °C for 30 min at top speed. Remove the largest volume of supernatant and carefully remove all remainder with a very fine pipet tip. Let the tubes stand open for ±10 min in the laminar flow to evaporate the remaining ethanol but avoid overdrying the pellet.
13. Add 10 μL of nuclease-free water to the RNA pellet. Do not resuspend in order to avoid loss of RNA in the pipet tip. Instead, incubate and vortex briefly. Spin down briefly and keep the RNA on ice from now on.
14. Measure the modified RNA concentration with a nanodrop spectrophotometer.
15. Modified RNA can be stored a couple of days at −80 °C, however, it is advised to continue immediately with primer extension due to increased stability of cDNA.

3.7 Primer Extension

3.7.1 [γ-^{32}P] Labeling of the Reverse Transcription Primer

The described labeling reaction will yield 30 μL of 2 pmol/μL [γ-^{32}P]-labeled RT primer.

1. If frozen, thaw the [γ-^{32}P] ATP at room temperature behind plexiglass screen.
2. Prepare a 1.5 mL tube and add the following components:

17.5 μL nuclease-free water

6 μL 10 pmol/μL RT primer.

3 μL 10× T4 reaction buffer A.

1 μL T4 polynucleotide kinase (10 U/μL).

2.5 μL [γ-^{32}P] ATP (3000 Ci/mmol).

3. Mix gently by pipetting a couple of times up and down and spin down briefly.
4. Incubate the reaction for 40 min in a 37 °C water bath.
5. Quench the reaction by incubating for 10 min at 65 °C in a heat block.
6. Store the [γ-^{32}P]-labeled primer in a lead box at −20 °C for up to a couple of weeks (*see* **Note 17**).

3.7.2 Reverse Transcription of the Modified RNA

1. Prepare the template RNA mix by adding to a RNase-free 1.5 mL tube (11 μL total volume):

   1–2 pmol of RNA sample in 10 μL nuclease-free water.

   1 μL 2 pmol/μL [γ-^{32}P]-labeled RT primer.
2. Hybridize the primer to the RNA by heating the template RNA mix for 5 min at 70 °C and putting it on ice for at least 5 min. Spin down briefly.
3. Prepare the reverse transcription master mix by adding the following to a tube (scale up accordingly for multiple samples) (9.2 μL total volume/reaction) (*see* **Note 18**):

   4 μL GoScript 5× reaction buffer.

   2 μL 25 mM $MgCl_2$.

   1 μL dNTP mix (10 mM each dATP, dGTP, dCTP, and dTTP).

   0.2 μL 0.1 M DTT.

   1 μL 100 mM RNaseOUT RNase inhibitor.

   1 μL GoScript reverse transcriptase
4. Combine 11 μL template RNA mix and 9.2 μL of reverse transcription master mix to an RNase-free PCR tube. Mix gently by pipetting and spin down briefly.
5. Prepare first-strand cDNA by incubating the reaction in a thermocycler using this program:

   25 °C for 5 min

   55 °C for 60 min

   70 °C for 15 min
6. Add 10 μL of 2× formamide dye to the 20 μL primer extension reactions.
7. Store the [γ-^{32}P]-labeled cDNA in a lead box at −20 °C for a couple of weeks.

3.7.3 Preparing the Dideoxy Chain-Termination Sequencing Ladder

For each transcript that will be analyzed on gel and each RT primer, a corresponding set of four sequencing ladders will have to be constructed in order to align the NMIA-modified nts with the sequence of the RNA species under study. The protocol below is based on the thermo sequenase cycle sequencing kit.

1. Prepare four PCR tubes labeled "A", "T", "G", and "C".
2. Add 4 μL of ddA termination mix to tube "A", 4 μL of ddT termination mix to tube "T", 4 μL of ddG termination mix to tube "G", and 4 μL of ddC termination mix to tube "C".
3. For each investigated RNA, prepare the following sequencing master mix (17.5 μL total volume):

   1 μg of midiprep plasmid with the cloned in vitro transcription construct.

   x μL nuclease-free water (up to 17.5 μL total volume).

   2 μL thermo sequenase reaction buffer.

   1 μL 2 pmol/μL [γ-^{32}P]-labeled RT primer.

   2 μL thermo sequenase DNA polymerase
4. Add 4 μL of the sequencing master mix to each of the four tubes "A", "T", "G", and "C".
5. Incubate the tubes in a thermocycler and perform the following program 45–55 times:

   30 s at 94 °C.

   30 s at 55 °C.

   60 s at 72 °C
6. Terminate the reaction by adding 4 μL of stop solution dye to each of the four 8 μL reactions. Mix well by pipetting.
7. Store the [γ-^{32}P]-labeled sequencing ladders in a lead box at −20 °C up to a couple of weeks.

3.8 Denaturing Acrylamide Gel Electrophoresis

3.8.1 Preparing the Denaturing Acrylamide Gel

1. Use a set of sequencing-type glass plates for gel electrophoresis. Make sure that the set of plates allows for a gel that fits perfectly in the electrophoresis system that will be used, in order not to create any buffer leaks. One plate should be slightly larger (large plate) than the other (small plate), to allow for the placement of the comb. Ensure the plates are assembled in pairs.
2. Clean both plates thoroughly with detergent and rinse with distilled water. Wipe dry. Degrease both plates with 70% ethanol and rinse with distilled water. Wipe dry.
3. Treat one side of the large plate with sticky solution by pouring about 5 mL of the solution on the plate and spreading the liquid over the complete surface with paper (*see* **Note 19**). Incubate the plate for at least 5 h (or overnight) at 30 °C.

4. Treat one side of the small plate with repellant by spraying the solution on the plate and spreading the liquid over the complete surface with paper (*see* **Note 20**). Let it evaporate for a couple of minutes.
5. Assemble the two plates and ensure that the treated sides are on the inside of the construction. Insert two spacers at the longest sides of the glass plates assembly. Temporarily secure clamps to the longest side of the plate/spacer assembly to facilitate its insertion in the rubber sealing system. Remove the clamps. Ensure that the small plate is on top (*see* Fig. 4a).
6. Prepare the 6% acrylamide solution (*see* **Note 21**). Mix 60 mL 6% polyacrylamide stock solution and 600 μL 0.1 g/mL APS in a measuring cylinder. Filter the solution on a Buchner filter and degas for 10 min while keeping the bottle on ice, thereby avoiding that the polymerization proceeds too fast. Transfer the solution to a beaker and add 30 μL TEMED (*see* **Note 22**).
7. Pour the gel immediately after addition of TEMED. Take up the acrylamide solution in a 50 mL syringe and inject the solution in between the glass plates. Start injecting in one corner of the plates and exert constant and even pressure in order to avoid formation of air bubbles. If needed, keep the plates slightly more vertical or horizontal to let the solution fill up the complete surface of the plates. Fill up to the top of the small plate with some excess. Leave the assembly at a slightly tilted position (e.g., by putting a small object underneath the plates). Immediately insert a 20-well comb underneath the small plate and make sure not to introduce any air bubbles in the solution while doing so. Put four clamps spread over the width of the gel to exert pressure on the comb (*see* Fig. 4a). Let the gel polymerize for at least 30 min. Remove the clamps.
8. Use the gel immediately or store the assembly at 4 °C (up to a few days) with wet paper on top of the comb and packed in plastic (do not remove the rubber sealing system or the comb) (*see* Fig. 4a).

3.8.2 Acrylamide Gel Electrophoresis

1. Prepare the acrylamide gel by removing the rubber sealing system and the comb from the gel and by removing all excess gel particles above the comb by passing a spacer above the small plate. Rinse off the plate assembly with distilled water. Dry the plates with paper and remove the distilled water from the wells by gently pressing a paper on top of the small plate.
2. Use a flat-tip gel-loading pipet tip to clean out all wells and ensure that no pieces of acrylamide are left, as small chunks of acrylamide could affect the migration of the DNA through the gel, and thus distort the formation of sharp bands.

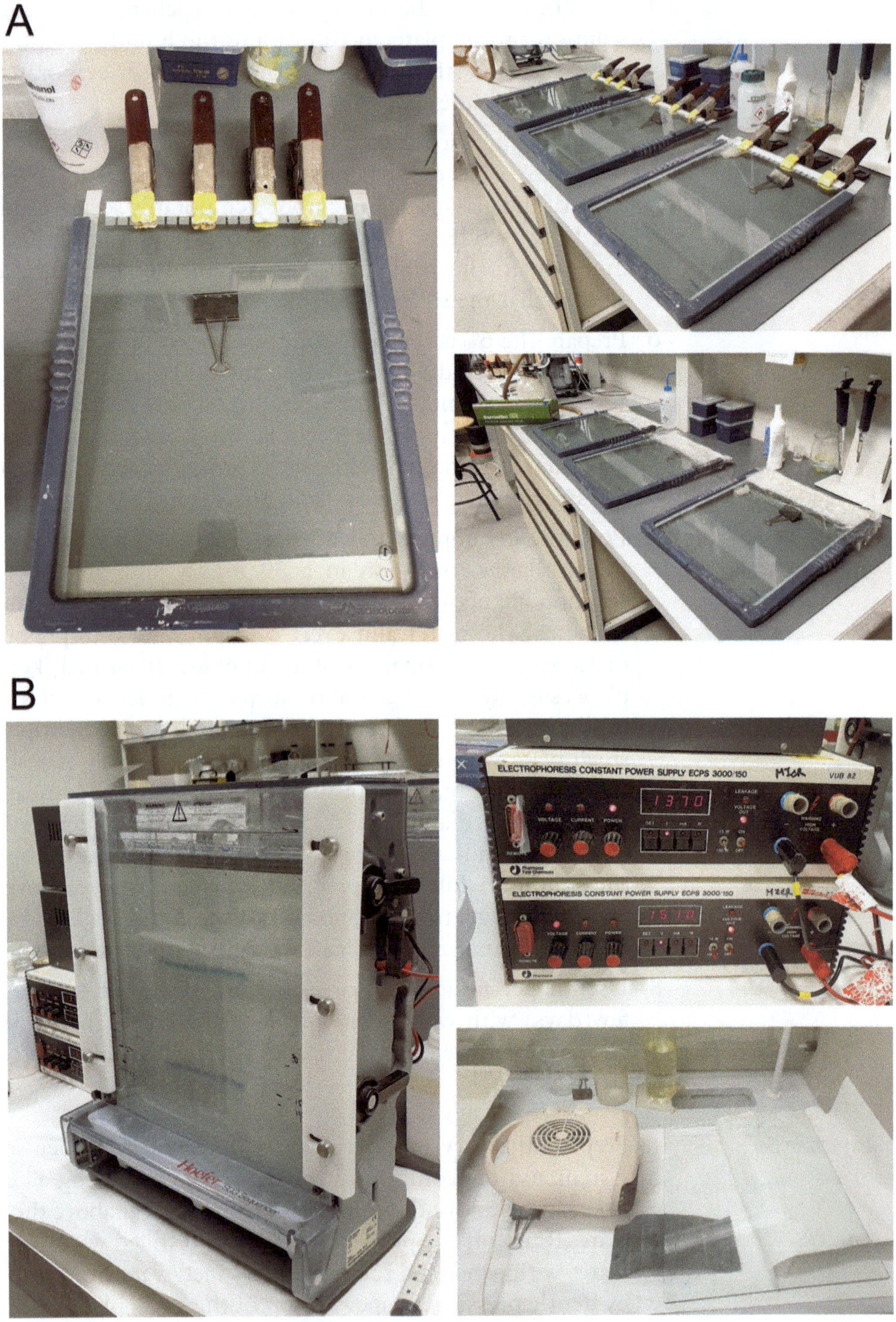

Fig. 4 (**a**) Assembly of the glass plate sandwich with spacers, comb, and rubber sealing system during polymerization of the acrylamide gel (left and top right). Acrylamide gels can be stored up to a week at 4 °C when wrapping the top part with wet paper and plastic foil (bottom right). (**b**) Acrylamide gel electrophoresis (left and top right). The migration of the DNA through the gel can be followed by the xylene cyanol (light blue) and bromophenol blue (dark blue), present in the formamide loading dye. After electrophoresis and incubation of the gel in 10% acetic acid, the gel is dried by putting it in front of a hot air blower (on the "eco" setting) in a fume hood until the gel is thin and completely hardened (bottom right)

3. Trace the wells with a marker on the large plate. Label the plate if needed.
4. Put a bit of Vaseline on the small plate where the plate makes contact with the electrophoresis system (usually the top sides) to avoid buffer leaks.
5. Insert the gel in the electrophoresis system and pour 1× TEB buffer in the upper and lower compartments of the electrophoresis system. Remove any air bubbles in the wells by injecting 1× TEB buffer in the wells with a syringe. Prerun the gel for 30 min at 60 W and limit the voltage to 1500 V.
6. Prepare the samples which will be loaded on the gel and denature the samples for 3 min in a heat block at 90 °C. Use the following volumes as a guideline (*see* **Note 23**): 10 μL of the primer extension reactions (to which formamide dye was already added); 4 μL of sequence ladders (to which the stop solution was already added). For pipetting ease, one can add additional formamide dye (e.g., 2 μL) to these reactions.
7. Remove the diffused urea in the wells of the gel by injecting 1x TEB buffer in the wells with a syringe.
8. Load the samples on gel using flat-tip gel loading pipet tips or capillary tips (*see* **Note 24**). Repeat the removal of urea in the wells in each well separately just before loading.
9. Run the gel at 60 W and limit the voltage to 1500 V (*see* Fig. 4b). Empirically estimate the duration of the electrophoresis run based on the fastest-migrating bromophenol blue dye in the formamide dye mixture. The dye comigrates with a particular size of DNA fragment based on the percentage of acrylamide gel: that is, in a 6% gel, the dye migrates approximately as a 26 nts long DNA (in a 8% gel this is a 19 nts DNA and in a 10% a 12 nts DNA). Note that the primer will be the shortest DNA fragment that is useful for proper interpretation of the results and should still be present on the gel. Shorter fragments are not meaningful and can be run off from the gel, since this will allow longer runtime and better size separation of the cDNA bands of interest. As an estimation, in the case of a "regular" 20–30 nts primer, a 6% gel can run for as long as the bromophenol blue dye is still present on the gel.
10. After electrophoresis, remove the gel from the system and discard the 1x TEB buffer in the radioactive waste if needed.
11. Remove all remaining buffer from the wells by gently pressing a paper on top of the small plate. Gently open the plate assembly by inserting a small piece of plastic between both plates and using it as a lever. The gel will stick on the large plate. Remove the small plate and the spacers.

12. Incubate the large glass plate with the gel in a large tray with 10% acetic acid (± 2 L) for 10 min in a fume hood to fixate the DNA on the gel and remove the urea. Gently shake and ensure that the gel is completely covered with solution. Gently blot the remaining acetic acid from the gel by slowly rolling a toilet paper roll over the gel.
13. Dry the gel with a hot air blower in a fume hood (*see* **Note 25**) (*see* Fig. 4b). It usually takes about 25 min for each of the four sides of the gel. The gel should become very thin and hard. Gently use a piece of paper to verify whether the gel is dry (the piece of paper should not stick to the gel at any point). Let the dried gel cool down for 10 min.
14. Place plastic foil on top of the dried gel, which sticks to the glass plate, to avoid any future direct contact of the radioactive material in the gel with the autoradiography film. Smoothen the plastic, thereby avoiding wrinkles, and tape it down at the backside of the glass plate. Put a piece of paper on top of the foil and close off with the small glass plate.

3.9 Autoradiography

When working with undeveloped autoradiography film, be very careful not to expose the film to any light. Both before use and during exposure, the film, with or without the gel, should be thoroughly wrapped in at least two black plastic bags. Work in a well-sealed dark room, without windows, and with a closed door. Do not bring in computers, smartphones, smartwatches, etc. in this room and ensure that nobody enters the room when manipulating. Only red lights should be used when manipulating the film.

1. Bring the glass plate/gel/foil assembly to the dark room. In the complete dark and with the red lights switched on, take a sheet of autoradiography film, and if needed, cut to the appropriate size of the gel.
2. Remove the small glass plate of the assembly and put the film on top of the dried gel with the matt side toward the gel and the shiny side upward. Cut off a small corner of the film for orientation purposes (e.g., top left corner). Immediately put back the small glass plate on the film, in order to avoid unnecessary exposure to radiation. Secure the sandwich with four clamps, two on each of the longest sides. Put the assembly in at least two black plastic bags. Close off the last bag with two clamps. Store the wrapped sandwich construction in a dark cupboard and expose the film for 2–3 days (*see* **Note 26**).
3. In the complete dark and with the red lights switched on, open the wrapped sandwich construction and take out the piece of autoradiography film. Place the film in the developer solution (±1 L) and make sure that the film is completely covered. Shake the tray until bands on the film become visible and the film no longer changes (this usually takes about 5 min).

4. Rinse off the film with water, in order not to contaminate the fixator solution with the developer solution. Place the film in fixator solution (±1 L) and make sure that the film is completely covered. Shake the tray for about 5 min.
5. Rinse the film first with tap water and then with distilled water. From this point on, it is safe to expose the film to light.
6. Dry the film by hanging it for at least 10 min at room temperature or at 30 °C. Verify whether all bands are visible and distinguishable (*see* **Note 27**).
7. Scan the film as a high-quality file (e.g., at least 600 dpi .tiff). A regular office scanner should suffice.
8. When pleased with the exposure time of the film, clean the glass plates. Soak the largest plate with the dried gel in water for ±1 h. Scrape off the gel with a plastic tool and discard the radioactive waste. Clean off the plates as described above (*see* Subheading 3.8.1, **Step 2**) and store them in a pair with a sheet of paper in between the two glass plates.

3.10 Interpretation and Quantification of the SHAPE Results

The single-stranded nts can be identified from the autoradiography film by comparing the NMIA-treated lane to the DMSO-treated lane, either by eye or further quantified by software such as ImageJ.

In theory, modifications of unbound nts manifest themselves as RT stops, and in this case, a cDNA band will be present at the corresponding position only in the NMIA lane and not in the DMSO lane. However, differences between both lanes are often not that clear-cut in case of less-efficient SHAPE modification and/or a high background signal in the DMSO lane. The latter can be explained by RT stops taking place in absence of modifications, which can be due to the inefficient denaturation of a strong, GC-rich, stem structure in the RNA prior to RT or suboptimal RT conditions. Therefore, densitometric quantification of band intensities is often useful and necessary for proper interpretation of the results.

Keep in mind during interpretation of the film that (i) the unextended primer is the shortest "meaningful" cDNA band on the gel and often forms a thick band in all lanes at the bottom of the film (both for sequence ladder and samples); (ii) during primer extension, the reverse complement cDNA strand is synthesized ($5' \rightarrow 3'$) and the corresponding RNA sequence should be derived from the sequence ladders from bottom (shortest cDNA fragment) to top (longest cDNA fragment); (iii) during primer extension, the RT stops at the nt before the modification, therefore, the cDNA band present in the NMIA lane and absent in the DMSO lane marks the last incorporated nt (length n). The modified, unpaired nt is at position $n + 1$ in the $3' \rightarrow 5'$ direction (*see* Fig. 1).

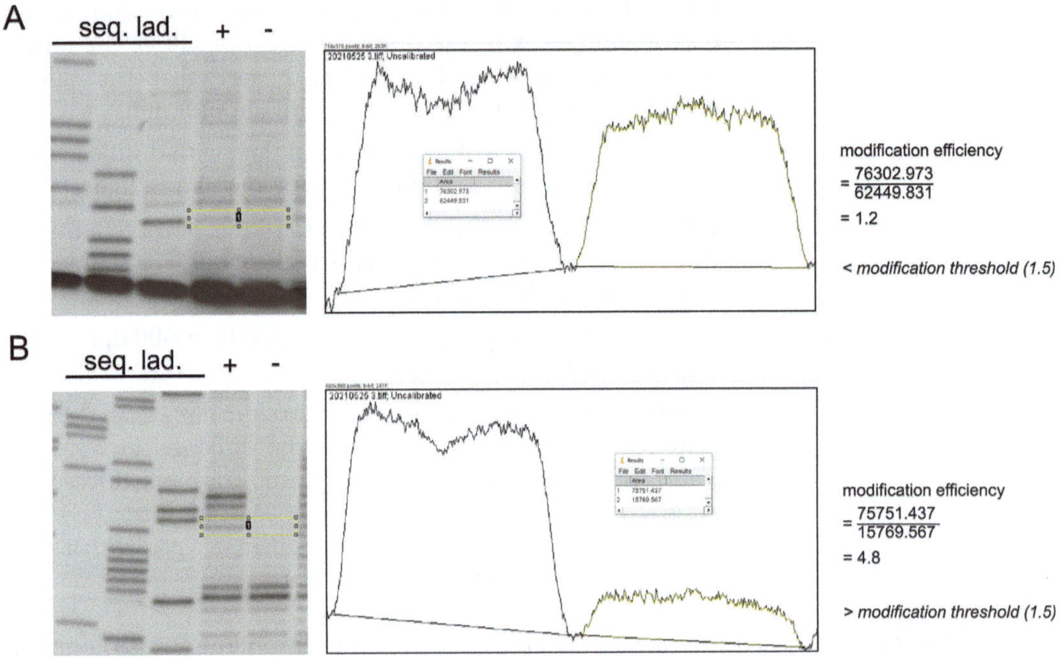

Fig. 5 Quantification of a SHAPE result with sequencing ladder (seq. lad.) and primer extension reactions from an NMIA-treated RNA sample (+) and DMSO mock control (−), using Image J software. By selecting a cDNA band of length n in both the NMIA- and DMSO-treated samples (left), the intensity of the corresponding bands can be plotted (right). The area under each band peak is calculated by the software and can be used to calculate the abundance of a cDNA with length n or the modification efficiency of nt $n + 1$. When this ratio approximately equals 1 (**a**), there is no considerable increase in abundance of a particular cDNA of length n in the NMIA-treated sample compared to the DMSO-treated sample, and thus, the nt +1 is not modified by NMIA and is most likely not in a single-stranded conformation. When the ratio exceeds an empirically set threshold (e.g., 1.5) (**b**), there is a considerable increase in abundance of a particular cDNA of length n in the NMIA-treated sample compared to the DMSO-treated sample, and thus, the nt +1 is modified by NMIA and is in a single-stranded conformation

1. Open Image J software, import scanned picture, and set image type to 8 bits.
2. Rotate the picture such that the bands of the NMIA-treated samples and DMSO-treated samples are completely horizontal.
3. Use the "rectangle" tool to draw a rectangle over the two bands of the NMIA-treated samples and DMSO-treated samples corresponding to the same nt. Use the "Select First Lane" option to select the drawn rectangle and then choose "Plot Lanes" (*see* Fig. 5, left).
4. A graph will be opened in a new window, in which the intensity is plotted of the signal over the selected rectangle. The intensities of both bands should be clearly distinguishable. Use the "straight" tool to draw a line at the bottom of each band peak, thereby cutting off the background signal. Use the "wand" tool to click on the two areas under the band peaks of the

intensity curve, the areas under the two band peaks will be calculated and will pop up in a new results window (*see* Fig. 5, right). Copy these results to an Excel file.

5. Repeat **steps 2-3-4** for other cDNA fragments.
6. When all raw area measurements are taken, calculate the abundance of a particular length (n) of cDNA, or, in other words, the modification efficiency of nt $n + 1$:

$$\begin{aligned}&\text{Abundance of a cDNA with length } n\\&\quad= \text{modification efficiency of RNA nt } n+1\\&\quad= \frac{\text{area under intensity curve of band } n_{\text{NMIA-treated sample}}}{\text{area under intensity curve of band } n_{\text{DMSO-treated sample}}}\end{aligned}$$

7. When the NMIA modification at nt $n + 1$ was efficient and induced a stop in RT, the abundance of a cDNA with length n will have a value ≥ 1. Due to the DMSO-background signal and in order to incorporate differences in RT efficiencies and gel loading between the NMIA- and DMSO-treated samples, it is beneficial to empirically set a "modification threshold" of, for example, 1.5 or 2. For nts of which this threshold is exceeded, the nts $n + 1$ are supposedly modified and in a single-stranded conformation in the RNA secondary structure (*see* Fig. 5).

As an illustration, a result is shown of an in vitro SHAPE experiment of *Sulfolobus acidocaldarius* 16S rRNA (helix 44–45), folded and modified at 37 °C with the NMIA SHAPE reagent (13 mM) (*see* Fig. 6). The cDNA bands are quantified with Image J and corresponding abundances of >1.5 are marked with an arrow next to the gel. In this way, a total of four modified regions (1–4) are identified (*see* Fig. 6a). Indeed, the corresponding unpaired nts map to known single-stranded regions on the secondary structure (*see* Fig. 6b).

4 Notes

1. Two classes of T7 promoters exist, each of which has been successfully used for in vitro transcription with little effect on transcriptional yield according to our experience: T7 promoter class II (5′ TAA TAC GAC TCA CTA TT<u>A</u> **G(G)** 3′) and T7 promoter class III (5′ TAA TAC GAC TCA CTA TA<u>G</u> **G(G)** 3′). Depending on the promoter class, in vitro transcription by the T7 RNA polymerase will start at the underlined A (class II) or underlined G (class III), followed by at least one G (a second G is not absolutely required for transcription); however, it might increase the RNA yield [13]. Depending on the first nts at the 5′ end of the RNA of interest, a well-considered

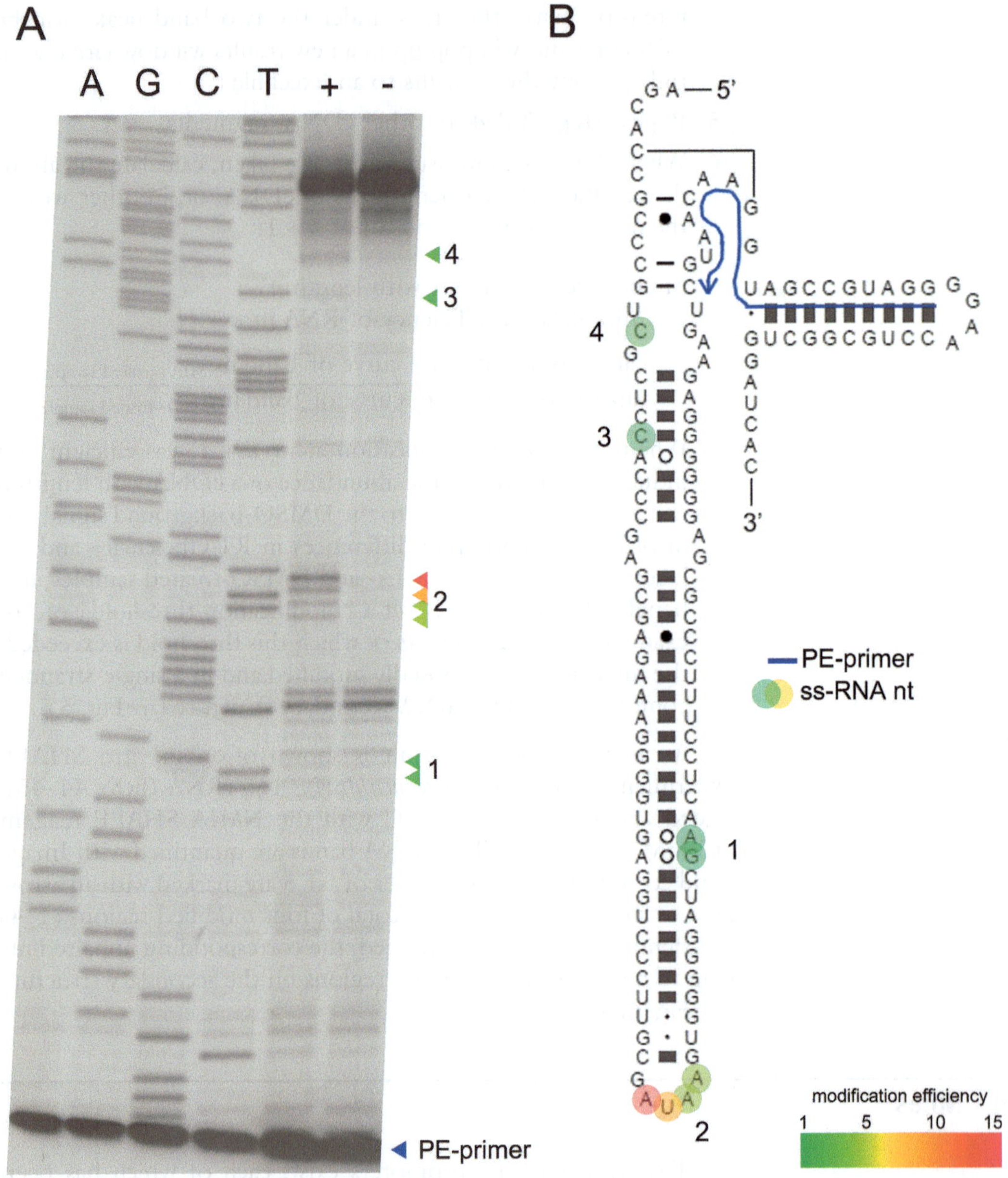

Fig. 6 Results from an in vitro SHAPE experiment of *Sulfolobus acidocaldarius* 16S rRNA (helix 44–45; Saci_1300). The RNA of interest was constructed with T7 in vitro transcription originating from the linearized plasmid template shown in Fig. 2. After purification, the RNA was folded at 37 °C and modified with 13 mM NMIA (+) or plain DMSO (mock control, −). The recovered, modified RNA was used as an input for primer extension using the ^{32}P-labeled PE-primer (5′-CCTACGGCTACCTTGTTACG-3′, blue line). RT will generate cDNAs of a specific length, due to a halt in elongation at position n when a modification is encountered at position $n + 1$ (3′ –> 5′ direction). Sequence ladders were constructed from the plasmid template encoding the RNA of interest and the same ^{32}P-labeled RT primer. Samples (10 μL of primer extension reactions and 4 μL of sequence ladders, both dyes included) were run on a 10% polyacrylamide gel and exposed to an autoradiography film for 3.5 days (**a**). The obtained cDNA signals were further quantified using Image J software by calculating the abundance of a particular length (n) of cDNA, or the modification efficiency of nt $n + 1$ as the ratio of the intensity of the NMIA band to the DMSO band for nt n (color-scale bottom right).

choice has to be made of which promoter will be used (in order not to include unnecessary nts at the 5′ end of the transcript which can influence the RNA secondary structure). For example, if the transcript starts with 5′ AG, the promoter class II seems a logical choice; in case the transcript starts with one G, it needs to be verified with RNA secondary structure software whether including an additional A or G affects the predicted structure.

2. To allow for specific transcription termination, a transcription terminator needs be included downstream of the sequence encoding the RNA of interest. A good alternative approach is to add a restriction site at the 3′ end of the sequence, preferably for a restriction enzyme yielding 5′ end sticky ends, which will allow for run-off in vitro transcription. The selection of an appropriate restriction site can again be based on the endogenous sequence in order not to add unnecessary nts to the transcript. For example, in case the last nt of the RNA of interest is an A, the MluI restriction site (A^CGCGT) can be incorporated.
3. In case the RNA of interest is longer than ±150 nts, it might be necessary to design a second RT primer and use this in separate primer extension reactions.
4. Other extraction methods are also possible. However, since the quality and purity of the template DNA are crucial for transcriptional yield, we recommend not using silica membranes, since purity is often less good.
5. In the case of nonefficient restriction, it could be beneficial to increase the incubation time of restriction enzymes, after which another equal volume of enzyme is added to the reaction and further incubated.
6. In case a longer RNA is transcribed (e.g., when analyzing a complete mRNA in combination with multiple primer extension reactions, with different RT primers), the amount of linearized input plasmid could be decreased and the incubation time shortened.
7. Make use of the following formula for dsDNA to calculate the amount of template that should be added (https://nebiocalculator.neb.com/#!/dsdnaamt):

Fig. 6 (continued) Modification threshold was set at 1.5. Regions 1 until 4 with a significant increased abundance of a particular length (n) of cDNA in the NMIA lane are annotated on the gel (**a**), and the corresponding NMIA modifications and single-stranded nts (n + 1) are annotated on the RNA secondary structure (**b**). (Derived from the Comparative RNA Web (CRW) Site, crw-site.chemistry.gatech.edu)

$$\text{mass of ds template DNA} = \frac{\text{moles of dsDNA (mol)}}{\left(\frac{\text{length of dsDNA (bp)}*617.96\frac{g}{mol}}{\text{bp}}\right) + 36.04\frac{g}{mol}}$$

8. In our own experience, overnight incubation for transcripts of ±100 nts does not further improve the yield compared to an incubation of 6 h. It should also be noted that longer incubation poses the risk of (more) RNA degradation in case of RNase contamination.
9. Since the expected RNA yield in the reaction is high, it is often better to make a dilution of the reaction (e.g., 1:150–1:300). Also note that this measurement only gives a rough idea whether the reactions are efficient and does not accurately reflect the yield of the synthesized RNA, since any unincorporated dNTPs will increase the signal. However, quantifying the yield at this stage gives an idea whether it is useful to perform RNA purification.
10. Ensure that 20 mL 100% ethanol was added to the wash solution concentrate.
11. For pure RNA, both the A260/A280 and A260/A230 ratio should be about 2 (>1.8 for A260/A280).
12. Do not skip this cooling step! If formaldehyde would be added to the solution at a higher temperature, it could lead to its evaporation and a risk of inhalation of toxic fumes.
13. Whereas it can be useful to load an RNA molecular weight (MW) ladder to verify the length of the in vitro transcription product, it is often not necessary for sporadic RNA work. Instead, incorporate a DNA MW ladder on the gel to follow the migration of the samples through the gel (do not use it for alignment).
14. In case multiple modification reactions are tested, scale up the volume of RNA sample. Count 2.5 pmol in 6 μL nuclease-free water per reaction. Make use of the following formula for ssRNA to calculate the amount of template that should be added (https://nebiocalculator.neb.com/#!/ssrnaamt):

$$\text{mass of ss RNA (g)} = \frac{\text{moles of ssRNA (mol)}}{(\text{length of ssRNA (bp)} * 321.47\ \text{g/mol/bp}) + 18.02\ \text{g/mol}}$$

15. In case multiple modification reactions will be tested, scale up the volume of RNA folding mix accordingly. Count 3 μL of 3.3× RNA folding mix per reaction.
16. Use the following formula to determine half-life: half-life (min) = 360 × e^[−0.102 × temperature(°C)] [8].
17. Note that the half-life of γ-^{32}P is 14 days, and therefore after 14 days, the signal will be only half as strong. Try to use the

[γ-^{32}P]-labeled primer as quickly as possible in the further reactions.

18. RNaseOUT is essential when working with low amounts of input RNA (<50 ng) and requires 1 mM DTT for activity.
19. Ensure that the complete surface of the large plate is treated with a sticky solution in order for the gel to thoroughly stick to the plate when disassembling the plates after electrophoresis. Ensure that each time the same side of the plate is treated, by putting a mark on the outside of the glass plate.
20. Ensure that the complete surface of the small plate is treated with repellant, otherwise the gel can tear when disassembling the plates after electrophoresis. Ensure that each time the same side of the plate is treated, by putting a marking on the outside of the glass plate.
21. The percentage of polyacrylamide stock solution is dependent on the expected size of the cDNA fragments and on the type of electrophoresis system. As a first trial, a 6% solution is a good starting point for cDNA of ±90–150 nt. Depending on the final resolution of the film, one can choose to increase the percentage (to 8–10%) in order to better separate small cDNAs, thereby increasing the resolution.
22. Polymerization will start immediately after TEMED is added. Ensure that the glass plates are assembled and that everything is ready for pouring the gel. When preparing multiple gels, only add TEMED to the fraction of acrylamide solution that will be poured first.
23. Mentioned volumes are only a guidance, optimal volumes will have to be determined experimentally and can depend on many factors, including date of [γ-^{32}P]-ATP used for labeling and radioactive signal at the time of running the gel/radiography, amount of template RNA used in the primer extension reactions, amount of DNA template used in construction of sequence ladders, and so on. Do not overfill the wells, because this will decrease sharpness of the bands (keep the volume per sample <15 μL).
24. Take especially care to load the same volume of NMIA-treated samples and DMSO-treated samples, since these two samples will have to be quantitatively compared.
25. Do not place the plate immediately next to the hot air blower and use the "eco" setting of the hot air blower to avoid blowing hot air over the gel, as otherwise the glass plates might crack.
26. The exact exposure time will depend on the signal of radioactivity and will have to be determined experimentally. Usually, 2–3 days are sufficient when using freshly labeled material.

27. In case the bands are too faint or the film is overexposed, repeat the autoradiography procedure with a new autoradiography film and expose the film for a longer or shorter time period, respectively.

Acknowledgments

This work was supported by the Vrije Universiteit Brussel and funded by Research Foundation Flanders (FWO-Vlaanderen) (FWO PhD fellowship 1134419N to Rani Baes). We are grateful to Sébastien Ferreira-Cerca for sharing his insights and knowledge.

References

1. Miao Z, Westhof E (2017) RNA structure: advances and assessment of 3D structure prediction. Annu Rev Biophys 46:483–503
2. Li B, Cao Y, Westhof E et al (2020) Advances in RNA 3D structure modeling using experimental data. Front Genet 11:1–19
3. Gruber AR, Lorenz R, Bernhart SH et al (2008) The Vienna RNA websuite. Nucleic Acids Res 36:70–74
4. Zuker M (2003) Mfold web server for nucleic acid folding and hybridization prediction. Nucleic Acids Res 31:3406–3415
5. Bellaousov S, Reuter JS, Seetin MG et al (2013) RNAstructure: web servers for RNA secondary structure prediction and analysis. Nucleic Acids Res 41:471–474
6. Mathews DH, Disney MD, Childs JL et al (2004) Incorporating chemical modification constraints into a dynamic programming algorithm for prediction of RNA secondary structure. Proc Natl Acad Sci 101:7287–7292
7. Kwok CK, Tang Y, Assmann SM et al (2015) The RNA structurome: transcriptome-wide structure probing with next-generation sequencing. Trends Biochem Sci 40:221–232
8. Wilkinson KA, Merino EJ, Weeks KM (2006) Selective 2′-hydroxyl acylation analyzed by primer extension (SHAPE): quantitative RNA structure analysis at single nucleotide resolution. Nat Protoc 1:1610–1616
9. Busan S, Weidmann CA, Sengupta A et al (2019) Guidelines for SHAPE reagent choice and detection strategy for RNA structure probing studies. Biochemistry 58:2655–2664
10. Knüppel R, Fenk M, Jüttner M et al (2020) In vivo RNA chemical footprinting analysis in archaea. In: Heise T (ed) RNA chaperones. Methods in molecular biology. Humana, New York, pp 193–208
11. Smola MJ, Weeks KM (2018) In-cell RNA structure probing with SHAPE-MaP. Nat Protoc 13:1181–1195
12. Schneider CA, Rasband WS, Eliceiri KW (2012) NIH image to ImageJ: 25 years of image analysis. Nat Methods 9:671–675
13. Beckert B, Masquida B (2011) Synthesis of RNA by in vitro transcription. In: Nielsen H (ed) RNA: methods and protocols methods in molecular biology. Clifton, pp 29–41

Chapter 15

Microscale Thermophoresis to Study RNA–RNA Binding Affinity

Britta Jordan, Lisa Nickel, and Ruth A. Schmitz

Abstract

Evaluation of RNA–RNA binding is crucial for in vitro studying of molecular mechanisms, for example, the interaction of noncoding RNAs (ncRNAs) with their respective targets. In recent years, the method of microscale thermophoresis (MST) has been developed, which is based on the physical phenomenon of thermophoresis (Ludwig–Soret Effect), defined as the migration of a molecule in a solution in response to a macroscopic temperature gradient. The method enables the fast detection and characterization of biophysical interaction between molecules, with the fundamental advantage that only small amounts of target and ligand are required. Here, we describe the characterization of RNA–RNA binding affinity using the example of the $sRNA_{41}$ from *Methanosarcina mazei* and its native target, the 5' UTR of mRNA-*MM2089*, the first gene of the operon encoding the acetyl–CoA decarboxylase/synthase complex.

Key words Microscale thermophoresis, MST, sRNA, RNA–RNA binding affinity

1 Introduction

RNA molecules play an important role in a wide range of biological processes, including for example transcription, translation, RNA processing, and different regulatory mechanisms on posttranscriptional level [1, 2]. This posttranscriptional regulation via regulatory RNAs is frequently/mainly mediated via RNA–RNA binding. The respective RNA–RNA interactions can be studied by numerous biophysical and biochemical methods, all of which are designed to distinguish specifically interacting RNAs from non-interacting ones and can detect those interactions on a molecular level. RNA–RNA complexes harboring a larger molecular mass than the single RNAs consequently migrate slower during gel electrophoresis, which can be studied via so-called electrophoretic mobility shift assays (EMSAs). Another approach to study RNA–RNA interaction is using surface plasmon resonance (SPR) or Forster resonance energy transfer (FRET) experiments, using one immobilized

Eveline Peeters and Indra Bervoets (eds.), *Prokaryotic Gene Regulation: Methods and Protocols*,
Methods in Molecular Biology, vol. 2516, https://doi.org/10.1007/978-1-0716-2413-5_15,

RNA and one mobile RNA in solution [3, 4]. Those commonly used methods however require complex experimental setups (particularly for SPR and FRET analysis), rely on high amounts of RNA and sometimes radioactive labeling, and are usually time-consuming.

The technology of microscale thermophoresis (MST) makes use of the physical phenomenon of thermophoresis, which describes the directed movement of molecules in a temperature gradient. This molecule movement depends on the respective size, charge, conformation, or hydrate shell of the molecule. MST enables the examination of diverse molecule interactions in a sensitive and highly flexible way. The interaction of two molecules results in a change in size, charge, conformation, or the respective hydrate shell and consequently in a change of movement within a temperature gradient. For MST, the temperature gradient is induced by an infrared laser (a schematic representation of the MST is shown in Fig. 1) and the molecular movements are tracked by covalently attached or intrinsic fluorophores (reviewed in [5]). The labeled ligand is used in a constant concentration, and the target is titrated in different concentrations. In principle, only small quantities of molecules are required for the MST measurement, and the analysis can be carried out particularly fast; consequently, several measurements with different targets are possible in only a short amount of time. Furthermore, the required reciprocal experiments, in which the fluorescence-labeled molecule is changed and the respective other is titrated, are simple and quick to conduct. Moreover, those experiments are highly reproducible and the interactions of a wide range of various combinations of interacting molecules can be studied, such as proteins, peptides, small molecules, DNA, and RNA, and also interaction of virus particles can be studied via MST (reviewed in [6]). The exceptionally high sensitivity of the MST technique enables measurements of affinity constants in the nanomolar and micromolar range. Thus, it is particularly suitable for the examination of RNA interactions and was already shown for a wide range of biologically relevant RNA interactions, such as RNA-protein [7, 8], RNA-small molecule [9], or RNA–RNA [10] interactions, in consequence representing a great tool for the investigations within the field of RNA biology.

Here we describe the method to study the RNA–RNA binding affinity of the 120 nucleotides (nt) long $sRNA_{41}$ from *M. mazei* and an 80 nt cyanine-labeled (Cy3) part of the 5′ UTR of *MM2089* which has been identified as the interacting partner. $sRNA_{41}$ with a mutated interaction site ($sRNA_{41}$*) was used as a control together with the 5′ UTR of *MM2089*, as well as with a compensatory mutated 5' UTR of *MM2089*. Thereby, significant binding between $sRNA_{41}$ and 5' UTR MM2089 was determined with a calculated K_D of 631 ± 133 nM (*see* Fig. 2, [10]).

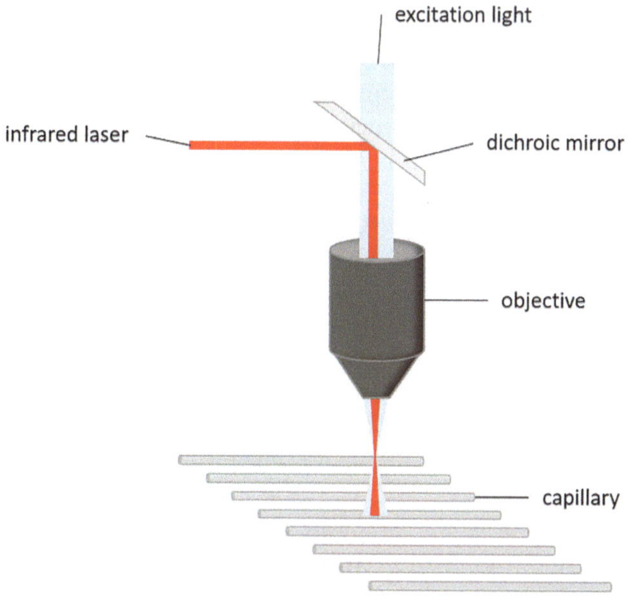

Fig. 1 Schematic representation of Microscale Thermophoresis. The fluorescence of the capillary is excited and detected by the same objective. The infrared laser locally heats the sample in the capillary

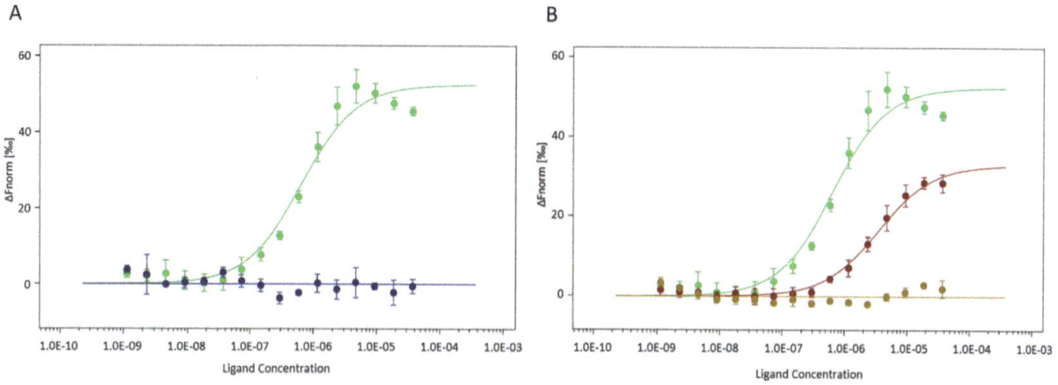

Fig. 2 (**a**) Interaction between Cy3-labeled wild type (wt) 5′UTR MM2089 mRNA and wt sRNA41 (green) and mutated sRNA41 (sRNA41*, blue). The calculated K_D for the interaction of 5′UTR MM2089 and $sRNA_{41}$ is 631 ± 133 nM. (**b**) Interaction between 5′UTR MM2089 mRNA mutated in four nucleotides (5′UTR MM2089*) and wt sRNA41 (brown) and compensatory mutated sRNA41 (sRNA41*, red). The calculated K_D for the interaction of 5′UTR MM2089* and $sRNA_{41}$* is 3.5 ± 0.43 μM. Green graph is identical to section **a**. The concentration of Cy3-labeled 5′UTR MM2089 and 5′UTR MM2089* was constant at 15 nM for all measurements, whereas the concentration of the nonlabeled sRNA41-mutants varied between 1.14 nM and 37.5 μM

2 Materials

Prepare all solutions with ultrapure water and sterilize them by autoclaving or filtration. Used materials like tips and reaction tubes need to be autoclaved twice.

2.1 Interaction Partners

1. Unlabeled RNA, generated by in vitro transcription (*see* **step 3.1**). We used wild type $sRNA_{41}$ and $sRNA_{41}$*, which is mutated in four nucleotides in the predicted binding region to *MM2089*. As a template for in vitro transcription, we used a linearized plasmid containing the T7 promoter followed by the sequence of the RNA of interest.
2. Fluorescently labeled RNA, generated by in vitro transcription (*see* **step 3.1**) and 3′ end labeled with Cyanine3 (*see* **step 3.3**). We used an 80 nt part of the 5′ UTR of *MM2089* and the compensatory mutated part (to $sRNA_{41}$*) of the 5′ UTR of *MM2089* (*MM2089**) (*see* **Note 1**). As a template for in vitro transcription, we used a linearized plasmid containing the T7 promoter followed by the sequence of the RNA of interest.

2.2 Buffers and Reagents

1. TranscriptAid T7 High Yield Transcription Kit (*see* **Note 2**).
2. 2x Ambion structure buffer: 20 mM Tris-HCl pH 7, 200 mM KCl, 20 mM $MgCl_2 \times 6\ H_2O$, 0.04 % Tween-20 (*see* **Note 3**).
3. DEPC-treated water at molecular biology grade, RNase-free.
4. DNase I and 10x DNase I buffer.
5. Phenol:Chloroform:Isoamylalkohol, P:C:I (25:24:1), RNA grade
6. 3 M NaAc pH 5.2: For 100 ml, weigh 24.6 g of sodium acetate (anhydrous) and add to a 100 ml bottle. Solve the sodium acetate in 70 ml water by stirring. Adjust the pH to 5.2 by adding concentrated acetic acid and fill up with water to 100 ml. Sterilize the solution by filtration with a 0.2 nm filter. Store at room temperature.
7. About 96% and 70% Ethanol (EtOH), molecular biology grade. Store the 96% EtOH at room temperature and the 70% EtOH at −20 °C.
8. T4 RNA Ligase and 10x ligation buffer.
9. 10 mM ATP solved in water. Store at −20 °C (*see* **Note 4**). The ATP is needed by the T4 RNA Ligase, if it is not provided by the ligation buffer.
10. BSA (1 mg/ml): solve 10 mg BSA in 10 ml Water. Sterilize the solution by filtration with a 0.2 nm filter. Store at −20 °C.

11. Cytidine-5′-phosphate-3′-(6-aminohexyl) phosphate, labeled with Cy3 (pCp-Cy3) (*see* **Note 5**).
12. Microspin G-25 column.

2.3 Instruments and Software

1. Thermocycler with a temperature range of 4–95 °C
2. Monolith NT.115
3. Centrifuge for 2 ml reaction tubes (cooled) with the ability to run at 4 °C.
4. Microvolume spectrophotometer (*see* **Note 6**).
5. MO.Control 2 software
6. MO.Affinity Analysis software

2.4 Materials

1. 1.5 ml reaction tubes, sterile and RNase-free
2. 0.1 ml reaction tubes, sterile and RNase-free
3. Monolith NT.115 capillaries, standard version, 16 per assay plus two for capillary scan.
4. 2 ml phase lock gel tubes
5. Dark 1.5 ml reaction tubes for labeled RNA to protect the fluorescence from light, sterile, and RNase-free (*see* **Note 7**).
6. Pipet tips: sterile and RNase-free.

3 Methods

Working with RNA needs to be performed under sterile conditions and in an RNase- free environment. It can be useful to work under an airflow cabinet and clean everything with RNase removal agents. It is recommended to wear gloves all the time working with RNA to protect the samples from contaminations.

3.1 In Vitro Transcription of RNA

1. Follow the instructions of the TranscriptAid T7 High-Yield Transcription Kit for high-yield T7 in vitro transcription.

   4 μl 5x TranscriptAid Reaction Buffer

   8 μl Nucleotides Mix (ATP, GTP, CTP, UTP; equal amount)

   2 μl TranscriptAid Enzyme Mix

   1 μg DNA Template

   ad. 20 μl H_2O
2. Incubate the mix at 37 °C for at least two hours (*see* **Note 8**).
3. Perform a DNase treatment with the in vitro transcribed RNA. For this, add 250 μl H_2O to the reaction and 30 μl 10x DNase buffer (*see* **Note 9**). Add 2 μl DNase I and mix well.
4. Incubate for 30 min at 37 °C.

3.2 PCI Clean-up to Remove the DNase from Generated RNA

1. Add 300 μl Phenol:Chloroform:Isoamylalkohol, P:C:I (25: 24:1) to the DNase treatment to stop the reaction and mix well.
2. Add the mix to a phase lock reaction tube (*see* **Note 10**).
3. Use a tabletop centrifuge to separate the phases at room temperature for 10 min at 10,000 rpm
4. Transfer the upper phase to a new reaction tube and add 40 μl 3 M NaAc pH 5.2 and 1 ml 100% EtOH.
5. Let the samples precipitate at −70 °C for 1–2 h or overnight at −20 °C.
6. Spin down the precipitates for 30 min at 4 °C and 13,000 rpm
7. Discard the supernatant (*see* **Note 11**).
8. Wash the pellet with 1 ml 70% EtOH (cold)
9. Centrifuge for 5 min at 13,000 rpm and 4 °C
10. Discard the supernatant completely (*see* **Note 12**).
11. Let the pellet dry for 20–30 min at RT (*see* **Note 13**).
12. Resolve the pellet in 30 μl H_2O (*see* **Note 14**).
13. Store the RNA at −70 °C or go to the next step.

3.3 3′ End Labeling of Generated RNA

1. The complete labeling procedure needs to be done without light exposure.
2. Dilute 100–200 pmol RNA (from in vitro transcription) in a volume of 20 μl in a 0.1 ml reaction tube and denaturize the RNA for 5 min at 95 °C (*see* **Notes 15** and **16**). Add 4 μl 10x ligation buffer, 4 μl 10 mM ATP, 4 μl BSA (1 mg/ml), 100–200 pmol pCp-Cy3 (equal molar amounts to RNA) (*see* **Note 17**) and 2 μl T4 RNA Ligase to the denaturized RNA. Fill up to 40 μl with DEPC-H_2O (*see* **Note 18**).
3. Run the following program over night in a thermocycler: 1 h at 37 °C, 3 h at 16 °C, and 10–12 h at 4 °C.
4. Deactivate the ligase at 70 °C for 10 min.
5. Purify the labeled RNA with a Microspin G-25 column by centrifugation, eluate in a dark reaction tube.
6. Measure the concentration at the Microvolume spectrophotometer (Microarray program, RNA, Cy3) (*see* **Note 19**).
7. Store the RNA at −70 °C or move on with the experiment.

3.4 Plan the Experiment

1. For planning the experiment, you can take the help of the MO. control software. Here you can set your parameters and the software will guide you through the pipetting steps.
2. Prepare the 2x Ambion structure buffer as shown in Subheading 2.2, **step 2**.

3. Start the Monolith and the computer (*see* **Note 20**).
4. Open the program MO. Control (NanoTemper Technologies).
5. Start a new session.
6. If you are working with new samples, it is recommended to run a pretest to check the fluorescence of the samples. If not, move on to **step 3.6**.
7. Choose the filter. The excitation of the green filter is 600–640 nm, and the emission is 660–725 nm (*see* **Note 21**). The red filter has its excitation at 510–550 nm and the emission at 560–590 nm (*see* **Note 22**).
8. Add the information in the green boxes (*see* **Note 23**). At the beginning, you can keep the settings on "auto-detect." If you want to compare the data with other results, you need to set a constant value at MST power and excitation power.
9. Press "go to instructions."
10. Dilute the fluorescently labeled samples as shown in the instructions and fill two capillaries (*see* **Notes 24** and **25**). Put the capillaries at position 1 and 2 of the rack and fix them with the magnets. Close the door and start the measurement.
11. The intensity of the fluorescence should be in a range of 300–400 counts. If you have a lower intensity, use a higher concentration of your labeled RNA. If your intensity is higher, use a lower concentration.
12. Before running the experiment, it is recommended to do a binding check. The binding check is a quick and easy method to see if a binding event is detectable. Here it is possible to check up to six buffers at the same time.
13. Start the binding check and follow the instructions of the software.

3.5 Run the Experiment

1. Start a new session and go to "binding affinity."
2. Fill all green boxes and set the MST power at 60% and the excitation power at 100%. Use 25 °C for the measurement (*see* **Note 26**).
3. When you press "next," the software will guide you through pipetting steps.
4. Dilute unlabeled RNA to 75 μM in H_2O (*see* **Note 27**).
5. Dilute labeled RNA to 30 nM in 200 μl 2x Ambion structure buffer in a dark reaction tube (*see* **Notes 28** and **29**).
6. For the dilution series of the unlabeled RNA, prepare 16 reaction tubes (0.1 ml) and add 10 μl H_2O in vial 2–16. In vial 1, add 20 μl of your unlabeled RNA (75 μM). Perform a 1:

2 dilution series by pipetting 10 μl from vial 1 to vial 2, mix, 10 μl from vial 2 to vial 3, mix, etc. Discard 10 μl of vial 16. Now you should have 10 μl of different RNA concentrations in your vials starting with 37.5 μM in vial 1 to 1.14 nM in vial 16 (*see* **Note 30**).

7. Mix thoroughly and spin down briefly.
8. Incubate the reaction tubes in a thermocycler for 1 min at 95 °C to break up homo-dimers followed by 5 min at 4 °C to cool down the samples.
9. Add 10 μl of your labeled RNA to each vial for a final concentration of 15 nM in each vial (*see* **Note 28**).
10. Mix thoroughly and spin down briefly.
11. Incubate the mix for 15 min at 37 °C (*see* **Note 31**). In this step, the interaction partners are binding.
12. Spin down briefly to make sure that everything is at the bottom of the tube.
13. Open the vials and place one capillary in each vial. Load them at least ¾ by capillary forces.
14. Load the loaded capillaries in the instrument. Start with your highest ligand concentration in slot 1 and go up to the lowest ligand concentration in slot 16. Take care that you keep the right order. Close the monolith and start the measurement.

3.6 Analyze the Data

1. The first part of the measurement is a capillary scan. During the capillary scan, the fluorescence of each capillary is measured and the quality is checked by the software. Here, the software determines whether the same amount of fluorescence-labeled ligand is present in all capillaries. See Fig. 3 for an exemplary capillary scan (*see* **Notes 32** and **33**).
2. The MST traces show ligand-induced changes in MST signal depending on the ligand concentration. See Fig. 4 for an example (*see* **Note 34**).
3. The dose response should be a sigmoidal curve as shown in Fig. 5.
4. Check the three parts for errors. If everything is positive, the program should tell you the K_D of this assay. You can find this in the conclusion part in the section of the interface (*see* **Note 35**).
5. Use at least three individual experiments for calculation of K_D using MO.Affinity Analysis software (Nanotemper). The software will guide you through the important steps. See Fig. 2 for the results of the binding affinity assay of $sRNA_{41}$ and 5′ UTR of *MM2089* (*see* **Note 36**).

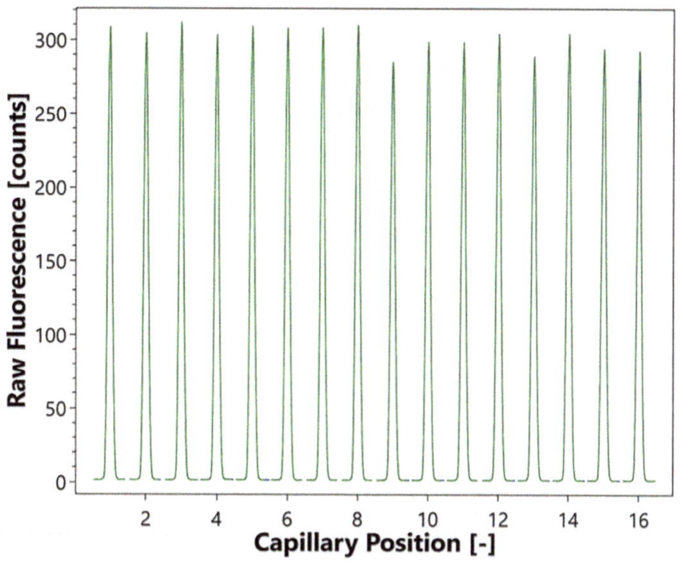

Fig. 3 Capillary scan. The raw fluorescence is measured for each capillary and shown as a peak. The peak height shows the intensity of raw fluorescence. The homogeneity of the samples is shown by identical peak height. The peak shape shows the adsorption to the capillary. Ligand-induced changes in fluorescence can be observed during capillary scan

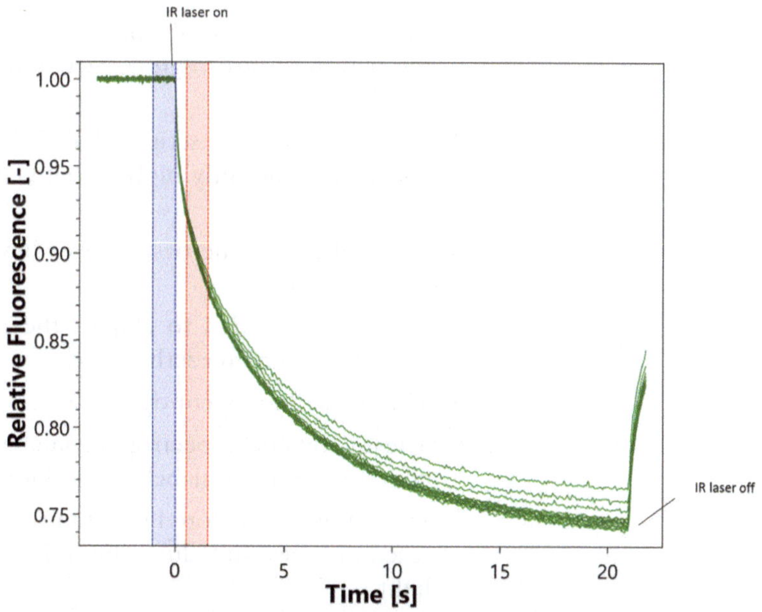

Fig. 4 MST traces. The relative fluorescence is measured over time. Each green line stands for a single capillary. The difference between the relative fluorescence of each capillary is due to the influence of the ligand. Blue box indicates F_0, red box indicates F_1, which are used for calculation of thermophoresis

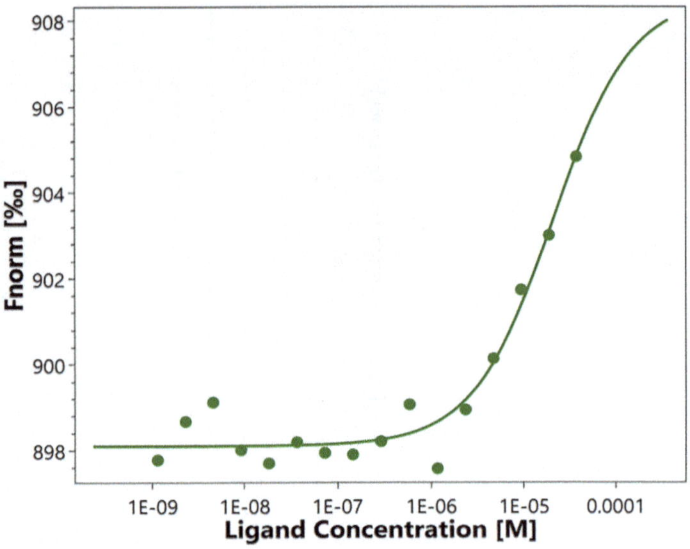

Fig. 5 Dose response. The signal of each capillary is plotted against the ligand concentration. The dose response shows the concentration dependency of the measurement parameters used for K_D calculation

4 Notes

1. In case of unsuccessful labeling of the RNA of interest, there is the option to buy a synthesized and 5′ end labeled version of the RNA.
2. The use of the TranscriptAid T7 High Yield Transcription Kit results in sufficiently high amounts and quality of mRNA for further processing.
3. This buffer was the best for our binding assay. Others might work well, too.
4. It is recommended to aliquot the ATP in small volumes to avoid too many freezethaw cycles.
5. This is the fluorescent dye for 3′-labeling of RNA.
6. The microvolume spectrophotometer is used for measuring the concentration of unlabeled and fluorescently labeled RNA.
7. Instead of dark-colored reaction tubes, you can wrap standard reaction tubes with aluminum foil to protect the samples from light.
8. For transcripts ≤100 bp, increase the incubation time to 4–8 h.
9. Depending on the DNase, the treatment may work better in a smaller volume. In this case, we recommend doing more than one reaction and pooling them before PCI clean up.

10. The phase lock gel helps to separate the organic and the aqueous phase
11. Sometimes the pellet is invisible. Move on with the purification, even if you cannot see the pellet.
12. Start with the blue tips to discard most of the supernatant, go over to the yellow tips and remove all you get. Let the tube sit for a few seconds and discard the rest with white tips.
13. Make sure you let the pellet dry in an RNase-free environment like in the sterile airflow cabinet.
14. For resolving, keep the tube at room temperature for around 20 min until the pellet disappears. You can mix carefully in between.
15. By using 200 pmol RNA, you have 5 μM RNA in total.
16. The amount of RNA to use in this part of the experiment needs to be calculated from the molecular weight of the RNA of interest.
17. To have a high labeling efficiency, it is important to use the same amount of fluorescent molecules as RNA molecule.
18. Fill up to 40 μl including the RNA from **step 1**.
19. To protect the RNA from light exposure and contaminations, use an aliquot for measurement.
20. It does not matter if you start first the computer or the monolith.
21. This filter is suitable for Cy3.
22. This filter is suitable for Cy5.
23. If possible, use nM as a unit for all concentrations in the software. Otherwise, there can be a problem with the data during the test.
24. The capillaries load faster when you turn the vial horizontally with the capillary inside.
25. Fill the capillaries at least ¾. Touch them only at the end, not in the middle.
26. 25 °C is the standard temperature for MST measurements. By changing the MST power, you adjust the intensity of the infrared laser. The higher the MST power, the lower is the temperature gradient in your experiment. With the excitation power, you set the LED lamp, which is used for the excitation of the fluorophores. By changing the settings, you can adjust the strength of the fluorescence signal. For this, you can use the auto-detect mode in MO.control as well.
27. 75 μM is the concentration we used in our assays. For your RNA of interest, it can be necessary to adjust the concentration.

28. To protect the fluorescently labeled samples from light, it is better to work in a darker environment.
29. It is important to have exactly the same concentration of the labeled interaction partner in every vial. Otherwise, you cannot calculate the binding affinity correctly.
30. Make sure that you mix well before you take the aliquot for the next vial.
31. In case you are working with RNA from an organism with a different growth temperature, you should try to do the incubation at the optimal temperature of your organism.
32. In the capillary scan, the initial fluorescence of each capillary is shown by the peak height. The peak shape shows the absorption. You can see the homogeneity of the fluorescence in your samples.
33. The peaks should have the same height and shape. If this is not the case, there are some points to check. It is important to have the same amount of labeled interaction partner in each capillary. Mix well before pipetting or filling the capillaries because changes in concentration of the labeled interaction partner influence the results. It is possible that the unlabeled interaction partner has an influence on the initial fluorescence either by changing the fluorescent signal of the other partner or by bringing its own fluorescence in the assay. If there are changes in the peak shape, aggregations to the capillary could be a reason. In this case, it is recommended to try premium-coated capillaries or to adjust the assay buffer. However, this problem is more known from binding assays containing proteins.
34. The higher the changes in MST signal, the higher the binding affinity. If there is only one line visible or only a small distance between the highest and the lowest lines, the binding affinity is low or not detectable. Changing the ligand concentration may improve the measurement. Changing the buffer system or incubation time and temperature could help as well.
35. To have three technical replicates, run the experiment three times with the same samples.
36. To have three biological replicates, perform the whole experiment three times, starting from in vitro transcription.

Acknowledgement

We thank the Schmitz–Streit working group for helpful discussions. The work was partially funded by the Deutsche Forschungsgemeinschaft (DFG) (SCHM1052/9-2, SCHM1052/21-1, SCHM1052/20-1).

References

1. Morris KV, Mattick JS (2014) The rise of regulatory RNA. Nat Rev Genet 15(6):423–437. https://doi.org/10.1038/nrg3722
2. Waters LS, Storz G (2009) Regulatory RNAs in bacteria. Cell 136(4):615–628. https://doi.org/10.1016/j.cell.2009.01.043
3. Bizebard T, Arluison V, Bockelmann U (2018) Single-molecule FRET assay to observe the activity of proteins involved in RNA/RNA annealing. In: Arluison V, Valverde C (eds) Bacterial regulatory RNA: Methods and protocols. Springer, New York, pp 301–319
4. Di Primo C, Dausse E, Toulmé J-J (2011) Surface plasmon resonance investigation of RNA aptamer–RNA ligand interactions. In: Goodchild J (ed) Therapeutic oligonucleotides: Methods and protocols. Springer Science+Business Media LLC, Totowa, pp 279–300
5. Jerabek-Willemsen M, André T, Wanner R et al (2014) MicroScale thermophoresis: Interaction analysis and beyond. J Mol Struct 1077:101–113. https://doi.org/10.1016/j.molstruc.2014.03.009
6. Asmari M, Ratih R, Alhazmi HA et al (2018) Thermophoresis for characterizing biomolecular interaction. Methods 146:107–119. https://doi.org/10.1016/j.ymeth.2018.02.003
7. Salze M, Muller C, Bernay B et al (2020) Study of key RNA metabolism proteins in Enterococcus faecalis. RNA Biol 17(6):794–804. https://doi.org/10.1080/15476286.2020.1728103
8. Sonnleitner E, Prindl K, Bläsi U (2017) The Pseudomonas aeruginosa CrcZ RNA interferes with Hfq-mediated riboregulation. PLoS One 12(7):e0180887. https://doi.org/10.1371/journal.pone.0180887
9. Moon MH, Hilimire TA, Sanders AM et al (2018) Measuring RNA-ligand interactions with microscale thermophoresis. Biochemistry 57(31):4638–4643. https://doi.org/10.1021/acs.biochem.7b01141
10. Buddeweg A, Sharma K, Urlaub H et al (2018) sRNA41 affects ribosome binding sites within polycistronic mRNAs in Methanosarcina mazei Göl. Mol Microbiol 107(5):595–609. https://doi.org/10.1111/mmi.13900

Chapter 16

Toeprint Assays for Detecting RNA Structure and Protein–RNA Interactions

Helen Yakhnin and Paul Babitzke

Abstract

Toeprint assays are primer extension inhibition assays that can detect the 3′ end of an RNA secondary structure, the position of a bound RNA binding protein, as well as the position of a bound 30S ribosomal subunit or a stalled ribosome. Here we describe how this assay was used to identify an RNA hairpin that sequesters a Shine–Dalgarno sequence, how the RNA-binding protein CsrA can alter RNA structure and affect 30S ribosomal subunit binding, and how the macrolide antibiotic tylosin can induce ribosome stalling.

Key words Toeprint, Primer extension inhibition, RNA structure, Protein–RNA interaction, Ribosome, Translational control

1 Introduction

The toeprint assay, also known as primer extension inhibition, is a method used in molecular biology for detecting RNA structure and protein–RNA interactions. This method is useful to identify the 3′ end of an RNA secondary structure, changes in RNA structure after interaction with an RNA-binding protein, and for determining the position of the 3′ edge of a bound protein [1, 2]. This assay can also provide mechanistic insight into the regulation of translation initiation by determining the position of the 3′ edge of a bound 30S ribosomal subunit in the presence or absence of an RNA-binding protein; bacterial 30S ribosomal subunits typically give an $tRNA^{f\text{-}Met}$-dependent toeprint band 16 nt downstream from the A of the AUG initiation codon [1–3]. Toeprint assays have also been used in coupled transcription-translation reactions to identify the position of antibiotic-induced ribosome stalling during polypeptide synthesis [4, 5]. It is important to empirically identify an appropriate buffer and temperature that is compatible with protein function, ribosome function, and reverse transcriptase activity. It is absolutely

Eveline Peeters and Indra Bervoets (eds.), *Prokaryotic Gene Regulation: Methods and Protocols*, Methods in Molecular Biology, vol. 2516, https://doi.org/10.1007/978-1-0716-2413-5_16,

critical to titrate reverse transcriptase to determine the optimal amount of enzyme that will allow you to identify the various molecular interactions. Too much reverse transcriptase can lead to full-size primer extension products and the absence of reverse transcription stops at the position(s) of bound proteins, ribosomes, and/or RNA structure. The utility of the toeprint assay is exemplified by demonstrating how binding of *Escherichia coli* CsrA protein disrupts an RNA structure that sequesters the *ymdA* Shine–Dalgarno (SD) sequence, how *E. coli* CsrA can affect ribosome binding, and how the antibiotic tylosin induces ribosome stalling during translation of the *Bacillus subtilis tlrB* leader peptide.

2 Materials

RNase-free conditions must be maintained throughout the procedure. All glassware, spatulas, and stir bars used for preparing solutions should be baked in an oven at 250 °C for 5 h to destroy RNases. Prepare all solutions with ultrapure water. All solutions that come into contact with RNA should be treated with diethylpyrocarbonate (DEPC) followed by autoclaving (*see* **Note 1**).

2.1 DNA Template for In Vitro Transcription (see Note 2)

1. Vent DNA polymerase.
2. 20 μM primers. Forward primer contains the T7 RNA polymerase (RNAP) promoter sequence in addition to complementarity to the target sequence.
3. Plasmid DNA or chromosomal DNA template.
4. Thermocycler.
5. PCR purification kit.

2.2 RNA Template Preparation (see Note 1)

1. DEPC-treated water: Add 1 mL of DEPC to 0.5 L of distilled water in a 1 L bottle. Mix, incubate overnight at room temperature in a hood, and autoclave for 30 min the next day. Store at room temperature.
2. *In vitro* transcription is carried out using the MEGAscript T7 or RNAMaxx transcription kit according to the manufacturer's specifications.
3. 5X TBE buffer for gel purification: Mix 465 mM Tris, 445 mM boric acid, 10 mM EDTA, pH 8.3. Weigh 54 g of Tris base, 27.5 g of boric acid, and 3.7 g of EDTA disodium salt in a 1 L glass beaker containing a stir bar. Add water to a volume of ~600 mL. Mix on a magnetic stirrer. Add water to a final volume of 1 L. pH does not require any adjustment. Filter and store at room temperature.
4. 6% polyacrylamide gel solution: Mix 75 mL of 40% acrylamide: bisacrylamide (19:1), 249.25 g urea, 100 mL 5X TBE buffer,

and add water to a final volume of 500 mL. Filter and store at 4 °C. Mix 12 mL of this solution with 12 μL of TEMED and 120 μL of fresh 10% ammonium persulfate. Cast gels using 10 × 10 cm gel plates with 0.5 mm spacers. Insert a 10 well comb. Allow the gel to polymerize for at least 30 min.

5. 2X RNA loading buffer (stop solution): For 10 mL mix 74.4 mg EDTA disodium salt, 48.4 mg Tris base, 10 mg sodium dodecyl sulfate (SDS), 3 mg bromophenol blue, and 3 mg xylene cyanol. Dissolve in 9.5 mL of formamide and store at -20 °C.
6. Shortwave UV light (254 nm) and a fluor-coated thin layer chromatography (TLC) plate for visualizing nucleic acids separated on polyacrylamide/urea gels.
7. Baked razor blades.
8. RNA elution buffer: 1.92 g ammonium acetate (0.5 M), 0.093 g EDTA disodium salt (5 mM), 0.1 g SDS (0.2%), and water to 50 mL. Treat solution with 50 μL DEPC and autoclave 30 min the next day.
9. Glycogen solution (5 mg/mL).
10. Phenol for RNA.
11. Chloroform.
12. 3 M sodium acetate, pH 5.2: Add 24.6 g of sodium acetate (anhydrous) to 70 mL distilled water. Dissolve the sodium acetate and adjust the pH to 5.2 with glacial acetic acid. Add water to a final volume of 100 mL.
13. 100% and 70% ethanol.
14. 100 mM dithiothreitol (DTT): dissolve 154 mg DTT in 10 mL of DEPC-treated water. Store at −20 °C.
15. TE buffer: 10 mM Tris-HCl, pH 7.5, 1 mM EDTA. Mix 0.5 mL of 2 M Tris-HCl, pH 7.5, 0.2 mL of 0.5 M EDTA, and 99.3 mL of DEPC-treated water.

2.3 5′ End-Labeled Primer

1. 18–25 nt DNA oligonucleotide with a melting temperature (Tm) of about 55 °C. The oligo should be complementary to a region near the 3′ end of the transcript that will be used in the toeprint reaction. The oligonucleotide should hybridize between 60 and 200 nt downstream of the expected toeprint position.
2. [γ-^{32}P] ATP (6000 Ci/mmol, 150 mCi/mL).
3. T4 polynucleotide kinase.
4. Mini quick spin oligo columns.

2.4 Toeprint Reaction

1. Reverse transcriptase (*see* **Note 3**).
2. 5X toeprint reaction buffer: 200 mM Tris-HCl, pH 8.0, 1 M KCl, 20 mM $MgCl_2$, 5 mM DTT. For 1 mL mix 0.1 mL 2 M Tris-HCl, pH 8.0, 0.25 mL 4 M KCl, 0.04 mL 0.5 M $MgCl_2$, 50 μL 100 mM DTT, and 560 μl of water. Store at −20 °C (*see* **Note 3**).
3. dNTP mixture.
4. RNasin RNase inhibitor.
5. 100 mM DTT.
6. 10 mg/mL BSA (Promega, NEB).
7. Protein dilution buffer (PDB): 10 mM Tris-HCl, pH 8, 1 mM DTT, 10% glycerol. Mix 10 μL 1M Tris, pH 8.0, 10 μL 100 mM DTT, 100 μL 100% glycerol, and 880 μL water. Store at −20 °C.
8. $tRNA^{fMet}$. We have been unable to identify a commercial source for this reagent.
9. 30S ribosomal subunits [6].
10. Yeast RNA. Dissolve 10 mg yeast RNA in 1 mL of DEPC-treated water. Store at −20 °C.
11. PURExpress *in vitro* protein synthesis kit.

2.5 Sequencing Gel

1. Sequenase version 2.0 DNA sequencing kit.
2. Gel casting system for sequencing gels.
3. Sigmacoat.
4. Whatman filter paper and plastic wrap.

3 Methods

3.1 DNA Template Containing a T7 RNAP Promoter

1. PCR reaction in 50 μL: 1 μL of Vent DNA polymerase, 5 μL of each 20 μM primers (forward primer contains the T7 RNAP promoter sequence), 5 μL of 10X buffer, 10–20 ng of plasmid DNA template or 100–300 ng of chromosomal DNA template, 1 μL of 100 mM $MgSO_4$ (supplied with Vent). 30–35 cycles with an annealing temperature 5 °C less than the melting temperature (Tm) of the primers, and extension at 72 °C for 2 min per kb. Examine a 5 μL aliquot of each reaction on an agarose gel.
2. Purify the PCR product with a PCR purification kit according to the manufacturer-provided manual.

3.2 RNA Preparation and Purification

1. Synthesize RNA with the MEGAscript T7 or RNAMaxx transcription kit. Use about 300 ng of the purified PCR product or 1 μg plasmid as a DNA template for each reaction according to the manufacturer's specifications.
2. Make 6% polyacrylamide gel solution with 8.3 M urea. Cast gel using 10 × 10 cm gel plates with 0.5 mm spacers. Insert a 10 well comb. Allow the gel to polymerize for at least 30 min. Transfer gel to a vertical gel running apparatus and fill the buffer reservoirs with 1X TBE buffer. Prerun the gel at 5 Watts for 10 min. Mix the RNA sample with the same volume of 2X loading buffer. Run gel at 3–5 Watts (*see* **Note 4**).
3. After electrophoresis, transfer gel from plates to plastic wrap. Cover the gel with a second sheet of plastic wrap. In a darkened room, place the plastic-wrapped gel on top of the fluor-coated TLC plate. Visualize nucleic acid bands by shining a hand-held UV light source (254 nm; shortwave) on the surface of the gel. Nucleic acid will appear as dark purple bands (see **Note 5**).
4. Mark the band to be removed. Cut out the band with a baked razor blade. Place the gel slice into a 2 mL microfuge tube with 300 μL of RNA elution buffer. Incubate at 37 °C overnight with shaking. The next day, centrifuge the tube at maximum speed for 5 min and then transfer the liquid to a new tube. Add glycogen to 50 μg/mL as a carrier for RNA precipitation. Add an equal volume of phenol and chloroform, vortex for 1–2 min, and then centrifuge for 5 min at maximum speed. Transfer the top aqueous layer to a new microfuge tube. Add 1/10 vol of 3 M sodium acetate, pH 5.3, and 2 vol of 100% ethanol. Incubate at -20 °C for 2 h and then pellet the RNA by centrifuging at 4 °C for 30 min at maximum speed. Wash the RNA pellet with 70% ethanol. Centrifuge at maximum speed for 5 min. Dry the RNA pellet and then dissolve in 30 μL of TE buffer. Determine the RNA concentration and then dilute to 300 nM. Store at -80 °C.

3.3 Preparation of 5' End-Labeled Primer

1. For a 10 μl reaction, mix 1 μL of 25 μM DNA oligonucleotide, 1 μL of 10X T4 polynucleotide kinase buffer, 2 μL of 25 μM [γ-^{32}P] ATP, 5 μL water, and then 1 μL (10 U) of T4 polynucleotide kinase.
2. Mix and incubate at 37 °C for 60 min.
3. Purify labeled primer using a mini quick spin oligo column. Centrifuge the column for 1 min at 3000 rpm, add 300 μL TE buffer and centrifuge for 2 min at the same speed.
4. Add 40 μL TE buffer to the 10 μL labeling reaction, then load onto the center of the column matrix, and centrifuge for 4 min.
5. Determine the concentration of the labeled primer and then dilute to 300 nM. Store at −80 °C.

3.4 Sequencing Ladder

1. Mix 1 μL of 1 μM plasmid DNA (*see* **Note 2**) or 1 μL of 1 μM PCR fragment and 3.33 μL of 300 nM of 5′ end-labeled oligonucleotide primer (*see* Subheading 3.3). Add 5.7 μL of water and then mix. Add 10 μL freshly prepared 2 N NaOH and incubate at 37 °C for 10 min. Add 2 μl 5 mg/mL glycogen, then add 6.7 μL 3 M sodium acetate, pH 5.2, and then add 175 μL of 100% ethanol. Mix and incubate at −20 °C for at least 1 h. Centrifuge at 4 °C for 30 min at maximum speed, wash with 70% ethanol, and then dry pellet. Dissolve pellet in 20 μL of 1X sequenase buffer (the buffer supplied in the sequenase kit is 5X).
2. Mix 4 μL of 100 mM DTT, 13.4 μL water, and 1 μL sequenase. Then add 11 μL of this mixture to the plasmid and primer in sequenase buffer. Transfer 5 μL of each ddNTP from the sequenase kit to separate tubes. Prewarm tubes at 37 °C, then add 7 μL of the sequenase-containing mixture and incubate at 42 °C for 5 min. Stop the reaction by adding 12 μl of 2X RNA loading buffer. Transfer 3 μL of each reaction to a new tube and incubate at 95 °C for 5 min prior to loading onto a sequencing gel.

3.5 Toeprint Reaction with *E. coli* CsrA and ymdA RNA to Demonstrate How Protein-Binding Can Alter RNA Structure

1. Hybridization reaction for 1 toeprint reaction (scale up as necessary): Prepare hybridization reaction by mixing 1 μL of 150 nM gel-purified RNA (*see* Subheading 3.2) and 1 μL of 150 nM of the 5′-end-labeled DNA primer (*see* Subheading 3.3) that is complementary to the 3′ end of the RNA. Incubate at 85 °C for 3 min, and then slowly cool down on the bench to room temperature (~10 min).
2. Reverse transcriptase mixture sufficient for 1 toeprint reaction (scale up as necessary): Mix 1 μL 10X SuperScript III reverse transcriptase buffer, 0.4 μL of 10 mM dNTP solution, 1 μL of 100 mM DTT, 1 μL of 2 μg/μL yeast RNA, 0.1 μL of RNasin, 0.2 μL 10 mg/mL BSA, and 2.3 μL water.
3. For each toeprint reaction mixture (10 μL) combine 2 μL of the hybridization mixture, 2 μL of 7.5 μM CsrA protein (1.5 μM final concentration) or the same volume of PDB, and 6 μL of the reverse transcriptase mixture.
4. Incubate at 37 °C for 20 min to allow CsrA–RNA complex formation.
5. Add 1 μL of SuperScript III reverse transcriptase (diluted in PDB) and continue incubation at 37 °C for 15 min (*see* **Note 6**).
6. Add an equal volume (11 μL) of 2X stop solution. Heat a 3 μL aliquot of each toeprint reaction and sequencing ladder samples at 95 °C for 5 min. Load onto a 6% sequencing gel that has been prerun for 20 min at 80 Watts. Continue running at 80 watts after loading the samples.

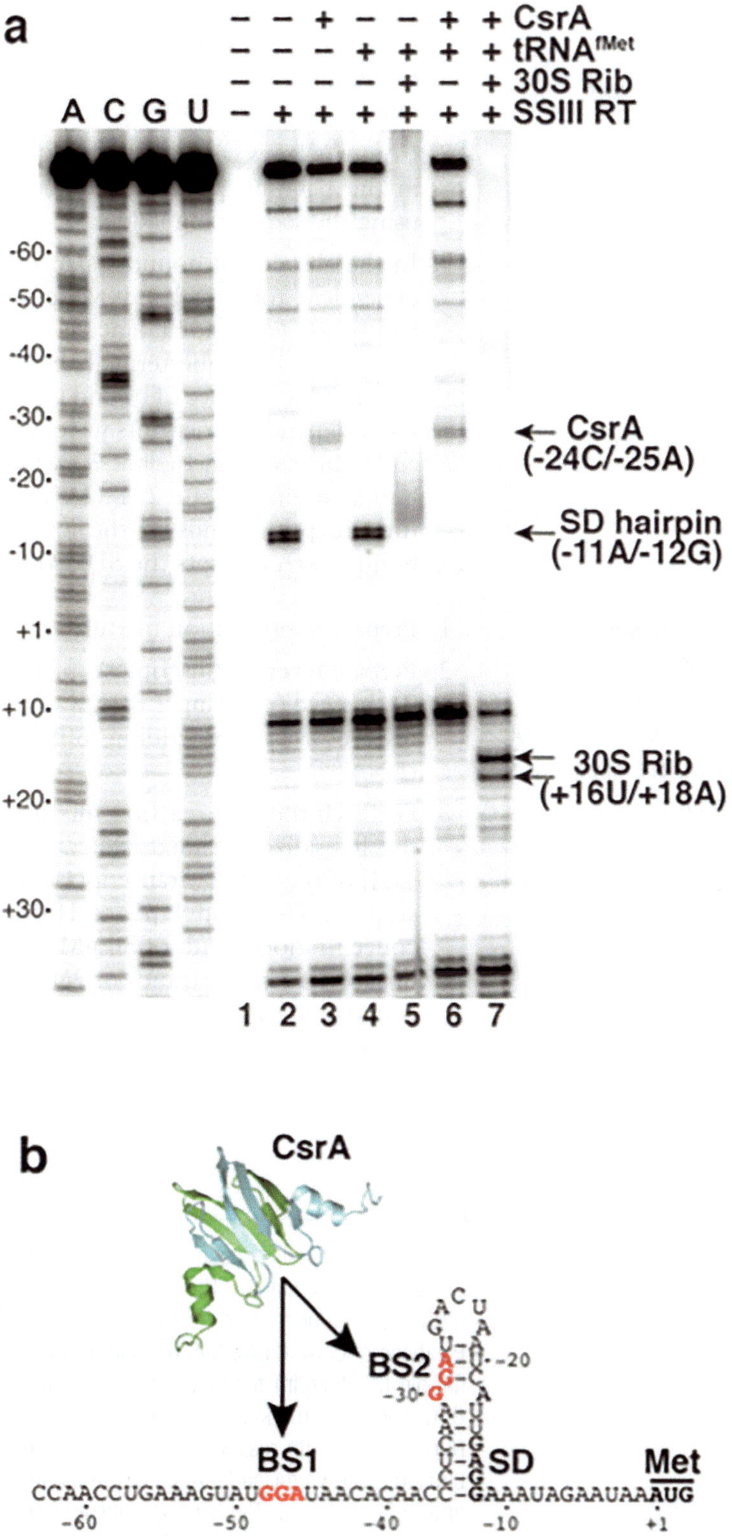

Fig. 1 RNA structure, CsrA, and ribosome toeprint analysis. (**a**) CsrA and 30S ribosome toeprint analysis of *ymdA* leader RNA. Bound CsrA facilitates 30S

7. Remove gel from glass plates and transfer to Whatman filter paper, cover with plastic wrap, and dry on a gel dryer at 80 °C for 30 min. Place completely dried gel in a phosphor imager cassette overnight or longer.
8. Compare lanes with and without protein to identify differences in the banding pattern and determine the precise positions using the sequencing ladder.
9. In the absence of bound CsrA, two adjacent RNA structure-dependent toeprint bands were identified at positions −11A and −12G (*see* Fig. 1a, lane 2). These bands are located at the base of the SD-sequestering hairpin, indicating that this hairpin forms in the absence of bound CsrA (*see* Fig. 1b). Two adjacent CsrA-dependent toeprints were observed at nucleotides −24C and −25A (*see* Fig. 1a, lane 3), which are at the 3′ edge of CsrA bound at binding site 2 (BS2). The absence of the RNA structural toeprint bands in the presence of CsrA indicates that bound CsrA disrupts the SD-sequestering hairpin.

3.6 30S Ribosomal Toeprint with *E. coli ymdA* RNA to Demonstrate How CsrA Can Affect Ribosome Binding

1. Prepare hybridization mixture (*see* Subheading 3.5, **step 1**).
2. Prepare reverse transcriptase mixture (*see* Subheading 3.5, **step 2**). Instead of adding 2.3 μL water, only add 0.3 μL, which will reserve 2 μL for adding 30S ribosomal subunits and/or $tRNA^{fMet}$.
3. For each toeprint reaction mixture (10 μL), combine 2 μL of the hybridization mixture (*see* Subheading 3.5, **step 1**) and 4 μL of the reverse transcriptase mixture (*see* Subheading 3.5, **step 2**). Then add 1 μL of 100 μM $tRNA^{fMet}$ (10 μM final concentration), 1 μL of 50 μM 30S ribosomal subunits (5 μM final concentration), and/or 2 μL of 7.5 μM CsrA (1.5 μM final concentration). Bring the total volume to 10 μL by adding PDB for reactions without CsrA or ribosomes and/or TE buffer for reactions without $tRNA^{fMet}$ as necessary (*see* **Notes 7** and **8**).

Fig. 1 (continued) ribosomal subunit binding by disrupting the *ymdA* SD-sequestering hairpin. The presence of CsrA, $tRNA^{fMet}$, 30S ribosomal subunit (30S Rib), and/or reverse transcriptase (SSIII RT) is indicated above each lane. Positions of CsrA (CsrA), 30S ribosomal subunit (30S Rib), and SD-sequestering hairpin (SD hairpin) toeprints are indicated with arrows. Sequencing lanes A, C, G, and U are marked, and lane numbers are shown at the bottom of the gel. Numbering is with respect to the start of *ymdA* translation. (**b**) Model for CsrA binding to the *ymdA* leader region. GGA motifs of CsrA binding sites 1 and 2 (BS1 and BS2) are in red. The SD sequence and *ymdA* translational start codon (Met) are marked. (This figure was adapted from Ref. 2. Copyright © American Society for Microbiology, mBio 11:e00849–20 (2020). DOI: 10.1128/mBio.00849-20)

3a. Add 2 μL of 7.5 μM CsrA and incubate at 37 °C for 20 min (*see* **Note 7**).

3b. Add 1 μL of 100 μM $tRNA^{fMet}$.

3c. Add 1 μL of 100 μM $tRNA^{fMet}$ and 1 μL of 50 μM 30S ribosomal subunits (*see* **Note 8**).

3d. Add 2 μL of 7.5 μM CsrA and incubate at 37 °C for 20 min. Then add 1 μL of 100 μM $tRNA^{fMet}$.

3e. Add 2 μL of 7.5 μM CsrA and incubate at 37 °C for 20 min. Then add 1 μL of 100 μM $tRNA^{fMet}$ and 1 μL of 50 μM 30S ribosomal subunits.

4. Then follow the procedure as written in Subheading 3.5, **steps 5–8**.
5. The addition of only $tRNA^{fMet}$ had no effect on the banding pattern (*see* Fig. 1a, compare lanes 2 and 4). The addition of $tRNA^{fMet}$ and 30S ribosomal subunits resulted in a smear near the base of the SD-sequestering hairpin but no toeprint band downstream of the start codon (*see* Fig. 1, lane 5). When CsrA was added prior to the addition of $tRNA^{fMet}$, the CsrA toeprint bands were observed (*see* Fig. 1a, compare lanes 3 and 6). When CsrA was added prior to the addition of $tRNA^{fMet}$ and 30S ribosomal subunits, two ribosomal toeprint bands were observed at +16U and +18A (*see* Fig. 1a, compare lanes 5 and 7). +16U is the expected position of a 30S ribosomal subunit toeprint. These results indicate that bound CsrA promotes 30S ribosomal subunit binding by destabilizing the *ymdA* SD-sequestering hairpin.

3.7 Toeprint Analysis of Tylosin-induced Ribosomal Stalling in the *tlrB* Leader Using PURExpress

1. Prepare DNA template with T7 promoter by PCR (*see* Subheading 3.1). Dilute template to 250 ng/μL.
2. Prepare 5′ end-labeled primer (*see* Subheading 3.3).
3. Transcription-translation mixture for 1 toeprint reaction (scale up as necessary): Mix 0.2 μL DNA template, 1.1 μL water, 2 μL solution A from the PURExpress kit, and 1.5 μL solution B from the PURExpress kit. Add 0.5 μL of tylosin at various concentrations (add water instead of tylosin as a negative control). Incubate at 37 °C for 1 h.
4. Hybridization mixture for 1 toeprint reaction (scale up as necessary): Mix 3 μL of 5X first strand (FS) buffer, 1.5 μL 100 mM DTT, 1.5 μL 2.5 mM dNTP, 0.375 μL RNasin, 1.125 μL water, 1 μL 5′ end-labeled primer (*see* Subheading 3.3).
5. Hybridization reaction for 1 toeprint reaction (scale up as necessary): Combine the 5.3 μL transcription-translational mixture (± tylosin) with 8.5 μL hybridization mixture. Incubate at 55 °C for 3 min. Chill on ice.

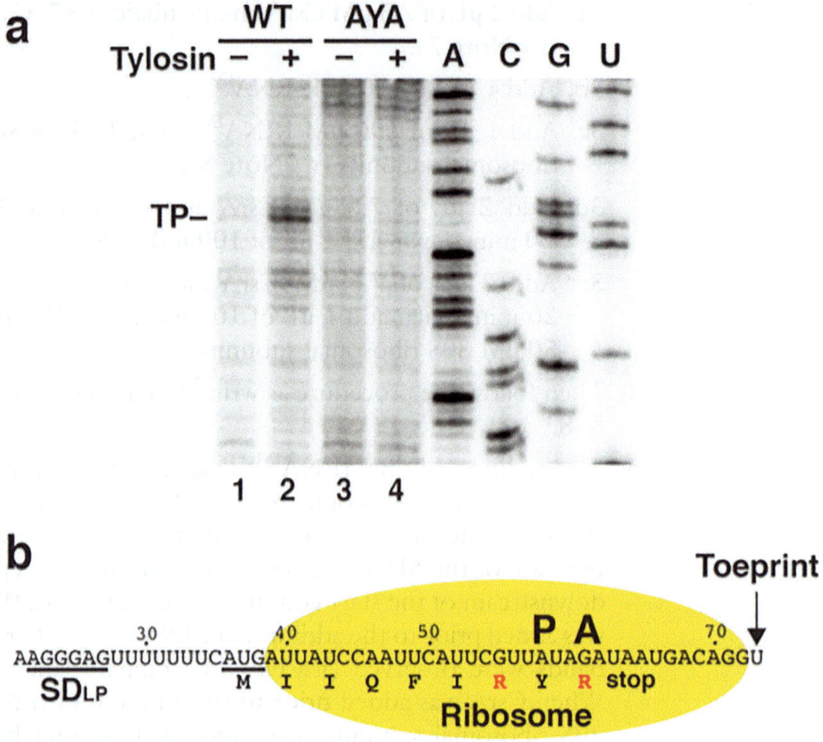

Fig. 2 Tylosin-dependent ribosome stalling in the *tlrB* leader peptide. (**a**) Toeprint analysis of tylosin-induced ribosome stalling during translation of the leader peptide using wild type (WT) and AYA mutant templates. The toeprint (TP) identified with the WT template in the presence of tylosin (+) is marked. Sequencing lanes A, C, G, and U are marked. The PURExpress kit containing T7 RNAP and *E. coli* ribosomes was used for this analysis. (**b**) *tlrB* leader region covered by the ribosome when tylosin induces stalling. Positions of the toeprint, and the ribosome peptidyl (P) and aminoacyl (A) sites are shown. The leader peptide coding sequence, and the leader peptide and *tlrB* SD sequences are shown. Numbering is with respect to start of *tlrB* transcription. (This figure was adapted from Ref. 5. Copyright © American Society for Microbiology, mBio 10:e02665–19 (2019). DOI: 10.1128/mBio.02665-19)

6. Add 0.5–2 μL of SuperScript III reverse transcriptase. Incubate at 37 °C for 1 h (*see* **Note 6**).
7. Then follow Subheading 3.5, **steps 6–8**.
8. A tylosin-dependent toeprint band was identified at position U74 with the WT *tlrB* leader template (*see* Fig. 2a, lane 2). This toeprint band was not observed in the RYR to AYA mutant template in which the tylosin-dependent ribosome stall motif was disrupted in the *tlrB* leader peptide (*see* Fig. 2a, lane 4 and 2b).

4 Notes

1. Maintaining RNase-free conditions is a vital consideration. Gloves should be worn at all times. Avoid using pipeters that have been used for procedures containing RNase A (e.g., plasmid purification). DEPC must be destroyed by autoclaving.
2. The use of a small plasmid vector is recommended to clone genes of interest (e.g., pTZ19R) to improve the generation of DNA templates via PCR and for DNA sequencing ladders. The concentration of Mg^{2+} in PCR using Vent DNA polymerase should be adjusted according to the manufacturer's suggestions.
3. Perform gel mobility assays to identify an appropriate buffer for toeprint reactions [7]. The buffer must be suitable for protein binding and reverse transcriptase activity. For example, we used 5X toeprint buffer (*see* Subheading 2.4, **step 2**) for *B. subtilis* TRAP protein toeprints (1) and avian myeloblastosis virus (AMV) or moloney murine leukemia virus (MMLV) reverse transcriptases (*see* Subheading 2.4, **step 1**). Alternatively, we used SuperScript III buffer and SuperScript III reverse transcriptase for *E. coli* CsrA toeprints (2).
4. 100 nt RNA runs approximately at the same place as xylene cyanol.
5. The xylene cyanol and bromophenol blue dyes may appear as dark bands during UV shadowing and are sometimes mistaken for nucleic acid. We recommend running an extra lane of loading buffer alone to easily distinguish between these dyes and a shadow caused by RNA.
6. It is vital to titrate RT to identify an appropriate amount of enzyme that maximizes the amount of toeprint signal at the expense of full-length products.
7. Subheading 3.6, **steps 3a–3e** are separate reactions containing various combinations of CsrA, $tRNA^{fMet}$, and/or 30S ribosomal subunits. Always add CsrA or other RNA-binding protein first to allow protein–RNA complex formation. Then add $tRNA^{fMet}$ and/or 30S ribosomal subunits.
8. Ribosomal subunits need to be activated by incubating at 37 °C for 15 min before adding to the toeprint reaction.

Acknowledgments

This work was supported by NIH grants GM059969 and GM098399.

References

1. Du H, Tarpey R, Babitzke P (1998) The *trp* RNA-binding attenuation protein regulates TrpG synthesis by binding to the *trpG* ribosome binding site of *Bacillus subtilis.* J Bacteriol 179: 2582–2586
2. Renda A, Poly S, Lai YJ, Pannuri A, Yakhnin H, Potts AH, Bevilacqua PC, Romeo T, Babitzke P (2020) CsrA-mediated translational activation of *ymdA* expression in *Escherichia coli.* MBio 11:e00849–e00820
3. Hartz D, McPheeters DS, Traut R, Gold L (1988) Extension inhibition analysis of translation initiation complexes. Methods Enzymol 164:419–425
4. Davis AR, Gohara DW, Yap MN (2014) Sequence selectivity of macrolide-induced translational attenuation. Proc Natl Acad Sci U S A 111:15379–15384
5. Yakhnin H, Yakhnin AV, Mouery BL, Mandell ZF, Karbasiafshar C, Kashlev M, Babitzke P (2019) NusG-dependent RNA polymerase pausing and tylosin-dependent ribosome stalling are required for tylosin resistance by inducing 23S rRNA methylation in *Bacillus subtilis.* MBio 10:e02665–19
6. Mehta P, Woo P, Venkataraman K, Karzai AW (2012) Ribosome purification approaches for studying interactions of regulatory proteins and RNAs with the ribosome. Methods Mol Biol 905:273–289
7. Yakhnin AV, Yakhnin H, Babitzke P (2012) Gel mobility shift assays to detect protein-RNA interactions. Methods Mol Biol 905:201–211

Index

A

B

C

D

E

G

H

I

L

M

N

O

Eveline Peeters and Indra Bervoets (eds.), *Prokaryotic Gene Regulation: Methods and Protocols*, Methods in Molecular Biology, vol. 2516, https://doi.org/10.1007/978-1-0716-2413-5,

MIX
Papier aus verantwortungsvollen Quellen
Paper from responsible sources
FSC® C105338

If you have any concerns about our products,
you can contact us on
ProductSafety@springernature.com

In case Publisher is established outside the EU,
the EU authorized representative is:
Springer Nature Customer Service Center GmbH
Europaplatz 3, 69115 Heidelberg, Germany

Printed by Libri Plureos GmbH
in Hamburg, Germany